Sustainable Developments by Artificial Intelligence and Machine Learning for Renewable Energies

Sustainable Developments by Artificial Intelligence and Machine Learning for Renewable Energies

Edited by

Krishna Kumar
UJVN Ltd., India

Ram Shringar Rao
NSUIT (East Campus), New Delhi, India

Omprakash Kaiwartya
Nottingham Trent University, United Kingdom

M. Shamim Kaiser
IIT Jahangirnagar University, Bangladesh

Sanjeevikumar Padmanaban
Aarhus University, Herning, Denmar

ELSEVIER

Elsevier
Radarweg 29, PO Box 211, 1000 AE Amsterdam, Netherlands
The Boulevard, Langford Lane, Kidlington, Oxford OX5 1GB, United Kingdom
50 Hampshire Street, 5th Floor, Cambridge, MA 02139, United States

Copyright © 2022 Elsevier Inc. All rights reserved.

No part of this publication may be reproduced or transmitted in any form or by any means, electronic or mechanical, including photocopying, recording, or any information storage and retrieval system, without permission in writing from the publisher. Details on how to seek permission, further information about the Publisher's permissions policies and our arrangements with organizations such as the Copyright Clearance Center and the Copyright Licensing Agency, can be found at our website: www.elsevier.com/permissions.

This book and the individual contributions contained in it are protected under copyright by the Publisher (other than as may be noted herein).

Notices
Knowledge and best practice in this field are constantly changing. As new research and experience broaden our understanding, changes in research methods, professional practices, or medical treatment may become necessary.

Practitioners and researchers must always rely on their own experience and knowledge in evaluating and using any information, methods, compounds, or experiments described herein. In using such information or methods they should be mindful of their own safety and the safety of others, including parties for whom they have a professional responsibility.

To the fullest extent of the law, neither the Publisher nor the authors, contributors, or editors, assume any liability for any injury and/or damage to persons or property as a matter of products liability, negligence or otherwise, or from any use or operation of any methods, products, instructions, or ideas contained in the material herein.

ISBN: 978-0-323-91228-0

> For information on all Elsevier publications
> visit our website at https://www.elsevier.com/books-and-journals

Publisher: Charlotte Cockle
Acquisitions Editor: Lisa Reading
Editorial Project Manager: Alice Grant
Production Project Manager: Surya Narayanan Jayachandran
Cover Designer: M les Hitchen

Typeset by STRAIVE, India

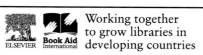

Dedication

Dedicated to the memory of
**The late Shri Bajilal Prasad (Grandfather) and the
late Smt. Guleshwari Devi (Grandmother),
who were the epitome of dedication and
humanity.**
Krishna Kumar

Contents

Contributors	xv
About the editors	xix
Preface	xxiii

1. Application of alternative clean energy

Adarsh Gaurav, Sujeet Kesharvani, Sakshi Sarathe,
Gaurav Dwivedi, Gaurav Saini, Anuj Kumar, and
Kamaraj Nithyanandhan

1.1	**Introduction**	1
1.2	**Solar energy**	2
	1.2.1 Photovoltaic systems	2
	1.2.2 Solar thermal energy systems	2
	1.2.3 Solar water heating (SWH) systems	2
	1.2.4 Solar cooker	3
	1.2.5 Solar water pumps	4
	1.2.6 Solar space heating	4
1.3	**Geothermal energy**	6
	1.3.1 Geothermal power generation	6
	1.3.2 Direct uses of geothermal energy	7
1.4	**Wind energy**	8
	1.4.1 Horizontal Axis wind turbine	8
	1.4.2 Vertical axis wind turbine	8
	1.4.3 Wind turbine applications	9
1.5	**Biomass energy**	10
	1.5.1 Method of biomass energy extraction	10
	1.5.2 Gasification	11
	1.5.3 Anaerobic digestion	11
	1.5.4 Biofuels	12
	1.5.5 Bioethanol production	12
	1.5.6 Biodiesel	13
1.6	**Ocean and tidal energy**	14
	1.6.1 Wave energy	14
	1.6.2 OTEC	15
	1.6.3 TIC	15
1.7	**Small, micro, and mini hydro plants**	15
1.8	**Case study**	16
1.9	**Conclusion**	17
	References	17

viii Contents

2. Optimization of hybrid energy generation
Poonam B. Dhabai and Neeraj Tiwari

2.1	Introduction	21
2.2	RES data and uncertainty statistical analysis	23
	2.2.1 Wind source analysis	24
	2.2.2 Solar source analysis	25
2.3	Test case modifications and solution methodology	27
	2.3.1 Test case modifications	27
	2.3.2 Configuration of cases	28
	2.3.3 Solution methodology	32
	2.3.4 Sensitivity factors	33
	2.3.5 Locational marginal pricing (LMP)	34
	2.3.6 Reliability parameters	34
2.4	Results	34
	2.4.1 Impact of probabilistic nature and location of RES on sensitivity factors	35
	2.4.2 Impact of probabilistic nature and location of RES on LMP	38
	2.4.3 Impact of the probabilistic nature and location of RES on TTC and TRM	39
2.5	Discussion and conclusion, future scope	41
	2.5.1 Discussion	41
	2.5.2 Conclusion	43
	2.5.3 Future scope	44
	Acknowledgment	44
	References	44

3. IoET-SG: Integrating internet of energy things with smart grid
M. Shahidul Islam, Md. Mehedi Islam, Sabbir Ahmed, Md. Sazzadur Rahman, Krishna Kumar, and M. Shamim Kaiser

3.1	Introduction	49
3.2	Traditional grid	50
3.3	Smart grid	51
3.4	Internet of energy things (IoET)	51
3.5	IoET-SG system	55
3.6	Research challenges and future guidelines	57
3.7	Conclusion	60
	References	60

4. Evolution of high efficiency passivated emitter and rear contact (PERC) solar cells
Sourav Sadhukhan, Shiladitya Acharya, Tamalika Panda, Nabin Chandra Mandal, Sukanta Bose, Anupam Nandi, Gourab Das, Santanu Maity, Susanta Chakraborty, Partha Chaudhuri, and Hiranmay Saha

4.1	Introduction	63
4.2	Photon absorption and optical generation	66

Contents   **ix**

4.3	**Loss mechanisms in PERC solar cells**	69
	4.3.1 Optical losses	70
	4.3.2 Electrical losses	70
4.4	**Carrier transport equations**	79
	4.4.1 Solar cell parameters	80
4.5	**PERC technology**	83
	4.5.1 PERC process flow	85
	4.5.2 Surface passivation	85
	4.5.3 LBSF and rear local contact	92
	4.5.4 Rear polishing	93
	4.5.5 PERC performance	93
	4.5.6 Improvements of PERC solar cells	95
	4.5.7 Further improvements	96
	4.5.8 Bifacial PERC	97
4.6	**Fabrication of PERC solar cells**	98
	4.6.1 Saw damage removal, texturization, and cleaning	99
	4.6.2 Diffusion and oxidation	101
	4.6.3 Reactive ion etching	103
	4.6.4 Plasma-enhanced chemical vapor deposition (PECVD)	104
	4.6.5 Atomic layer deposition (ALD)	106
	4.6.6 Laser ablation	107
	4.6.7 Metallization	109
4.7	**Characterization equipment**	111
	4.7.1 Scanning electron microscopy (SEM)	111
	4.7.2 Four point probe measurement	111
	4.7.3 Thickness profilometer	111
	4.7.4 I-V and C-V measurement	114
	4.7.5 X-ray photo electron spectroscopy (XPS)	114
	4.7.6 Lifetime and Suns-V_{oc} measurement	115
	4.7.7 Reflectance and external quantum efficiency (EQE) measurement	116
	4.7.8 Current-voltage (I-V) measurement	118
4.8	**Conclusion**	121
	References	121

5. Online-based approach for frequency control of microgrid using biologically inspired intelligent controller

Bhola Jha, Manoj Kumar Panda, and Yatindra Kumar

5.1	**Introduction**	131
5.2	**Test system description**	133
	5.2.1 Photovoltaic model	133
	5.2.2 Wind energy	135
	5.2.3 Diesel engine generator (DEG) model	136
	5.2.4 Fuel cell, BESS, and FESS	136
5.3	**Fuzzy logic controller**	137
5.4	**Particle swarm optimization (PSO)**	140
5.5	**Gray wolf optimization (GWO)**	142

x Contents

5.6	Results analysis	145
5.7	Conclusion	145
	References	146

6. Optimal allocation of renewable energy sources in electrical distribution systems based on technical and economic indices

Mohamed Zellagui, Samir Settoul, and Heba Ahmed Hassan

6.1	Introduction	149
	6.1.1 Motivation	149
	6.1.2 Literature review	150
	6.1.3 Contribution and chapter organization	151
6.2	Problem formulation	152
	6.2.1 Multiobjective function	152
	6.2.2 Equality constraints	153
	6.2.3 Inequality constraints of distribution line	153
	6.2.4 Inequality constraints of DG units	154
6.3	Cosine adapted whale optimization algorithm (CAWOA)	154
6.4	Results and discussion	155
	6.4.1 Test systems	155
	6.4.2 Analysis of optimal results	158
	6.4.3 Comparison results	166
	6.4.4 Impact of DG on branch currents	171
	6.4.5 Impact of loadability variation on EDS	172
6.5	Conclusions	176
	References	182

7. Optimization of renewable energy sources using emerging computational techniques

Aman Kumar, Krishna Kumar, and Nishant Raj Kapoor

7.1	Introduction	188
7.2	Sources of renewable energy	190
	7.2.1 Bioenergy (BE)	193
	7.2.2 Geothermal energy (GE)	193
	7.2.3 Hydropower energy (HPE)	197
	7.2.4 Hydrogen energy (HE)	200
	7.2.5 Solar energy (SE)	203
	7.2.6 Wind energy (WE)	206
	7.2.7 Ocean energy (OE)	206
7.3	Artificial intelligence (AI)	206
	7.3.1 Artificial intelligence in bioenergy	213
	7.3.2 Artificial intelligence in geothermal energy	214
	7.3.3 Artificial intelligence in hydro energy	215
	7.3.4 Artificial intelligence in hydrogen energy	221

Contents **xi**

	7.3.5 Artificial intelligence in solar energy	221
	7.3.6 Artificial intelligence in wind energy	221
	7.3.7 Artificial intelligence in ocean energy	221
7.4	Conclusion	221
	References	229

8. Advanced renewable dispatch with machine learning-based hybrid demand-side controller: The state of the art and a novel approach

Yuekuan Zhou

8.1	Introduction	237
8.2	Building energy demand forecasting with machine learning	238
	8.2.1 Predictions on cooling/heating/electrical loads	239
	8.2.2 Machine learning modeling techniques	239
8.3	Flexible demand-side management strategies	242
	8.3.1 Smart appliances	247
	8.3.2 HVAC systems	247
	8.3.3 Plug-in loads and storages	248
8.4	Machine learning-based advanced controllers	250
	Acknowledgment	252
	References	252

9. A machine learning-based design approach on PCMs-PV systems with multilevel scenario uncertainty

Yuekuan Zhou

9.1	Introduction	257
9.2	Overview on PCMs-PV systems and operations	259
	9.2.1 Passive PCMs-PV systems	259
	9.2.2 Active PCMs-PV systems	261
	9.2.3 Combined passive/active PCMs-PV systems	262
9.3	Mechanism for machine learning on performance prediction of nonlinear systems	263
9.4	Application of machine learning in PCMs-PV systems	264
	9.4.1 Surrogate model for performance prediction	264
	9.4.2 System optimization	265
	9.4.3 Robust optimization with multilevel scenario uncertainty	267
9.5	Challenges and outlooks	268
	9.5.1 Uncertainty quantification and probability density function	268
	9.5.2 Stochastic sampling size and uncertainty-based optimization function	268

xii Contents

	9.5.3 Hybrid learning and advanced optimization algorithms	270
	9.5.4 Multicriteria decision-marking for trade-off solutions	270
	Acknowledgment	270
	References	270

10. Agent-based peer-to-peer energy trading between prosumers and consumers with cost-benefit business models

Yuekuan Zhou and Jia Liu

10.1	Introduction	273
10.2	Agent-based peer-to-peer energy trading with dynamic internal pricing	274
	10.2.1 P2P energy trading modes with different energy forms	274
	10.2.2 Mechanisms and mathematical models for dynamic internal pricing	276
10.3	Blockchain and machine learning technologies in P2P energy trading	282
	10.3.1 Blockchain in P2P energy trading	282
	10.3.2 Machine learning technologies in P2P energy trading	283
10.4	Electricity market and techno-economic incentives for P2P energy market	284
	10.4.1 Decentralized electricity market design	285
	10.4.2 Techno-economic incentives	285
10.5	Challenges and outlook	286
	Acknowledgment	286
	References	286

11. Machine learning-based hybrid demand-side controller for renewable energy management

Padmanabhan Sanjeevikumar, Tina Samavat, Morteza Azimi Nasab, Mohammad Zand, and Mohammad Khoobani

11.1	Introduction	291
	11.1.1 Renewable and hybrid energy system	293
	11.1.2 Demand-side management	294
11.2	Machine learning at a glance	295
	11.2.1 Machine learning meets model-based control	296
	11.2.2 The application of machine learning in hybrid demand-side controllers	297
	11.2.3 Support vector machine	299
	11.2.4 K-means clustering	302

Contents **xiii**

	11.2.5 Extreme learning machine	303
	11.2.6 Linear regression	303
	11.2.7 Partial least squares	304
	11.2.8 Challenges and future research direction	304
11.3	Conclusion	304
	References	305

12. Prediction of energy generation target of hydropower plants using artificial neural networks

Krishna Kumar, Gaurav Saini, Narendra Kumar, M. Shamim Kaiser, Ramani Kannan, and Rachna Shah

12.1	Introduction	309
12.2	Artificial neural network (ANN)	310
12.3	Performance measurement parameters	313
12.4	Modeling and analysis	314
12.5	Conclusion	319
	References	319

13. Response surface methodology-based optimization of parameters for biodiesel production

Pijush Dutta, Bittab Biswas, Biplab Pal, Madhurima Majumder, and Amit Kumar Das

13.1	Introduction	321
13.2	Problem formulation	324
13.3	Mathematical model of biodiesel production	324
	13.3.1 Optimization of the mathematical model	326
	13.3.2 Proposed methodology	327
	13.3.3 Basic elephant swarm water search algorithm (ESWSA)	327
13.4	Methodology	328
13.5	Reaction conditions by RSM	329
13.6	Surface plot by different combinations in RSM model	329
13.7	Conclusion	331
	References	336

14. Reservoir simulation model for the design of irrigation projects

Siva Ramakrishna Madeti, Gaurav Saini, and Krishna Kumar

14.1	Introduction	341
14.2	System description	343
14.3	Cost-benefit functions	343
14.4	Methodology	345

xiv Contents

	14.4.1 Linear programming model (LP model)	345
	14.4.2 Reservoir simulation	349
14.5	Simulation computations	350
14.6	Results and discussion	352
14.7	Response of Harabhangi irrigation project	353
	14.7.1 Support for the use of simulation	353
14.8	Conclusion	355
	References	357

15. Effect of hydrofoils on the starting torque characteristics of the Darrieus hydrokinetic turbine

Gaurav Saini, Anuj Kumar, and R.P. Saini

15.1	Introduction	359
15.2	Investigated parameters for the Darrieus hydrokinetic turbine	362
15.3	Numerical simulation analysis	362
	15.3.1 Turbine model development	363
	15.3.2 Grid generation	364
	15.3.3 Boundary conditions and turbulence modeling	365
15.4	Results and discussion	366
	15.4.1 Performance characteristics	366
	15.4.2 Flow contours	370
15.5	Conclusions	374
	References	374

Index   377

Contributors

Numbers in parentheses indicate the pages on which the authors' contributions begin.

Shiladitya Acharya (63), School of Advanced Materials, Green Energy and Sensor Systems (SAMGESS), Howrah, India

Sabbir Ahmed (49), Institute of Information Technology, Jahangirnagar University, Dhaka, Bangladesh

Bittab Biswas (321), Department of Electronics and Communication Engineering, Global Institute of Management and Technology, Krishnagar, West Bengal, India

Sukanta Bose (63), School of Advanced Materials, Green Energy and Sensor Systems (SAMGESS), Howrah, India

Susanta Chakraborty (63), Department of Computer Science and Technology (CST), Indian Institute of Engineering Science and Technology (IIEST) Shibpur, Howrah, India

Partha Chaudhuri (63), School of Advanced Materials, Green Energy and Sensor Systems (SAMGESS), Howrah, India

Amit Kumar Das (321), Department of Physics, Global Institute of Management and Technology, Krishnagar, West Bengal, India

Gourab Das (63), School of Advanced Materials, Green Energy and Sensor Systems (SAMGESS), Howrah, India

Poonam B. Dhabai (21), Poornima University, Jaipur, India

Pijush Dutta (321), Department of Electronics and Communication Engineering, Global Institute of Management and Technology, Krishnagar, West Bengal, India

Gaurav Dwivedi (1), Energy Centre, Maulana Azad National Institute of Technology, Bhopal, India

Adarsh Gaurav (1), Energy Centre, Maulana Azad National Institute of Technology, Bhopal, India

Heba Ahmed Hassan (149), Electrical Power Engineering Department, Cairo University, Giza, Egypt

M. Shahidul Islam (49), Institute of Information Technology, Jahangirnagar University, Dhaka, Bangladesh

Md. Mehedi Islam (49), Department of Electronics & Communication Engineering, Hajee Mohammad Danesh Science & Technology University, Dinajpur, Bangladesh

Bhola Jha (131), G. B. Pant Institute of Engineering and Technology, Pauri, Uttarakhand, India

xvi Contributors

M. Shamim Kaiser (49, 309), Institute of Information Technology, Jahangirnagar University, Dhaka, Bangladesh

Ramani Kannan (309), Electrical and Electronics Engineering Department, University of Technology PETRONAS (UTP), Seri Iskandar, Malaysia

Nishant Raj Kapoor (187), CSIR-CBRI, AcSIR—Academy of Scientific and Innovative Research, Roorkee, India

Sujeet Kesharvani (1), Energy Centre, Maulana Azad National Institute of Technology, Bhopal, India

Mohammad Khoobani (291), Department of Mechanical and Mechatronics Engineering, Shahrood University of Technology, Shahroud, Iran

Aman Kumar (187), CSIR-CBRI, AcSIR—Academy of Scientific and Innovative Research, Roorkee, India

Anuj Kumar (1, 359), School of Mechanical Engineering, Vellore Institute of Technology, Vellore, India

Krishna Kumar (49, 187, 309, 341), Department of Hydro and Renewable Energy, Indian Institute of Technology Roorkee, Roorkee; Research & Development Unit, Uttarakhand Jal Vidyut Nigam (UJVN) Ltd., Dehradun, Uttarakhand, India

Narendra Kumar (309), School of Computing, DIT University, Dehradun, Uttarakhand, India

Yatindra Kumar (131), G. B. Pant Institute of Engineering and Technology, Pauri, Uttarakhand, India

Jia Liu (273), Department of Building Environment and Energy Engineering, Faculty of Construction and Environment, Hong Kong Polytechnic University, Kowloon, Hong Kong, China

Siva Ramakrishna Madeti (341), University of Santiago Chile, Santiago, Chile

Santanu Maity (63), School of Advanced Materials, Green Energy and Sensor Systems (SAMGESS), Howrah, India

Madhurima Majumder (321), Department of Electrical & Electronics Engineering, Mirmadan Mohanlal Government Polytechnic, Gobindapur, West Bengal, India

Nabin Chandra Mandal (63), School of Advanced Materials, Green Energy and Sensor Systems (SAMGESS), Howrah, India

Anupam Nandi (63), School of Advanced Materials, Green Energy and Sensor Systems (SAMGESS), Howrah, India

Morteza Azimi Nasab (291), Department of Business Development and Technology, CTIF Global Capsule, Aarhus University, Herning, Denmark

Kamaraj Nithyanandhan (1), Department of Automobile Engineering, Kongu Engineering College, Erode, India

Biplab Pal (321), Department of Electronics and Communication Engineering, Global Institute of Management and Technology, Krishnagar, West Bengal, India

Manoj Kumar Panda (131), G. B. Pant Institute of Engineering and Technology, Pauri, Uttarakhand, India

Tamalika Panda (63), School of Advanced Materials, Green Energy and Sensor Systems (SAMGESS), Howrah, India

Md. Sazzadur Rahman (49), Institute of Information Technology, Jahangirnagar University, Dhaka, Bangladesh

Sourav Sadhukhan (63), School of Advanced Materials, Green Energy and Sensor Systems (SAMGESS), Howrah, India

Hiranmay Saha (63), School of Advanced Materials, Green Energy and Sensor Systems (SAMGESS), Howrah, India

Gaurav Saini (1, 309, 341, 359), School of Advanced Materials, Green Energy and Sensor Systems, Indian Institute of Engineering Science and Technology Shibpur, Howrah, West Bengal, India

R.P. Saini (359), Department of Hydro and Renewable Energy, Indian Institute of Technology Roorkee, Roorkee, India

Tina Samavat (291), Department of Mechanical and Mechatronics Engineering, Shahrood University of Technology, Shahroud, Iran

Padmanabhan Sanjeevikumar (291), Department of Business Development and Technology, CTIF Global Capsule, Aarhus University, Herning, Denmark

Sakshi Sarathe (1), Energy Centre, Maulana Azad National Institute of Technology, Bhopal, India

Samir Settoul (149), Department of Electrotechnic, Mentouri University of Constantine 1, Constantine, Algeria

Rachna Shah (309), Indian Institute of Information Technology, Guwahati, Assam, India

Neeraj Tiwari (21), Poornima University, Jaipur, India

Mohammad Zand (291), Department of Business Development and Technology, CTIF Global Capsule, Aarhus University, Herning, Denmark

Mohamed Zellagui (149), Department of Electrical Engineering, University of Quebec, Montreal, QC, Canada; Department of Electrical Engineering, University of Batna 2, Batna, Algeria

Yuekuan Zhou (237, 257, 273), Sustainable Energy and Environment Thrust, Function Hub, The Hong Kong University of Science and Technology, Guangzhou, China; Department of Mechanical and Aerospace Engineering, The Hong Kong University of Science and Technology, Clear Water Bay, Hong Kong SAR, China

About the editors

Er. Krishna Kumar is presently working as a research and development engineer at UJVN Ltd. Before joining UJVNL, he worked as Assistant Professor at BTKIT, Dwarahat (Uttarakhand), India. He received his BE (Electronics and Communication Engineering) from Govind Ballabh Pant Engineering College, Pauri Garhwal, Uttarakhand, India, and MTech (Digital Systems) from Motilal Nehru National Institute of Technology, Allahabad, India. He is also pursuing his PhD from the Indian Institute of Technology, Roorkee, India. He has more than 12 years of industrial and teaching experience and has published numerous research papers in international journals like IEEE, Elsevier, Taylor & Francis, Springer, and Wiley. His research area includes renewable energy, artificial intelligence, cloud computing, and the Internet of things.

Dr. Ram Shringar Rao received his PhD (Computer Science and Technology) from the Jawaharlal Nehru University, New Delhi, India. He obtained his MTech (IT) and BE (CSE) in 2005 and 2000, respectively. He worked as an Associate Professor in the Department of Computer Science, Indira Gandhi National Tribal University (a Central University in Madhya Pradesh) from April 2016 to March 2018. He is currently working in the Department of Computer Science and Engineering of Netaji Subhas University of Technology, East Campus, Delhi, India. He has more than 18 years of teaching, administrative, and research experience. He has published more than 100 research papers, including edited books with good impact factors in reputed International journals and conferences, including IEEE, Elsevier, Springer, Wiley & Sons, Taylor & Francis, IERI Letters, American Institute of Physics, etc. His current research interests include mobile ad hoc networks, vehicular ad hoc networks, flying ad hoc networks, and cloud computing.

Dr. Omprakash Kaiwartya is currently working at the School of Science & Technology, Nottingham Trent University (NTU), United Kingdom, as a Senior Lecturer and Course Leader for MSc Engineering (Electronics, Cybernetics, and Communications). Previously, he was a Research Associate (equivalent to Senior Lecturer) in the Department of Computer and Information Science at Northumbria University, Newcastle, United Kingdom, where, he was involved in the gLINK European Union project. Prior to this, he was a Post-Doctoral Fellow (equivalent to Lecturer) in the Faculty of Computing,

xx About the editors

University of Technology (UTM), Malaysia. Before moving to Malaysia, he completed his BSc in Computer Science from Guru Ghansidas Central University, Bilaspur Chhattisgarh, India, and combined master's degree and PhD from the School of Computer and Systems Science, Jawaharlal Nehru University (JNU), New Delhi, India. Overall, he has authored/co-authored over 100 international publications including journal articles, conference proceedings, book chapters, and books. Dr. Kaiwartya's research interest focuses on the IoT-centric smart environment for diverse domain areas, including transport, healthcare, and industrial production. His recent scientific contributions are to the Internet of connected vehicles (IoV), E-mobility, electronic vehicles charging management (EV), Internet of healthcare things (IoHT), smart use case implementation of sensor networks, and next generation wireless communication technologies (6G and beyond).

Furthermore, Omprakash is a Fellow of the Higher Education Academy (FHEA), IEEE Senior member, and BCS Professional member. He has served as a TPC member or reviewer in over 100 international conferences and workshops including IEEE Globecom, IEEE ICC, IEEE CCNC, IEEE ICNC, IEEE VTC, IEEE INFOCOM, ACM CoNEXT, ACM MobiHoc, ACM SAC, and many more. Furthermore, he has reviewed papers for more than 30 international journals including IEEE Magazines on Wireless Communications, Networks, Communications, IEEE Communications Letters, IEEE Sensors Letters, IEEE Transactions on Industrial Informatics, Vehicular Technologies, Intelligent Transportation Systems, Big Data, and Mobile Computing. Moreover, Dr. Kaiwartya has been an editorial member of various special issues of the top-ranked IEEE Open Journal of the Communication Society, and served as an Associate Editor of IET Intelligent Transport Systems, IEEE Internet of Things Journal, Springer, EURASIP Journal on Wireless Communication and Networking, MDPI Electronics, Ad Hoc and Sensor Wireless Networks, and KSII Transactions on Internet and Information Systems.

Dr. M. Shamim Kaiser is currently working as a Professor at the Institute of Information Technology of Jahangirnagar University, Savar, Dhaka, Bangladesh. He received his bachelor's and master's degrees in Applied Physics, Electronics, and Communication Engineering from the University of Dhaka, Bangladesh, in 2002 and 2004, respectively, and PhD in Telecommunication Engineering from the Asian Institute of Technology (AIT) Pathumthani, Thailand, in 2010. His current research interests include data analytics, machine learning, wireless networks and signal processing, cognitive radio networks, big data and cyber security, and renewable energy. He has authored more than 100 papers in different peer-reviewed journals and conferences and his Google Scholar citation count is more than 1020.

He is an Associate Editor of the IEEE Access Journal and Guest Editor of Brain Informatics Journal and Cognitive Computation Journal. Dr. Kaiser is a life member of the Bangladesh Electronic Society and Bangladesh Physical

About the editors **xxi**

Society. He is also a senior member of IEEE, United States, and IEICE, Japan, and an active volunteer of the IEEE Bangladesh Section. He is the founding Chapter Chair of the IEEE Bangladesh Section Computer Society Chapter.

Dr. Sanjeevikumar Padmanaban (Member 2012, Senior Member 2015, IEEE) received his PhD in Electrical Engineering from the University of Bologna, Bologna, Italy, in 2012. He was an Associate Professor with VIT University from 2012 to 2013. In 2013, he joined the National Institute of Technology, India, as a Faculty Member. In 2014, he was invited as a Visiting Researcher at the Department of Electrical Engineering, Qatar University, Doha, Qatar, funded by the Qatar National Research Foundation (Government of Qatar). He continued his research activities with the Dublin Institute of Technology, Dublin, Ireland, in 2014, and served as an Associate Professor with the Department of Electrical and Electronics Engineering, University of Johannesburg, Johannesburg, South Africa, from 2016 to 2018. Since 2018, he has been a Faculty Member with the Department of Energy Technology, Aalborg University, Esbjerg, Denmark. He has authored over 300 scientific papers.

Dr. Padmanaban was the recipient of the Best Paper cum Most Excellence Research Paper Award from IET-SEISCON 2013, IET-CEAT 2016, IEEE-EECSI 2019, and IEEE-CENCON 2019, and five best paper awards from the ETAEERE 2016 sponsored Lecture Notes in Electrical Engineering, Springer, book. He is a Fellow of the Institution of Engineers, India, the Institution of Electronics and Telecommunication Engineers, India, and the Institution of Engineering and Technology, United Kingdom. He is an Editor/Associate Editor/Editorial Board for refereed journals, in particular the IEEE Systems Journal, IEEE Transaction on Industry Applications, IEEE Access, IET Power Electronics, IET Electronics Letters, and Wiley International Transactions on Electrical Energy Systems, Subject Editorial Board Member—Energy Sources—Energies Journal, MDPI, and the subject editor for the IET Renewable Power Generation, IET Generation, Transmission and Distribution, and FACTS journal (Canada).

Preface

Energy is a basic need for the development of human beings and will always be part of the context of our daily life. An increase in energy demand due to an increase in population and utilization has put extra pressure on the existing energy generation infrastructures. This burden can only be resolved by installing new power plants based on renewable energy resources. Energy generation through renewable energy sources is sustainable in nature, which minimizes the environmental effect. However, an increase in the renewable energy generation share destabilizes the grid. Among all the available renewable energy sources, solar and wind energy sources are the dominant emerging forms, and these power plants have short gestation periods. Hydro and wind energy generation systems convert mechanical power to electrical power, whereas solar energy generation systems convert solar radiation into electricity. The selection of a particular energy source is generally made based on the available energy density along with the techno-economic feasibility of the conversion.

The development in capacity building for renewable energy generation has sparked a paradigm change in the energy sector. Due to changes as the sources of energy generation shift, issues of grid stability have been affected. The energy demand market accelerates to focus on innovative technologies integrated with renewable energy systems. Various complex nonlinear interactions among different parameters drive the integration of renewable energy with the grid. Artificial intelligence (AI), the Internet of things (IoT), and cloud computing techniques are being utilized to produce more reliable energy generation and to optimize system performance.

The major topics in this book are covered in depth. Chapter 1 is about the application of alternative clean energy in the transportation sector, energy production, and cooking. Chapter 2 presents a detailed analysis of a standard IEEE 30 bus system considering wind and solar as energy sources. Traditional methods like Monte Carlo, two-point estimation, and Taylor series are also discussed. Chapter 3 focuses mainly on IoET-SG, advantages, and future challenges, along with effective solutions. Chapter 4 discusses the evaluation of PERC/PERT/PERL solar cells. Chapter 5 addresses the issues of installing a feasible and reliable micro-hybrid grid system in remote regions. Chapter 6 presents a new optimization algorithm to solve the techno-economic problem of optimal location and sizing of RES-based distributed generators (DG) in the EDSs. Chapter 7 is an overview of available renewable energy generation technologies.

xxiii

xxiv Preface

Chapter 8 presents an overview of machine learning (ML) applications in building energy demand prediction for cooling, heating, and electrical energy systems. Chapter 9 discusses an overview of solar cell cooling techniques, including passive, active, and combined passive/active strategies. The mechanism of machine learning for the performance prediction of nonlinear systems is studied in terms of training, validation, and testing processes. Chapter 10 presents agent-based, peer-to-peer (P2P) energy trading, with dynamic internal pricing in terms of various energy trading forms, underlying mechanisms, and mathematical models. The applications and prospects of blockchain and machine learning technologies in P2P energy trading have been reviewed to show recent progress and advances. Chapter 11 discusses machine learning-based hybrid demand-side controllers. Chapter 12 discusses the classifications of daily power generation data of a hydropower plant using the unsupervised self-organizing maps (SOM) clustering technique to decide the energy generation target for individual power plants.

Chapter 13 covers biodiesel energy generation optimization techniques. Equation response surface methodology is utilized to build the numerical model while the advanced elephant swarm water search algorithm (ESWSA) is used to test the nonlinear model. Chapter 14 presents a simulation model for optimizing a reservoir to determine the best, and correct quantity of, future irrigation facilities that can be provided in the proposed scheme. It goes on to look at the applicability of simulation in reservoir planning. Chapter 15 discusses the performance of hydrokinetic turbines under various angles of attack (AoA) of blade profile.

The target audience of this book are researchers, academicians, technical institutes, R&D laboratories, data scientists working in the fields of renewable energy, AI, cloud computing, IoT, microgrid, smart grid, and water resources, and investors who want to analyze and understand the challenges, benefits, and possibilities available for the improvement in system performance.

Chapter 1

Application of alternative clean energy

Adarsh Gaurav[a], Sujeet Kesharvani[a], Sakshi Sarathe[a], Gaurav Dwivedi[a], Gaurav Saini[b], Anuj Kumar[c], and Kamaraj Nithyanandhan[d]

[a]*Energy Centre, Maulana Azad National Institute of Technology, Bhopal, India,* [b]*School of Advanced Materials, Green Energy and Sensor Systems, Indian Institute of Engineering Science and Technology Shibpur, Howrah, West Bengal, India,* [c]*School of Mechanical Engineering, Vellore Institute of Technology, Vellore, India,* [d]*Department of Automobile Engineering, Kongu Engineering College, Erode, India*

1.1 Introduction

Fossil fuels are nonrenewable energy that emits quantities of greenhouse gases and other pollutants, which cause climate change and global warming. According to WHO-2018, 8 million people die (4.2 million due to ambient air +3.8 million due to household air) every year due to air pollution (Ierodiakonou et al., 2016). The IPCC (2018), sr15_chapter1, says that in 2017 global warming had reached approximately 1°C (0.8–1.2°C) above preindustrial (1750–1900) levels, increasing at a rate of 0.2°C per decade (2018). This is causing the melting of glaciers, extinction of several species, increased threat of wildfire, storms, and drought, and effects on agriculture and cultivation, which are very susceptible to climate change and temperature. Poverty may increase drastically because many people depend on agriculture. Furthermore, fossil fuels are depleting rapidly. Many researchers predict that, with the present consumption rate, coal will likely be exhausted by the year 2090, oil will have run out by 2052, and gas by 2060. According to the WHO-2018 report, the amounts of CO_2, N_2O, and CH_4 in 2018 were 407.8 ± 0.1 ppm, N_2O at 331.1 ± 0.1 ppb, and 1869 ± 2 ppb, respectively; these values having increased by 147%, 123%, and 259%, respectively, from preindustrial (before 1750) levels (World Meteorological Organization and Global Atmosphere Watch, 2019). The human population has increased to more than 7.5 billion, so the energy demand has also increased. In order to mitigate pollution and release greenhouse gases to meet our energy demand, the need for clean sources of energy is required.

Sustainable Developments by Artificial Intelligence and Machine Learning for Renewable Energies.
https://doi.org/10.1016/B978-0-323-91228-0.00004-5
Copyright © 2022 Elsevier Inc. All rights reserved.

2 Sustainable developments by artificial intelligence & machine learning

Alternative, clean sources of energy (ACE), such as hydro, solar, wind, biomass, geothermal, and ocean energy, emit a smaller quantity of pollutants and greenhouse gases. They are generally renewable forms of energy. They have the possibility to replace fossil fuels. The use of solar and wind, hydro, biomass, and other clean energy sources is growing rapidly. Solar, biogas, small-scale hydro, wind, and other source-based applications help rural areas to meet their energy demand. Every ACE source has a wide variety of applications, such as use in the transportation sector, existing buildings, energy production, heating, cooking, etc., which will be discussed in the following chapter.

1.2 Solar energy

Solar energy can be utilized in two ways, either by converting it into electrical energy by using photovoltaic cells or by conversion into thermal energy.

1.2.1 Photovoltaic systems

Photovoltaic cells are generally made up of semiconductor materials of silicon. When sunlight strikes these semiconductor materials, they lose electrons and, by joining negative and positive terminals to complete the external circuit, electricity can be generated. These cells are connected to a form module. When several modules are connected, either in parallel or series combination, an array is formed. A typical cell generally produces 1.5 watts of power (Bhawan & Puram, 2006a).

Some applications for photovoltaic systems are lighting commercial buildings, street lighting systems, lighting rural areas, etc. (Bhawan & Puram, 2006b).

1.2.2 Solar thermal energy systems

Solar thermal energy devices convert solar energy into thermal energy. On the basis of temperature, these can be classified as:

Low-grade temperature devices: Temperature up to 100°C obtained;
Medium grade temperature devices: Temperature between 100°C and 300°C;
High-grade temperature devices: Temperature above 300°C.

1.2.3 Solar water heating (SWH) systems

One of the traditional methods of water heating is solar water heating (SWH). The majority of SWHs are built without a solar energy concentrator (Abed, 2021). A typical SWH consists of a rectangular box with tubes through which liquid (either water or another liquid) to be heated flows, a solar collector (usually flat plate collectors), a clear glass cover, an absorber plate attached to tubes to absorb heat, and an insulated storage tank to store heated water. The system is available in either active or passive form. The active system includes a pump for liquid flow, whereas the passive system is without a pump and the liquid

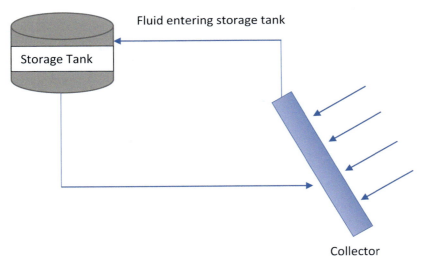

FIG. 1.1 Solar water heater.

flows naturally due to gravity. A schematic diagram of a simple SWH is shown in Fig. 1.1.

A few industrial applications of solar water heaters are listed below (Bhawan & Puram, 2006b):

- Used in chemical/bulk drug units for fermentation of mix and boiler feed applications.
- Used in the kitchen, for bathing, washing, and laundry applications in hotels.
- Used in the pulp and paper industries for soaking of pulp and also used in boiler feed pumps.
- Used in electroplating/galvanizing units for heating of plating baths, cleaning, and degreasing applications.
- Used in clarified butter production, cleaning, sterilizing, and pasteurization. Used in breweries and distilleries for bottle washing, wort preparation, boiler feed heating.
- Used in textiles for bleaching, boiling, printing, dyeing, curing, aging, and finishing purposes.

1.2.4 Solar cooker

Solar cookers are used to cook food, such as cereals, rice, etc., by utilizing solar energy. On the basis of design, these can be classified as:

1.2.4.1 Box type solar cooker

The box-type solar cooker consists of an insulated box with a glass covering and a reflecting mirror placed in such a position that it concentrates the rays of the

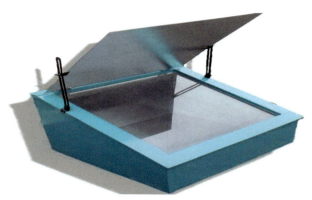

FIG. 1.2 Box type solar cooker. *(Adopted from Ghodake, D. (2016). A review paper on utilization of solar energy for cooking. Imperial International Journal of Eco-friendly Technologies.)*

sun. Fig. 1.2 shows a box type solar cooker (Bhawan & Puram, 2006b; Ghodake, 2016).

The payback period for the cooker is 3 to 4 years, and the cost of the cooker in India is around ₹3000 to ₹4000.

1.2.4.2 Parabolic concentrating type solar cooker

A parabolic concentrating type solar cooker consists of a paraboloid mirror with the cooking vessel placed on the focus point of the mirror (Kumaresan, Raju, Iniyan, & Velraj, 2015). It is used in baking and cooking food at high temperatures, as the cooker produces a higher temperature—up to 300°C—as compared to the box-type solar cooker. Fig. 1.3 shows a parabolic collector (Asif, 2017; Bhawan & Puram, 2006b).

The payback period is about six years and the cost of the cooker is around ₹3300 to ₹5000.

1.2.5 Solar water pumps

Solar water pumps use electricity produced by solar energy instead of conventional electricity to run the pumps. In this type of system, the photovoltaic array is connected to a motor-pump set. Applications include pumping water for agriculture, drinking, and other daily uses.

1.2.6 Solar space heating

The energy consumed by buildings is very high and still corresponds to about 40% of the final energy demand in most developed countries, out of which 22% is utilized by residential buildings and 18% by commercial ones. Residential buildings consume a large part of space heating worldwide, being responsible

FIG. 1.3 Parabolic collector. *(Adopted from Asif, M. (2017). Fundamentals and application of solar thermal technologies (Vol. 3, pp. 27–36). Elsevier.)*

for more than 50% in International Energy Agency (IEA) countries Directive 2002/91/EC of the European Parliament and of the Council, 2002; Directive, 2010; Laustsen, 2008).

Solar space heating can be categorized as active or passive:

1.2.6.1 Active space heating

In active solar space-heating systems, solar energy is used to heat a fluid or gas (liquid or air), this liquid (antifreeze liquid) is then stored in a storage tank for further use, or the heated air can be used directly for space heating. When liquid is used as a working fluid, then a pump is used, whereas a fan is used in the case of air.

1.2.6.2 Passive space heating

This type of system may be divided into several categories. In a direct-gain passive system, the floors or walls behave as a storage system and windows act as solar collectors, which means they are part of the occupied space. Thermal masses are used to absorb the solar radiation during the daytime and slowly release it during the night. Thermal masses are generally kept insulated from the outside environment and the ground to reduce heat losses (Sârbu, 2007).

In indirect-gain passive systems, the wall facing south or the roof is used to absorb solar radiation, thereby increasing the temperature, thus conveying heat into the building in various ways. The heat loss to the atmosphere from the wall is reduced by glazing and hence it improves the overall system efficiency (Sârbu, 2007).

1.3 Geothermal energy

1.3.1 Geothermal power generation

Geothermal energy is ecofriendly and has a wide range of applications. It is found in three forms: wet steam, dry steam, and hot water. There are four basic methods of extracting power from geothermal resources: direct steam power plants, single flash systems, double flash systems, and binary cycle power plants.

1.3.1.1 Direct steam power plants

Direct steam power plants generally require a very high geothermal temperature reservoir—greater than 455°F. Steam is taken out via production wells that are 3280 ft to two and a half miles underground to run the turbine (Zoet, Bowyer, Bratkovich, Frank, & Fernholz, 2011). Various impurities and particulates need to be removed from the steam.

1.3.1.2 Single flash system power plants

This type of system is used when the production well consists of a mixture of steam and water; a cylindrical cyclone separator is used to separate the steam and water. According to DiPippo (2015) and Valdimarsson (2021), the word "single" indicates a single flashing process in a mixture of steam and water: due to lowering the pressure of the mixture, some liquids convert into vapor, which occurs either in the production wells, reservoir, or cyclone inlet. The steam obtained is used to run the turbine. The remaining liquid, which does not flash into steam, is transferred to the reservoir for further use.

1.3.1.3 Double flash steam power plants

According to DiPippo (2015), double flash is an improvement on the single flash system and produces 25% more power output than a single flash for the same fluid conditions. In double flash, the liquid that remains in the first separator is converted to steam in the second separator. Generally, flash systems (single or double) require high-temperature geothermal reservoirs ranging from 300°F to 700°F (ZOET et al., 2011).

1.3.1.4 Binary cycle power plants

Binary cycle power plants utilize geothermal reservoirs ranging from 212°F to 302°F (ZOET et al., 2011). Binary cycle power plants use secondary fluids. The hot liquid from the reservoir vaporizes the secondary fluid at a lower temperature than the hot liquid in a heat exchanger, which runs the turbine.

1.3.2 Direct uses of geothermal energy

Geothermal energy is generally used as an indirect form, however, according to Lund and Boyd in the year 2015, out of the total worldwide installed capacity: 70.9% (55.15% annual energy use) used for ground-source heat pumps, 12.9% (20.18% annual usage) for swimming and bathing purposes (including balneology), 10.72% (14.9% annual use) for space heating, 2.78% (4.5% annual usage) for purpose of greenhouses and open heating, 0.98% (2.02% annual usage) for purpose of raceway and agricultural pond heating, 0.87% (1.76%) for industrial use applications, 0.51% (0.44%) for snow melting and cooling, 0.23% (0.34%) for agricultural drying, and 0.1% (0.24%) for other uses (Lund & Boyd, 2016). According to Lund, 350 million barrels of equivalent oil per year can be saved by direct usage, which reduces CO_2 by emissions about 148 million tonnes (Lund & Boyd, 2016).

Figs 1.4 and 1.5 shows direct application installed and usage worldwide, respectively (Lund & Boyd, 2016).

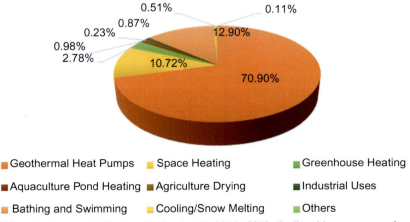

FIG. 1.4 Geothermal direct application (worldwide) in 2015, distributed by percentage of total installed capacity (MWt).

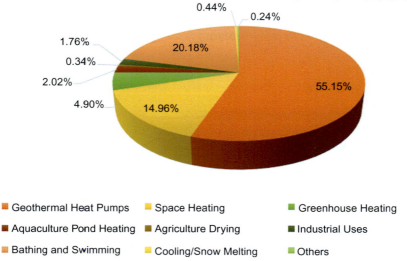

FIG. 1.5 Geothermal direct applications (worldwide) in 2015, distributed by percentage of total energy used (TJ/year).

1.4 Wind energy

The wind is a source of energy that humanity has utilized since ancient times. Historically, it was used to power sailing ships or spin windmills to pump water or air for blacksmiths, or run grinders to mill cereals (Mathew, 2017). The wind turbine can be broadly classified into two categories:

1.4.1 Horizontal Axis wind turbine

In this type of turbine, the axis of rotation is parallel to the ground. Most commercial wind energy uses this type of turbine, as they are more efficient and generate more power. The major advantage of this type of turbine is that we can control the rotor speed and power output by regulating blade pitch. In addition, blade pitch control protects the wind turbine against over speed when the wind speed is too high (Dewek, 2017). Based on the position of the rotor with respect to the tower, the rotors are classified as upwind (the rotor faces the wind and nacelle is at the back side relative to the wind directions) and downwind (the nacelle faces the wind and the rotor is at the back). This type of turbine requires a mechanism to move the rotor in the direction of the wind.

1.4.2 Vertical axis wind turbine

In this type of turbine, the axis of rotation is perpendicular to the ground. These are various types, like Darrius, Savonious, H-type, and V-shaped type. They generally have a very low tip speed ratio and power coefficient and are hence used where wind speed is low (Dewek, 2017). They are simple in design. Since the shaft is vertical, the generator is mounted on the ground. The shaft needs wires and cables for support and the power output cannot be controlled by pitching the rotor blades (Dewek, 2017). These are the reasons why they are not used often. However, these types of turbines generally do not have a yaw mechanism, as they can capture wind from any direction.

Fig. 1.6 shows the types of wind turbine (horizontal axis and vertical axis) and their components.

1.4.3 Wind turbine applications

- The wind turbines which are utility interconnected produce electricity that is synchronous with the grid, reducing utility bills by displacing the utility power used by households and by selling the excess power back to the electricity company.
- The DC current generated by wind turbines is used to charge batteries, thus, it helps remote homes (off the grid).

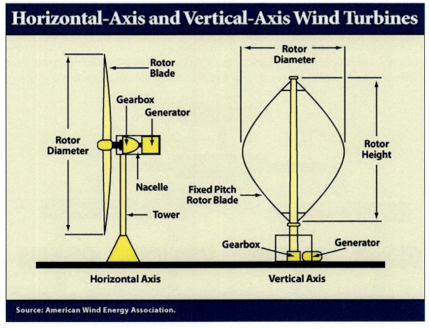

FIG. 1.6 Types of wind turbine. *(From Turbine Options: Vertical Axis vs. Horizontal Axis— UrbanWind. (2021). Retrieved May 30, 2021, from https://sites.google.com/a/temple.edu/ urbanwind/services/turbine-options-and-specifications.)*

- The three phase AC current generated by wind turbines is used in remote water pumping, which is suitable for running an electrical submersible pump directly. Residential or village scale wind turbines are suitable in a power range from 480 watts to 52 kW (Bhawan & Puram, 2006b).

1.5 Biomass energy

Biomass is an organic source of material having carbonaceous elements which are derived from plants and animals like animal residue, plant residue, agricultural crops and residue, sewage, municipal solid waste, etc.

1.5.1 Method of biomass energy extraction

- By direct combustion of biomass like wood.
- By thermochemical or biochemical methods, as shown in Fig. 1.7 (Ratho, 2020).

1.5.1.1 Pyrolysis

Pyrolysis is a Greek word in which the word "pyro" means fire and "lysis" means isolating or breaking. The process is carried out in the absence of air or partial presence of air and the biomass is heated at a very high temperature (normally above 430°C). All types of organic materials can be used as input, including rubber and plastics. The product obtained can be a gas mixture, solid residues like biochar, or liquid/oil form. Pyrolysis of wood is done in a restricted airflow, which yields charcoal, and the process is known as carbonization.

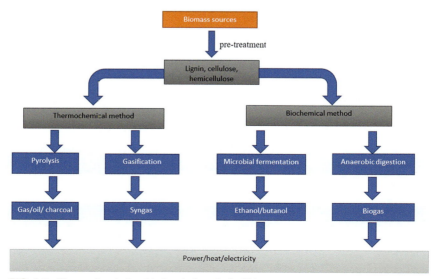

FIG. 1.7 Thermochemical or biochemical method of biomass energy extraction.

Applications of pyrolysis include:

- The process is utilized in various synthetic industries for the production of methanol, carbon, charcoal and other materials, and other organic materials (Ratho, 2020).
- Waste from the pyrolysis process, like a blend of stone, soil, pottery, and glass acquired can be utilized as a structural material (Ratho, 2020).
- It is used in some cooking techniques like barbecuing and caramelizing (browning of sugar) (Ratho, 2020).
- Biochar is used to increase soil fertility and recycling of agricultural waste. It can also be used in catalytic utilization, hydrogen energy storage, and protection of the environment, and is a sustainable platform for other applications where carbon is used as it has a very high content of carbon, Sometimes used as a sorbent for the removal of pollutants in water, biochar is also used as a sorbent for the removing pollutants in flue gas, such as SO_2 and NO_x (Ahmad et al., 2014).

1.5.2 Gasification

Biomass gasification is a process in which biomass is combusted with controlled air/oxygen/nitrogen/steam or a mixture of them to form producer gases (a mixture of N_2, CO, H_2, CO_2, and CH_4).

Applications of gasification include:

- These gases are combustible fuel, therefore, used to produce heat and electricity.
- Used for production of hydrogen and liquid fluid after further processing.
- It is burnt in boilers to get steam and heat.
- Can be used in dual-fuel mode, to run pumps for irrigation purposes, or coupled with a generator to produce electricity.
- Used in a gas turbine (Brayton cycle).
- The heat obtained from producer gas is used as a dryer for drying materials like flowers, tea, pottery, spices, etc.

1.5.3 Anaerobic digestion

This is a process in which biomass like cow dung, green plants, sewage sludge, etc., is decomposed by bacteria in the absence of oxygen to get biogas (mainly a mixture of CH_4 and CO_2). The slurry (mixture of biomass and water) is kept in a biogas plant for many months for decomposition.

Applications of biogas from anaerobic digestion include:

- Used in cooking.
- Lightning.

12 Sustainable developments by artificial intelligence & machine learning

- Converts organic wastes into usable products and emits less pollutants than fossil fuels, thus helping in environmental protection.
- Used in cogeneration cycle (CHP) to produce heat and electricity.
- Biogas has a high octane and low cetane value. The auto ignition temperature of biogas is higher than LPG and natural gas. Biogas has a much lower flame speed and a lower heating value than CNG and LPG fuel. These properties influence the application of biogas in SI engines (Qian, Sun, Ju, Shan, & Lu, 2017).
- Also used in CI engines.
- Used in gas turbine cycles.

1.5.4 Biofuels

Biofuels are liquid fuels that are clean and renewable sources of energy derived from biomass. Biofuels emit fewer pollutants and are mostly used in the transportation sector. Biofuel may be classified as first generation, second generation, and third generation, as shown in Table 1.1 (Dragone, Fernandes, Vicente, & Teixeira, 2010).

1.5.5 Bioethanol production

1.5.5.1 Sugar or starch fermentation

This is a process in which biomass (vegetable matter, agricultural waste, crops, sawdust and straw, pulp, molasses, etc.) is decomposed by yeast in the absence of

TABLE 1.1 Biofuels classifications.

First generation	Second generation	Third generation
Ethanol or butanol by fermentation process of raw material rich in sugar (sugarcane, sugar beet) or starch (barley, wheat, potato, corn, etc.). Biodiesel production by transesterification of oil crops (soyabeans, vegetable oils, palm, coconuts, animal fats, rapeseeds)	Bioethanol and biobutanol are produced from sugar, starch, and oil (jatropha, pongamia, cassava, miscanthus); the production method is the same as that of first generation. Lignocellulose (grass, straw, wood bark) fermentation, which is a different process from the conventional, is used to produce bioethanol, biobutanol, and syndiesel	Microalgae is used as raw material for the production of bioethanol and biobutanol. Bioethanol is also produced from seaweed. Hydrogen is produced from microbes and microalgae

Application of alternative clean energy **Chapter | 1** **13**

air to get ethyl alcohol (ethanol) as the product. Bioethanol is a clean, colorless, and flammable liquid. The basic conversion of glucose is as shown below:

$$C_6H_{12}O_6 \rightarrow 2C_2H_5OH + 2CO_2$$

$$Glucose \rightarrow Ethanol + Byproduct$$

1.5.5.2 Bioethanol from lignocellulose fermentation

Lignocellulose is found in grass, tree bark, and straw, and consists of sugar and hemicellulose. In order to separate sugar from hemicellulose requires pretreatment and hydrolysis, so it's a very complex process to get sugar from bioethanol, unlike simple sugar or starch fermentation. It is a four-step process: pretreatment, hydrolysis, fermentation, and purification. The basic steps are shown below:

$$Pretreatment \rightarrow Hydrolysis \rightarrow Fermentation \rightarrow Purification$$

Bioethanol is a clean and renewable source of energy and is mostly used in the transport sector, and emits less pollution. Bioethanol contains approximately 35% oxygen. The addition of bioethanol to gasoline improves combustion by reducing particulates, nitrogen oxides (NOx), CO, and unburned hydrocarbons from exhaust emissions, thereby reducing the consumption of fuel and GHG (Balat & Balat, 2009; Gavrilescu & Chisti, 2005). The blending of bioethanol with gasoline is very economical because 1 L of bioethanol is equivalent to 0.72 liters of gasoline (Kim & Dale, 2004). Mostly the blending is done in five categories: E0, E10 (90% gasoline + 10% ethanol), E20 (80% gasoline + 20% ethanol), E30 (70% gasoline + 30% ethanol), and E85 (15% gasoline + 85% ethanol). However, bioethanol blending decreases the viscosity and increases the acid number, thereby increasing the wear and corrosion. E10 has less effect on the wear and lubricating properties of engine oil (Khuong et al., 2017).

1.5.6 Biodiesel

Biodiesel is alkyl esters (methyl or ethyl esters) made from vegetable oils (soya bean, sunflower, peanuts) and animal fats. It is made by a process known as transesterification, in which vegetable oil or animal fats get modified with alcohols (methanol or ethanol) in the presence of an acid catalyst to produce esters and glycerine. Methanol is the most widely used alcohol. The reaction is shown in Fig. 1.8 (Koberg & Gedanken, 2013).

Biodiesel is mostly used in diesel engines. It can reduce the emissions of various pollutants like CO, CO_2, SO_2, and hydrocarbons, thus providing a clean source of energy. However, NOx increases slightly if maintenance of the engine is not performed properly.

The national biofuel policy of India—2018, has an aim of reaching 20% ethanol blending and 5% biodiesel blending by the year 2030.

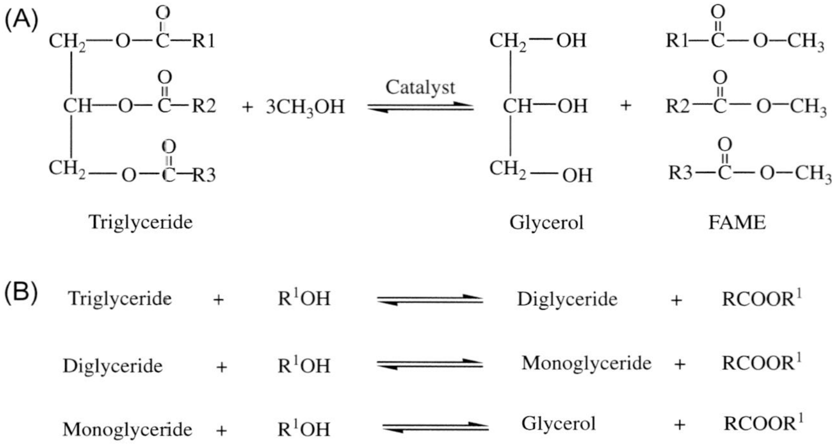

14 Sustainable developments by artificial intelligence & machine learning

(A)

$$
\begin{array}{c}
CH_2-O-\overset{O}{\overset{\|}{C}}-R1 \\
| \\
CH-O-\overset{O}{\overset{\|}{C}}-R2 \quad + \quad 3CH_3OH \quad \underset{}{\overset{Catalyst}{\rightleftharpoons}} \quad
\begin{array}{c} CH_2-OH \\ | \\ CH-OH \end{array} \quad + \quad
\begin{array}{c} R1-\overset{O}{\overset{\|}{C}}-O-CH_3 \\ R2-\overset{O}{\overset{\|}{C}}-O-CH_3 \end{array} \\
| \\
CH_2-O-\overset{O}{\overset{\|}{C}}-R3 \quad\quad\quad CH_2-OH \quad\quad R3-\overset{O}{\overset{\|}{C}}-O-CH_3
\end{array}
$$

Triglyceride   Glycerol   FAME

(B)

Triglyceride + R^1OH $\rightleftharpoons$ Diglyceride + $RCOOR^1$

Diglyceride + R^1OH $\rightleftharpoons$ Monoglyceride + $RCOOR^1$

Monoglyceride + R^1OH $\rightleftharpoons$ Glycerol + $RCOOR^1$

FIG. 1.8 Transesterification reaction. *(From Koberg, M. K., & Gedanken, A. G. (2013). Using Microwave Radiation and SrO as a Catalyst for the Complete Conversion of Oils, Cooked Oils, and Microalgae to Biodiesel. In New and Future Developments in Catalysis: Catalytic Biomass Conversion. Elsevier B.V. https://doi.org/10.1016/B978-0-444-53878-9.00010-2.)*

1.6 Ocean and tidal energy

The ocean accounts for nearly 70% of the surface area of the earth and also contains a large amount of energy in the form of thermal and mechanical. Thermal energy is due to heat absorbed by the ocean from the sun and mechanical energy is stored in the form of tidal and ocean currents, due to the interaction of the Earth's moon and the solar system. To harness power from ocean energy, basically, three types of conversion systems are used:

- OTEC (ocean thermal energy converter);
- WEC (wave energy converter);
- TIC (tidal energy converter).

1.6.1 Wave energy

The sun is responsible for the wind blowing; the uneven heating of the ocean surface and the ground creates differences and thus causes the wind to flow. This wind, when flowing over the ocean surface, causes ocean waves. In addition, there are some external forces like earthquakes, and the gravitational force of attraction of the moon and sun, which cause ocean waves. These ocean waves can be harnessed to generate electricity. The energy density formed by wave farms is higher than solar parks and wind farms. For harnessing power from waves, four main types of devices are used: attenuators, point absorbers, terminators or oscillating water columns, and overtopping devices (Irrizary-Rivera, Colluci-Rios, & O'Neill-Carrillo, 2009).

1.6.2 OTEC

The temperature difference between warm surface ocean water (26–29°C) and deep ocean water (5–7°C), at around 800 to 1000 m deep, should be around the 20°C required to run a heat engine under the Rankine cycle. In open cycle OTEC, the warm water is utilized to evaporate a low boiling temperature substance like ammonia, and the vapor obtained is utilized to run the turbine for power generation. The cold deep seawater is used to condense the fluid and recirculate, whereas, in an open cycle system, a vacuum chamber is used to make hot surface water flash evaporate. The low-pressure steam generated is used to run a large diameter turbine and is condensed by circulating the cold deep seawater.

1.6.3 TIC

The difference between high tide and low tide, which should be a minimum of 5 m, is used to get power from tidal energy. For this, a dam or barrage should be made in a shallow area like a gulf, bay, etc. Power is generated when the water flows through turbines located in the barrage separating an estuary from the sea or by running the turbines freely on the tidal current, which is located in creeks or channels. Turbines with casings/shrouds generate three to four times more power than an open flow or free stream turbine (Ravindran & Raju, 2015).

Apart from electricity generation, ocean energy is also used for many purposes, like desalination of water, refrigeration, and conditioning, pumping of water. Ocean waves can be used for the desalination of water. This can be achieved either directly or indirectly. In an indirect system, the ocean wave energy is converted into electrical power to run the desalination plant, whereas the direct system converts ocean wave energy into pressure energy and directly runs the process (Leijon & Boström, 2018).

OTEC is also used for refrigeration and conditioning by supplying deep sea cold water. OTEC is used to desalinate water. The freshwater can be obtained when steam after expansion in the turbine is made to condense in the condenser by cold seawater, thus helping tropical areas where scarcity of water is present and ocean surface temperature is high by producing both power and desalinated water.

The OTEC also helps in the extraction of minerals.

1.7 Small, micro, and mini hydro plants

A small hydropower plant (SHP) is a clean (Kumar & Saini, 2021), reliable, and green source of alternative energy. A major application of SHP is for eco-friendly electricity generation. According to The World Small Hydropower Development Report (WSHPDR) 2019 report, the total installed capacity of SHP was around 78GW. WSHPDR further stated the contribution of SHP

16 Sustainable developments by artificial intelligence & machine learning

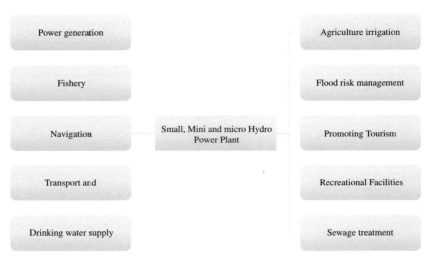

FIG. 1.9 The wide applications of hydropower projects. *(Data from Pollitt, D. (2018). The Uses of Hydropower Energy. https://bizfluent.com/info-8419828-uses-hydropower-energy.html (Original work published 2018).)*

was 1.5% for world electricity generation (United Nations Industrial Development Organization (UNIDO), 2019, World Small Hydropower Development Report 2019—Case Studies, 2019). Apart from power generation, SHP has some other applications such as irrigation, flood management, recreation facilities, and tourism (Pollitt, 2018; Salian, 2021).

The wide variety of applications for small, mini, and micro-hydropower plants has been represented in the Fig. 1.9 (Pollitt, 2018; Salian, 2021).

1.8 Case study

Bhandari, Basnet, Pokharel, Shrestha, and Baral (2020) studied the multipurpose small hydro plant at Padhu Khola River in Nepal. The purpose of this project was to understand the other applications of small hydro apart from power generation. The major concern for their study was to understand the possibility of the development of drinking water plants and freshwater aquaculture, which was found to have a benefit to cost ratio of 1.97 and 1.98, respectively. They further found that other facilities can be developed, such as tourism, local employment, and forestry.

Kucukali, Al Bayatı, and Maraş (2021) conducted screening on an existing irrigation dam to mark a potential site for future small hydro development in Turkey. The study was evaluated by selecting criteria such as spatial characteristics, grid connection, and dam characteristics with a geographical information system. Karadere dam was found to be the best potential site for future development.

Worku (Bezabih, 2021) evaluated a potential site for the development of small hydro for rural electrification on the Ribb dam in Ethiopia. The purpose of his study was to raise awareness among researchers, scientists, and governments so that they can establish a small hydro plant for the development of rural areas in countries such as Ethiopia, which is targeting to become a middle-income country by 2025.

Raczyński (2020) studied a multipurpose dam regarding the incidence of low flow and flood on the Wieprz River in Poland. The investigation was carried out in the hydrological period 1976–2014. The result showed that small multipurpose hydro has very little effect on floods.

Shao et al. conducted a study to find a potential site for the development of a multipurpose project using the dam suitability stream model (DSSM). Every input constraint was weighted and the investigation carried out using a geographical information system (GIS). The objective of their study was to find a potential site for a variety of applications including reducing floods, household applications, etc., at the local and regional levels (Shao, Jahangir, Yasir, Atta-Ur-Rahman, & Mahmood, 2020).

García, Díaz, Morillo, and McNabola (2021) studied the utilization of microhydro as sewage treatment and power generation in private and municipal industries in Spain. Munaaim, Hamidin, Ayob, Sani, and Salleh (2016) also studied the co-utilization of microhydro in Malaysia. This technology can help to generate power and reduce greenhouse gases simultaneously.

1.9 Conclusion

Fossil fuels are depleting day by day and are disastrous for our ecosystem. The GHG content in the environment increases rapidly, and CO_2 and CH_4 content increases by 147% and 259% above the preindustrial phase, therefore, it is necessary to increase the proportion use of clean energy rapidly. Furthermore, fossil fuels are not present uniformly all over the entire world, many countries do not have enough resources of fossil fuels, so energy is not equally distributed. This is not the case with clean energy sources, as they is found almost everywhere, which strengthens socioeconomic life. This chapter consists of various applications of clean sources of energy such as solar, wind, biomass, hydro, etc. These various applications to help us to lower the dependency on conventional fuel as well emissions aspects. ACEs possess wide application, yet they have some limitations, such as not being consistently available, such as solar and wind. The electricity generation capacity is still not large enough. These energies can be unreliable, have low-efficiency levels, require a huge upfront capital outlay, take a lot of space to install, have expensive storage costs, etc.

References

Abed, Q. A. (2021). *Some solar energy technologies and applications.* Retrieved June 2, 2021, from: https://www.researchgate.net/publication/304784549_Some_Solar_Energy_Technologies_and_Applications.

18 Sustainable developments by artificial intelligence & machine learning

Ahmad, M., Rajapaksha, A. U., Lim, J. E., Zhang, M., Bolan, N., Mohan, D., et al. (2014). Biochar as a sorbent for contaminant management in soil and water: A review. *Chemosphere*, *99*, 19–33. https://doi.org/10.1016/j.chemosphere.2013.10.071.

Asif, M. (2017). *Fundamentals and application of solar thermal technologies. Vol. 3* (pp. 27–36). Elsevier.

Balat, M., & Balat, H. (2009). Recent trends in global production and utilization of bio-ethanol fuel. *Applied Energy*, *86*(11), 2273–2282. https://doi.org/10.1016/j.apenergy.2009.03.015.

Bezabih, A. W. (2021). Evaluation of small hydropower plant at Ribb irrigation dam in Amhara regional state, Ethiopia. *Environmental Systems Research*, *10*(1). https://doi.org/10.1186/s40068-020-00196-z.

Bhandari, A., Basnet, K., Pokharel, N., Shrestha, K. K., & Baral, N. P. (2020). Assessment of the multi functionality of a small hydropower project at Padhu Khola, Kaski, Nepal. *Technical Journal*, 30–39. https://doi.org/10.3126/tj.v2i1.32827.

Bhawan, S. B., & Puram, R. K. P. (2006a). *Energy performance assessment for equipment and utility systems: Chapter 12—Application of non-conventional & renewable energy sources*. Bureau of Energy Efficiency.

Bhawan, S., & Puram, R. K. (2006b). Energy performance assessment for equipment and utility systems: Chapter 12—Application of non-conventional & renewable energy sources. In *Bureau of energy efficiency* (pp. 147–161).

Dewek, D. (2017). *Wind energy wind energy. Vol. 10*. https://doi.org/10.1016/B978-0-12-804448-3/00004-9.

DiPippo, R. D. (2015). *Geothermal power plants: Principles, applications, case studies and environmental impact*. 4th ed. https://doi.org/10.1016/C2014-0-02885-7.

Directive. (2010). *31/EU of the European parliament and of the council of 19 may 2010 on the energy performance of buildings*.

Directive 2002/91/EC of the European Parliament and of the Council. (2002). *On the energy performance of buildings*.

Dragone, G., Fernandes, B., Vicente, A. A., & Teixeira, J. A. (2010). *Third generation biofuels from microalgae*.

García, A. M., Díaz, J. A. R., Morillo, J. G., & McNabola, A. (2021). Energy recovery potential in industrial and municipal wastewater networks using micro-hydropower in Spain. *Water (Switzerland)*, *13*(5). https://doi.org/10.3390/w13050691.

Gavrilescu, M., & Chisti, Y. (2005). Biotechnology—A sustainable alternative for chemical industry. *Biotechnology Advances*, *23*(7–8), 471–499. https://doi.org/10.1016/j.biotechadv.2005.03.004.

Ghodake, D. (2016). A review paper on utilization of solar energy for cooking. *Imperial International Journal of Eco-friendly Technologies*.

Ierodiakonou, D., Zanobetti, A., Coull, B. A., Melly, S., Postma, D. S., Boezen, H. M., et al. (2016). Ambient air pollution. *Journal of Allergy and Clinical Immunology*, *137*(2), 390–399. https://doi.org/10.1016/j.jaci.2015.05.028.

IPCC. (2018). Framing and context. In *Global warming of 1.5°C*. https://www.ipcc.ch/site/assets/uploads/sites/2/2019/05/SR15_Chapter1_Low_Res.pdf.

Irrizary-Rivera, A. A., Colluci-Rios, J. A., & O'Neill-Carrillo, E. (2009). *Achieveable renewable energy targets for Puerto Rico's renewable energy portfolio standard* (pp. 1–35).

Khuong, L. S., Masjuki, H. H., Zulkifli, N. W. M., Mohamad, E. N., Kalam, M. A., Alabdulkarem, A., et al. (2017). Effect of gasoline-bioethanol blends on the properties and lubrication characteristics of commercial engine oil. *RSC Advances*, *7*(25), 15005–15019. https://doi.org/10.1039/c7ra00357a.

Kim, S., & Dale, B. E. (2004). Global potential bioethanol production from wasted crops and crop residues. *Biomass and Bioenergy, 26*(4), 361–375. https://doi.org/10.1016/j.biombioe.2003.08.002.

Koberg, M. K., & Gedanken, A. G. (2013). Using microwave radiation and SrO as a catalyst for the complete conversion of oils, cooked oils, and microalgae to biodiesel. In *New and future developments in catalysis: Catalytic biomass conversion* Elsevier B.V. https://doi.org/10.1016/B978-0-444-53878-9.00010-2.

Kucukali, S., Al Bayatı, O., & Maraş, H. H. (2021). Finding the most suitable existing irrigation dams for small hydropower development in Turkey: A GIS-Fuzzy logic tool. *Renewable Energy, 172*, 633–650. https://doi.org/10.1016/j.renene.2021.03.049.

Kumar, K., & Saini, R. (2021). Application of artificial intelligence for the optimization of hydropower energy generation. In *EAI endorsed transactions on industrial networks and intelligent systems* (p. 170560). https://doi.org/10.4108/eai.6-8-2021.170560.

Kumaresan, G., Raju, G., Iniyan, S., & Velraj, R. (2015). CFD analysis of flow and geometric parameter for a double walled solar cooking unit. *Applied Mathematical Modelling, 39*(1), 137–146. https://doi.org/10.1016/j.apm.2014.05.010.

Laustsen, M. J. (2008). Energy efficiency requirements in building codes, energy efficiency policies for new buildings IEA information paper. In *Support of the G8 plan of action*.

Leijon, J., & Boström, C. (2018). Freshwater production from the motion of ocean waves—A review. *Desalination, 435*, 161–171. https://doi.org/10.1016/j.desal.2017.10.049.

Lund, J. W., & Boyd, T. L. (2016). Direct utilization of geothermal energy 2015 worldwide review. *Geothermics, 60*, 66–93. https://doi.org/10.1016/j.geothermics.2015.11.004.

Mathew, S. (2017). Wind energy. In *Vol. 10. 13th Ger. Wind Energy Conf* (pp. 1–158). https://doi.org/10.1016/B978-0-12-817012-0.00030-X.

Munaaim, M. A. C., Hamidin, N., Ayob, A., Sani, N. M., & Salleh, A. M. (2016). *Application of MICRO hydro electric system attached to effluent discharge point of sewerage treatment plant*. https://doi.org/10.5176/2251-189x_sees16.33.

Pollitt, D. (2018). *The uses of hydropower energy*. https://bizfluent.com/info-8419828-uses-hydropower-energy.html (Original work published 2018).

Qian, Y., Sun, S., Ju, D., Shan, X., & Lu, X. (2017). Review of the state-of-the-art of biogas combustion mechanisms and applications in internal combustion engines. *Renewable and Sustainable Energy Reviews, 69*, 50–58. https://doi.org/10.1016/j.rser.2016.11.059.

Raczyński, K. (2020). Influence of a multipurpose retention reservoir on extreme river flows, a case study of the Nielisz reservoir on the Wieprz River (Eastern Poland). *Water Resources, 47*(1), 29–40. https://doi.org/10.1134/S0097807820010091.

Ratho, B. (2020). Biomass extraction of energy transformation. *Journal of Advanced Research in Power Electronics and Power Systems, 07*(1 & 2), 1–6. https://doi.org/10.24321/2456.1401.202001.

Ravindran, M., & Raju, V. S. (2015). Ocean energy. *Proceedings of the Indian National Science Academy, 81*(4), 983–991. https://doi.org/10.16943/ptinsa/2015/v81i4/48306.

Salian, P. (2021). *Hydroelectric power plant—Classification, working & applications*. Retrieved May 31, 2021, from: https://electricalfundablog.com/hydroelectric-power-plant/.

Sârbu, I. (2007). *Solar water and space-heating systems*. https://doi.org/10.1016/B978-0-12-811662-3.00005-0.

Shao, Z., Jahangir, Z., Yasir, Q. M., Atta-Ur-Rahman, & Mahmood, S. (2020). Identification of potential sites for a multi-purpose dam using a dam suitability stream model. *Water (Switzerland), 12*(11). https://doi.org/10.3390/w12113249.

United Nations Industrial Development Organization (UNIDO). (2019). *World small hydropower development report 2019—Case studies*. United Nations Industrial Development Organization.

Valdimarsson, P. (2021). *Power, geothermal, plant, & components.*

World Meteorological Organization and Global Atmosphere Watch. (2019). *WMO greenhouse gas bulletin (GHG bulletin).* World Meteorological Organization.

Zoet, A., Bowyer, D. J., Bratkovich, D. S., Frank, M., & Fernholz, K. (2011). *Geothermal 101: The basics and applications of geothermal energy.*

Chapter 2

Optimization of hybrid energy generation

Poonam B. Dhabai and Neeraj Tiwari
Poornima University, Jaipur, India

2.1 Introduction

The consciousness of saving energy has been elevated among the public, thus building the utilization of solar, wind, biomass, and hydropower energies. India is experiencing an expansion of RES (renewable energy sources) in the energy sector, with an aim of 175 GW by 2022. Wind power, as the largest, and solar, as the fastest growing renewable energy source, have established their role in electricity production. RES possesses a number of advantages, but being the most intermittent resource of energy drills a hole in the bucket of advantages (Burke & Malley, 2008). Wind power production habitually includes uncertainties owing to the stochastic temperament of wind speeds, leading to disparity in the nonlinear power curve; day-night and seasonal variation in insolation and irradiance level also hamper solar output, to an extent (Karki, Hu, & Billinton, 2006). Therefore, (Celeska & Najdenkoski, 2005) analyzing these uncertainties for optimized hybrid generation is an imperative part of any assessment for the extensive-term energy production of any RES farm. Uncertainty scrutiny is a methodical progression to propagate uncertainty in a model, which involves a careful statistical and quantification approach (Sjodin, Gayme, & Topcu, 2012). Analysis of these uncertainties can be conducted through different approaches which embrace the established methods, such as Monte-Carlo, Taylor series, first-order second moment, two-point estimate methodology, PDF and CDF analogy, etc. (Ahmadi & Ghasemi, 2011; Morgan, 2010). A statistical approach establishes the ultimate footprint of the uncertainties. To analyze the uncertainties in the wind, PDF and CDF based on Weibull, Rayleigh, or lognormal parameters give the best fit (Ahmadi & Ghasemi, 2011); whereas modeling of PV module (Jamil, Zhao, Zhang, Rafique, & Jamil, 2019), and analyzing the power output based on insolation level best establishes the solar-based uncertainties (Albadi, El-Rayani, El-Saadany, & Al-Riyami, 2017).

Sustainable Developments by Artificial Intelligence and Machine Learning for Renewable Energies.
https://doi.org/10.1016/B978-0-323-91228-0.00001-X
Copyright © 2022 Elsevier Inc. All rights reserved.

As the percentage of renewable energy capacity on the stabilized grid is mounting exponentially, the issues linked to RES integration into an established network become greater (Ela et al., 2010). For assessing and maintaining an unbiased and balanced state among the generation, transmission, and the demand on the power grid, the operators have got to be alert on a prior basis of instantaneous availability of RES, the expectation of RES, and response to variation in RES (Dhabai & Tiwari, 2020a, 2020b, 2020c). This is an easier said than done task for the operators, due to the ample variation in location and size of RES across the power grid (Lowery & O'Malley, 2014). In the worst case scenario, an alteration in the power flow pattern of the transmission lines due to a discrepancy in generation perspective could possibly violate the system limits (Kane & Ault, 2014). OPF (optimal power flow), DCPF (distributed continuation power flow), and ACPF (alternating current power flow) methods are employed to record and verify the variation in power flow with respect to post-integration of stochastic and intermittent RES generation (Banerjee, Jayaweera, & Islam, 2012).

The occurrence of congestion in the system owing to changes in power flow of transmission lines and operation under stressed circumstances (Christie, Wollenberg, & Wangensteen, 2000) introduces the congestion cost element into the LMP (locational marginal pricing) (Morales, Conejo, & Pérez-Ruiz, 2011) of the bus nodes fluctuating the market options (Neuhoff, Boyd, & Grau, 2011), which leads to an assortment of challenges in the computation of LMP, including congestion cost. ACOPF (alternating current optimal power flow), 3-parameters, decision tree model, and P2P (peer-to-peer) methodologies are utilized to assess the LMPs of the system with or without transmission constraints (Nwulu, 2018). The network reliability parameters of the transmission lines ensure the secure and reliable operation of the network. The real-time market state of affairs directly relies on TTC (total transfer capacity), ATC (available transfer capability), and TRM (transmission reliability margin) values (in MW) (Bhesdadiya & Patel, 2014; Prabha & Venkataseshaiah, 2009). These parameters oscillate with respect to deviation in power flow of the lines, which in turn is adversely varied due to uncertain generation, and are a function of time and network state. Long-established methods like Monte-Carlo, PLF, ACPTDFs, CEED, parametric bootstrap, and auction-based methods like EDRP and DADRP, deterministic methods such as RPF, CPF, DCPF, and intelligent methods ANN and PSO contribute to attaining values of these reliability parameters. The insertion of RES into the lattice frequently leads to the congestion within the system. By analyzing the power flow of the lines, the LMPs of the buses (Daneshi & Srivastava, 2011; Morstyn, Teytelboym, Hepburn, & McCulloch, 2020; Umale & Warkad, 2016), and the reliability parameters we can supervise and manage the congestion within the system without jeopardizing the safe and sound operation mode of the network.

Accessibility of high RES attracts consumers to pay money for the electricity from the cheaper source, leading to congestion possibilities in transmission

Optimization of hybrid energy generation **Chapter | 2** **23**

lines (Bohn, Caramanis, & Schweppe, 1984; Shahidehpour, Yamin, & Li, 2002). This causes competition between producers as well as the consumers. At the point when the producers and consumers of electricity, respectively, want to generate and consume power in amounts that lead the transmission network to function at or ahead of one or more transfer limits, the network is then said to be congested. Therefore, congestion management is controlling the power flow of transmission lines of the network so that transfer limits are monitored and maintained within the limits. It is conceivably an elementary transmission management problem (Daneshi & Srivastava, 2011). The term congestion management relates to the economics of the power system. Hence, optimization of hybrid generation can be attained by managing congestion episodes in the system (Kothari & Dhillon, 2004; Shayesteh, Parsa Moghaddam, Haghifam, & Sheikh-EL-Eslami, 2009). Optimization of RES includes the complete framework, right from availability and assessment of RES, finding out the optimized location for integration, impact on the power flow of the lines, and variation in LMPs (Dhabai & Tiwari, 2020a), to estimating the ATC and TRM of the network (Ahmadi & Ghasemi, 2011; Fangxing & Bo, 2008; Othman & Musirin, 2011).

The optimization of hybrid generation (wind + solar + conventional) framework is subdivided into three segment analyses carried out on a standard IEEE 30 bus system. A brief statistical analysis dealing with the uncertainties in wind and solar is also presented. The first segment of optimization focuses on the power flow patterns of the transmission lines post RES integration with the help of three linear sensitivity factors, that is, GSDF (generation perspective), PTDF (transmission perspective), and LODF (consumer/fault perspective). The second segment deals with the LMPs (lambda) of the buses, including the congestion cost component to it, if any. The third segment estimates the reliability parameters (TTC, TRM) useful to determine the ATCs that can be further extended to market policies. These segments are studied together for the optimized location of hybrid generation to manage congestion.

2.2 RES data and uncertainty statistical analysis

To examine a realistic, practical, and real-time case, the authentic wind and solar data of the Pune (India) region is considered. The windowpane of the data inward is from 1st January to 31st December. The data was provided by the Indian Meteorological Department (IMD), Pune, which is the nationwide meteorological sector of the country and the prime government group in all divisions including weather, meteorology, seismology, etc. It provides real-time scenario and also forecasts meteorological data for the most constructive process of the weather-sensitive recital. It conducts and promotes research in meteorology and associated disciplines. The data and uncertainty analysis of wind and solar is carried out statistically.

24 Sustainable developments by artificial intelligence & machine learning

2.2.1 Wind source analysis

Real-time wind speed data was recorded and collected at a regular interval period of 3 h for every day of a complete year. The recorded wind speed data was measured in Km/h and then utilized by converting it into m/s for analysis (Khatavkar, Mayadeo, Dhabai, & Dharme, 2017). The collected data was also scrutinized on a monthly and hourly basis. It was from the PDF that the wind speed follows the normal distribution. The mean and variance for every month are computed using:

$$\mu = \ln \left[\frac{U}{\sqrt{\left[1 + \left(\frac{S^2}{U^2} \right) \right]}} \right] \tag{2.1}$$

$$\sigma = \sqrt{\left(\ln \left[1 + \left(\frac{S^2}{U^2} \right) \right] \right)} \tag{2.2}$$

where μ and σ are the mean and the variance of the natural logarithm, respectively, of the U, and S is the standard deviation.

Table 2.1 shows the minimum and maximum speed and variance of wind data for every month.

TABLE 2.1 Wind speed data.

Month	Minimum wind speed (m/s)	Maximum wind speed (m/s)	Variance (σ)
January	3.04	12.56	23.66
February	3.00	11.58	21.45
March	3.04	12.01	23.02
April	3.01	13.04	24.23
May	3.03	15.08	25.64
June	3.09	20.02	52.33
July	3.07	19.85	48.05
August	3.01	18.53	46.21
September	3.00	14.58	37.65
October	3.03	12.33	23.09
November	3.04	11.06	20.87
December	3.04	13.33	25.22

Maximum and minimum wind speed data for 12 months with variance.

Optimization of hybrid energy generation **Chapter | 2   25**

TABLE 2.2 Technical data—GE 1.5 SLE wind turbine generator set.

Parameter	Value
Cut—in speed	3.5 m/s
Cut—out speed	25 m/s
Rated speed	14 m/s
Area Swept	4567 m^2
Rotor Diameter	77 m
Power Output	1.5 MW

Comparison of accessed data to technical set.

The corresponding wind speeds were equated to wind power output by means of the cubic power-speed relation obtained from the nonlinear power curve. The power output equations are:

$$P_W = \frac{1}{2}\rho V^3 A \qquad (2.3)$$

$$P_{W=} 0 \, V \le V_{in} \qquad (2.4)$$

$$P_W = P_{W(rated)} \, V_r \le V_{in} \qquad (2.5)$$

$$P_W = \frac{(V - V_{in})}{(V_r - V_{in})} P_{W(rated)} \, V_{in} \le V_r \qquad (2.6)$$

The obtained power outputs were then compared to a practical wind turbine generator set, GE 1.5 SLE. Table 2.2 shows the technical data of the set.

From the derived mean and variance, 1000 random samples of power output corresponding to 1000 wind speeds are generated every month for the analysis. The distribution considered for generated power output samples is normal.

2.2.2   Solar source analysis

Solar irradiance in real time and practical data was recorded from dawn, at 05.00h, to dusk, at 18.00h, at intervals of 1 h for the whole year. The accessible data was scrutinized on a monthly and hourly basis. The solar insolation data was recorded in MJ/m^2 and later it was converted to W/m^2 for the analysis. After the conversion, the power output was derived with the help of the equation relating to solar insolation and power output. It was observed from the PDF that the solar insolation nature follows the normal distribution (Sathya, 2014). The mean and variance of the power output was calculated. The standard and established power output equations are represented as below:

26 Sustainable developments by artificial intelligence & machine learning

$$I = N_P I_{PH} - N_P I_S \left[\exp\left(\frac{q(V/N_S + IR_S)/N_P}{kT_CA} \right) - 1 \right] - (N_PV/N_S + IR_S)/R_{SH}$$

$$(2.7)$$

$$I_{PH} = \left[I_{SC} + K_1 \left(T_C - T_{ref} \right) \right] \lambda \qquad (2.8)$$

$$I_S = I_{RS} \left(\frac{T_C}{T_{ref}} \right)^3 \exp\left[\frac{qE_G \left(\frac{1}{T_{ref}} - \frac{1}{T_C} \right)}{kA} \right] \qquad (2.9)$$

Table 2.3 shows the minimum and maximum solar irradiance levels of all the months.

To have a sensible rhythm to the system, a genuine solar generator model is taken into consideration to calculate power output at various insolation levels. The solar generator model is an "OEM 10 kit" (Solar Data Sheet, Omnisite, 2009). The technical data is shown in Table 2.4.

According to the obtained mean and variance, 1000 random samples of power output equivalent to 1000 solar insolations are derived for every month. The distribution of generated power output samples is normal. The PDF and CDF distribution of the data obtained was in a math-wave environment,

TABLE 2.3 Solar insolation data.

Month	Minimum solar insolation (KW/m^2)	Maximum solar insolation (KW/m^2)	Variance (σ)
January	13.09	17.84	11.51
February	13.11	18.55	10.45
March	13.85	18.65	13.10
April	14.56	22.35	22.41
May	19.59	30.55	35.12
June	07.12	24.25	30.65
July	04.54	18.69	18.25
August	05.00	14.65	14.11
September	10.89	16.22	07.08
October	12.96	12.10	05.20
November	12.88	11.02	02.03
December	13.00	15.09	02.68

Comparison of accessed data to technical set.

Optimization of hybrid energy generation Chapter | 2 **27**

TABLE 2.4 Technical parameters of OEM 10 Kit.

Parameter	Value
Rated voltage	16 V
Rated current	0.61 A
Rated power	10 W
Dimensions	15.18″ × 13.28″

Maximum and minimum solar insolation for 12 months with variance.

obtaining by the mean and variance of the data points. The random power output samples, which are normally distributed along the mean, reflect the uncertainty in the RES generation. These random samples are used as RES generation for further analysis and the impact of these uncertainties is further analyzed for different parameters. The uncertainty in RES is also guided by the location impact.

2.3 Test case modifications and solution methodology

2.3.1 Test case modifications

Power flow analysis is the bakbone of power system design, analysis, and operation (Dharmjit & Tanti, 2012). The major purpose of power flow investigation is to find the magnitude and phase angle of voltage on each bus and the real and reactive power flowing on each transmission line (Table 2.5). For the analysis, the IEEE 30 bus system is divided into two areas (Table 2.6) with six connecting tie lines. The tie lines are used for power exchange within the two areas. There are six tie lines in the interconnected system (Table 2.7). Single line diagram,

TABLE 2.5 Conventional generator data of IEEE 30 bus system.

Bus number	P_G	Q_G
1	23.54	0
2	60.97	0
22	21.59	0
27	26.91	0
23	19.2	0
13	37.0	44.7

Real and reactive power.

28 Sustainable developments by artificial intelligence & machine learning

TABLE 2.6 Area details.

Parameter	Area 1	Area 2
Buses	12,13,14,15,16,17,18,23,25, 26,27,29,30	1,2,3,4,5,6,7,8,9,10,11,19, 20,21,22,24,28
Generators	Bus number-27,23,13	Bus number-22,2,1
Total generation (MW)	83.11	106.1
Total Load(MW)	70.5	118.7

Two area briefing.

TABLE 2.7 Tie line details.

Tie line	From bus to bus	Line number
T-1	4–12	15
T-2	10–17	26
T-3	18–19	23
T-4	23–24	32
T-5	24–25	33
T-6	28–27	36

Details of tie lines connecting both the areas.

generator, bus (Table 2.8), and line data (Table 2.9) of a standard IEEE 30 bus system is as shown in Fig. 2.1.

2.3.2 Configuration of cases

Five different cases were cnsideredd with different locations of wind and solar generators. Table 2.10 reflects all the cases considered.

Case-1 is the base case, where only conventional generation is considered with no RES. The remaining cases include RES at different locations. First, the study is carried out with different locations of wind, and later the same analysis is done for different locations of solar. Table 2.11 gives the statistical input parameters for variation and distribution of generation and load for interarea intermediate exchange of power.

For inclusion of generation uncertainty, two wind farms and two solar farms of 10 MW each are added at diverse available locations. Each wind farm

Optimization of hybrid energy generation Chapter | 2   **29**

TABLE 2.8 Bus data of IEEE 30 bus system.

Bus number	P_D	Q_D	Base KVA	Bus number	P_D	Q_D	Base KVA
1	0	0	135	16	3.5	1.8	135
2	21.7	12.7	1	17	9	5.8	135
3	2.4	1.2	135	18	3.2	0.9	135
4	7.6	1.6	135	19	9.5	3.4	135
5	0	0	0	20	2.2	0.7	135
6	0	0	135	21	17.5	11.2	1
7	22.8	10.9	1	22	0	0	135
8	30	30	135	23	3.2	1.6	135
9	0	0	135	24	8.7	6.7	0
10	5.8	2	135	25	0	0	135
11	0	0	135	26	3.5	2.3	135
12	11.2	7.5	0	27	0	0	135
13	0	0	135	28	0	0	135
14	6.2	1.6	135	29	2.4	0.9	135
15	8.2	2.5	135	30	10.6	1.9	0

Details of bus data.

TABLE 2.9 Line data of IEEE 30 bus system.

Line no.	From bus	To bus	Flow limit
1	1	2	130
2	1	3	130
3	2	4	65
4	3	4	130
5	2	5	130
6	2	6	65
7	4	6	90
8	5	7	70
9	6	7	130

Continued

TABLE 2.9 Line data of IEEE 30 bus system—cont'd

Line no.	From bus	To bus	Flow limit
10	6	8	32
11	6	9	65
12	6	10	32
13	9	11	65
14	9	10	65
15	4	12	65
16	12	13	65
17	12	14	32
18	12	15	32
19	12	16	32
20	14	15	16
21	16	17	16
22	15	18	16
23	18	19	16
24	19	20	32
25	10	20	32
26	10	17	32
27	10	21	32
28	10	22	32
29	21	22	32
30	15	23	16
31	22	24	16
32	23	24	16
33	24	25	16
34	25	26	16
35	25	27	16
36	28	27	65
37	27	29	16
38	27	30	16
39	29	30	16
40	8	28	32
41	6	28	32

Details of line data.

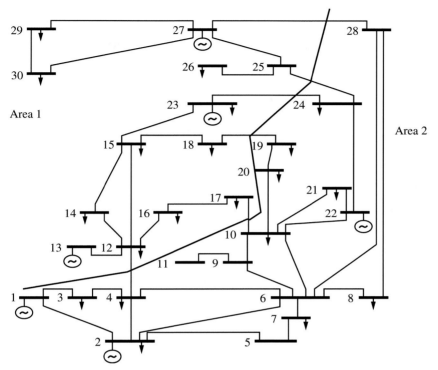

FIG. 2.1 One line diagram of Standard IEEE 30 Bus System in two areas. *(Modified from Attia, A.-F., Al-Turki, Y. A., & Abusorrah, A. M. (2012). Optimal Power Flow Using Adapted Genetic Algorithm with Adjusting Population Size. Electric Power Components and Systems, 40(11), 1285–1299. https://doi.org/10.1080/15325008.2012.689417. The two area representation of IEEE 30 Bus system.)*

TABLE 2.10 Details of cases.

Case number	Base Case	Case 1	Case 2	Case 3	Case 4	Case 5
Case description	Base Case. No RES	RES in Area1	RES in Area1	RES in Area1	RES in Area2	RES in Area2
RES generation included	NO	YES	YES	YES	YES	YES
Location of RES generation	–	Bus No. 23	Bus No. 27	Bus No. 13	Bus No. 22	Bus No. 2

Details of different scenarios considered.

32 Sustainable developments by artificial intelligence & machine learning

TABLE 2.11 Statistical variation in generation and load.

Generation/load	PDF	Variation (%)
Conventional generation	Normal	4–6
Load	Normal	4–6
Wind generation	Log-normal	20.00–55.00
Solar generation	Normal	20.00–33.56

Percentage variation considered in generation (Conventional + RES) and load.

consists of a wind turbine generator set of 2 MW, such that each farm has five units of wind turbine generator set. Wind and solar farms are added to both areas accordingly.

2.3.3 Solution methodology

The amalgamation of RES into the grid adversely affects the power flow of the transmission lines. "The set of optimization problems while taking into account the uncertainties in electric power systems engineering is known collectively as Probabilistic Power Flow (PPF)" (Li, Li, Yan, Yu, & Zhao, 2014; Madrigal, Ponnambalam, & Quintana, 1998). The DC-Probabilistic-Optimal Power Flow (DC-P-OPF) methodology is employed in this analysis in the MATPOWER package of MATLAB. The optimization problem is formulated as maximization of active power generation, subject to active power balance, reactive power balance, apparent power flow limit, from and to side, bus voltage limits, and active and reactive generation limits.

The following algorithm has been prepared and executed to optimize the hybrid (conventional + wind; conventional + solar) generation based on the congestion scenario within the system:

(a) Suitable wind and solar data is collected over a period of 1 year.
(b) The data is analyzed and the distribution of data is observed based on the mean and variance to obtain the power output of RES generators.
(c) Wind speed and solar insolation of 1000 random power output samples normally distributed are generated and 1000 samples of load for 12 months. That is a total of 12,000 samples of generation and 12,000 samples of load.
(d) Bus number 1 is kept as the slack bus.
(e) Base case is run with conventional generators only.
(f) Base case power flow, base case TTCs, and base LMP are recorded for each transmission line, tie lines, and bus node, respectively.

Optimization of hybrid energy generation **Chapter | 2** **33**

(g) The load is normally varied, with a variance of 4%–6% for initiating inter-area power transactions.
(h) The lowest conventional generation (here 19.2 MW) is first replaced with a wind farm. The DC-P-OPF is run to record (step-5) parameters.
(i) The wind farm is then shifted to area 1 and area 2 of the available locations.
(j) New power flow, TTCs and LMP are recorded for each different location of the wind farm (all five locations available) by DC-P-OPF run.
(k) Linear sensitivity factors (GSDF, PTDF, and LODF) are obtained based on base case power flow and new power flow for each location of RES, reflecting the stressed lines of the network.
(l) Change in LMP is obtained with base LMP and new recorded LMPs, reflecting the congestion cost factor.
(m) The difference between base TTC and new TTC results in TRM value, reflecting the congestion scenario.
(n) Based on the linear sensitivity factors, LMP, and TRM values, the optimized location for the wind farm is chosen within the grid.
(o) The wind farm is replaced with a solar farm and steps 7 to step 13 are repeated.

The DC-P-OPF (Madrigal et al., 1998; Sarkar & Khaparde, 2009) method converges faster as compared to the other methods. If the system is congested, the limits of transmission lines are violated, endangering the stability of the system. The parameters calculated are discussed below.

2.3.4 Sensitivity factors

The easiest way to offer a speedy calculation of promising power flows is to employ linear sensitivity factors. Three linear sensitivity factors are accessed in this analysis to encompass a brief study. The factors are GSDF, PTDF, and LODF. These three factors are highlighted here due to their comparative simplicity and ease in understanding, as well as computation with a touch of generation, transmission, and load/fault perspectives. These parameters are calculated as shown:

$$a_{li} = \frac{\Delta F_l}{\Delta P_i} = \text{GSDF} \qquad (2.10)$$

$$b_{li} = \frac{f_l}{P_{inj}} = \text{PTDF} \qquad (2.11)$$

$$d_{lk} = \frac{\Delta f_l}{f_k^{old}} = \text{LODF} \qquad (2.12)$$

These three factors mutually give a finer sight of the variation in power flow patterns of the transmission system.

34 Sustainable developments by artificial intelligence & machine learning

2.3.5 Locational marginal pricing (LMP)

LMP is the marginal price of supplying, at the minimum price and the subsequent addition of electric demand, at a definite site market price of generation in the controlled zone (Dharme, Khatavkar, Myadeo, & Dhabai, 2020). The objective function of LMPs is computed by maximizing the demand and supply bids. It is subject to the constraints of the existence of transmission restrictions and line losses. The LMP of any bus j can be computed as:

$$LMP_j = LMP_{ref} + LMP_{loss\,j} + LMP_{congestion\,j} \qquad (2.13)$$

The LMPs, which are also recognized as nodal prices, are the permutation of three parts: marginal price on the reference bus, marginal price dealing with transmission losses, and marginal price reflecting transmission system congestion.

2.3.6 Reliability parameters

To operate the transmission system within safe limits, reliability margins are assessed. These take into account TTC, ATC, CBM, ETC, and TRM (Shin, Lee, & Kim, 2003; Thomas & Padma, 2015; Venkatesh, Ganangdas, & Narayan, 2004).

Total transfer capability (TTC) is the sum of electric power that can be transferred in an interrelated network consistently within the system limits. Hence, TTC is the minimum of all the limits (thermal limit, voltage limit, and stability limit).

Transmission Reliability Margin (TRM) is the sum of transmission capacity reserved to assure that the interconnected power system will be secure. TRM incorporates inherent uncertainties of the system limits including the system changes.

The difference between the base TTC and the new TTCs obtained is the value of the TRM of the transmission lines. The flowchart to calculate TTC and TRM of the network is as shown in Fig. 2.2.

Repeated DC-P-OPF was carried out for all the cases with variation in the generation. The power flow variations were recorded and the parameters were computed accordingly.

2.4 Results

To recognize the discrepancy in sensitivity factors, LMPs, and TTC, TRM with the variation in wind farm power output and solar farm power output, six different cases were considered and variation graphs were plotted. Results of each parameter are shown separately with worst-case scenarios.

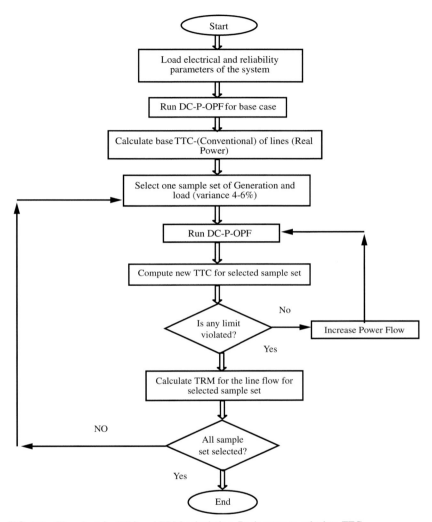

FIG. 2.2 Flowchart for TTC and TRM calculation. Basic steps to calculate TTC.

2.4.1 Impact of probabilistic nature and location of RES on sensitivity factors

First, the impact of wind generation is analyzed, followed by solar. The 19.2 MW conventional generator is replaced by equivalent wind and solar generation. Consequently, the variables are being studied. Later on, the RES is shifted in the system (i.e., at bus numbers: 23, 2, 22, 27, and 13) and optimal location is obtained for inclusion of RES. The three sensitivity factors are calculated for each generation value for each line. Tables 2.12 and 2.13 show all

36 Sustainable developments by artificial intelligence & machine learning

TABLE 2.12 LMP values for wind generation.

Bus number	Base LMP	Case 1 Bus 23	Case 2 Bus 2	Case 3 Bus22	Case 4 Bus 27	Case 5 Bus 13
1	3.635	3.3478	4.3693	4.5656	4.7666	2.3245
2	3.635	4.6752	4.5722	2.1123	4.5454	4.5672
3	3.635	4.8786	4.5762	4.6787	3.8768	4.6666
4	3.635	3.6612	4.5661	5.1231	3.9821	4.8675
5	3.635	4.2234	4.5872	4.3323	2.9803	3.4356
6	3.635	2.3345	4.5931	4.5679	4.9912	1.2378
7	3.635	4.6364	4.6364	3.4434	4.7856	8.4324
8	3.635	5.3432	4.6264	4.6659	2.7645	1.3244
9	3.635	4.1121	4.6414	1.2343	2.9754	3.4564
10	3.635	4.6454	5.1322	4.6758	3.3312	4.1209
11	3.635	6.5856	5.4189	1.2342	4.7682	8.6754
12	3.635	5.1322	5.1322	8.4564	4.569	9.8767
13	3.635	4.2213	4.4911	5.1233	4.2189	8.3456
14	3.635	2.3433	3.2214	3.6584	3.6715	3.5421
15	3.635	5.3539	4.2665	3.1111	4.1212	6.5454
16	3.635	8.4887	4.8981	5.4542	4.0987	6.0895
17	3.635	2.7614	5.2821	1.1209	4.0012	1.2232
18	3.635	3.6099	4.7065	2.3342	2.3354	8.7675
19	3.635	6.6638	4.9683	7.8989	3.4356	4.5467
20	3.635	8.8231	5.0371	4.4564	2.3498	9.6756
21	3.635	3.8764	8.1301	1.1287	2.4598	3.3453
22	3.635	7.5605	0.9897	0.8765	3.1145	8.8865
23	3.635	5.0569	3.4722	2.3476	4.5362	7.6686
24	3.635	8.8311	2.2864	1.3543	2.1114	6.1239
25	3.635	3.6074	3.1568	10.8712	3.4312	5.4365
26	3.635	3.3424	3.1693	3.4789	4.0876	7.7895
27	3.635	3.8773	3.5198	4.5687	4.5766	4.5675
28	3.635	0.8205	4.554	2.3487	4.0987	7.3412
29	3.635	7.0734	3.5335	0.1111	3.1212	7.3478
30	3.635	2.7014	3.5428	6.3432	3.1212	4.3312

LMPs of all 30 buses.

TABLE 2.13 LMP values for solar generation.

Bus number	Base LMP	bus 23 Case 1	Bus 2 Case 2	Bus 22 case 3	Bus 27 case 4	Bus 13 case 5
1	3.635	3.567	4.223	2.346	3.444	3.122
2	3.635	3.454	4.216	4.332	2.341	2.423
3	3.635	2.324	3.122	0.998	4.561	2.111
4	3.635	3.454	4.657	1.236	3.456	3.654
5	3.635	4.545	3.223	4.332	4.511	3.554
6	3.635	3.997	5.435	3.445	4.562	3.453
7	3.635	3.222	5.467	3.452	4.561	3.451
8	3.635	3.123	6.334	3.445	4.907	4.341
9	3.635	4.324	0.331	3.458	2.111	3.445
10	3.635	3.348	4.123	3.906	2.342	2.994
11	3.635	2.334	4.334	3.125	3.124	3.005
12	3.635	4.345	5.334	3.004	2.116	2.432
13	3.635	3.435	6.232	4.556	3.567	3.452
14	3.635	2.343	3.221	3.658	3.671	3.542
15	3.635	2.876	2.222	4.002	3.871	3.441
16	3.635	3.444	6.434	3.442	4.001	3.442
17	3.635	2.331	5.343	3.451	3.128	3.442
18	3.635	5.432	4.112	3.234	3.334	2.431
19	3.635	3.112	3.456	3.556	5.341	3.221
20	3.635	4.327	3.897	3.556	5.619	3.996
21	3.635	2.678	3.009	2.314	4.999	3.561
22	3.635	3.223	3.445	3.114	4.324	4.561
23	3.635	4.115	3.126	3.095	5.123	4.129
24	3.635	3.324	4.331	3.215	1.341	3.456
25	3.635	4.343	4.556	3.451	2.452	4.112
26	3.635	4.121	5.455	3.451	3.451	3.778
27	3.635	4.324	3.435	3.451	2.443	3.553
28	3.635	2.112	4.556	3.446	2.116	3.645
29	3.635	3.223	3.445	4.241	2.996	2.943
30	3.635	3.223	4.445	3.443	3.024	3.445

LMPs of 30 buses.

38 Sustainable developments by artificial intelligence & machine learning

the cases and the three sensitivity factors, the value of sensitivity factors, and the lines affected most in each case. For the worst circumstances, the line with the highest disparity in GSDF is preferred to be out of service and comparative LODF is evaluated for every case.

2.4.2 Impact of probabilistic nature and location of RES on LMP

The base LMP is used as a reference for the new LMPs. The LMPs obtained with RES inclusion reflect the congestion cost component. Tables 2.12 and 2.13 reflect the LMPs for wind as well as solar generation respectively for all five cases for 30 buses. The performance indices for both generations for each case are represented in Tables 2.14 and 2.15. The graphs of the performance index for wind and solar variation are represented in Figs. 2.3 and 2.4.

TABLE 2.14 Performance index representing the congestion scenario.

Cases	Case 1	Case 2	Case 3	Case 4	Case 5
Mean	4.70751	4.315917	3.959883	3.670673	5.759677
Median	4.4299	4.5742	4.39435	3.92945	5.763
Variance	4.219505	1.499285	6.193097	0.737845	6.596978
Performance Index	−30%	−19%	−9%	−1%	−58%

Performance of wind generation.

TABLE 2.15 Performance index representing congestion scenario for solar generation.

Case	Case 1	Case 2	Case 3	Case 4	Case 5
Mean	3.645	4.214	3.256	3.658	3.654
Median	3.235	4.125	3.546	3.658	3.658
Variance	0.641	1.765	0.652	1.564	0.345
Performance Index	5%	−15%	9%	2%	6%

Performance of solar generation.

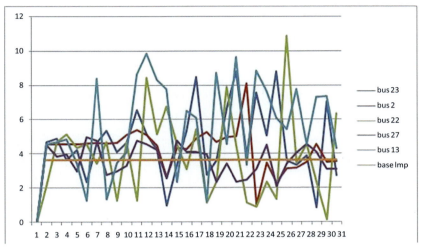

FIG. 2.3 LMP variation of all five cases with respect to base LMP and uncertain wind generation.

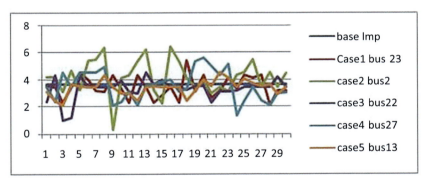

FIG. 2.4 Variation of all five cases with respect to base LMP and uncertain solar generation.

2.4.3 Impact of the probabilistic nature and location of RES on TTC and TRM

To understand the variation in TTCs with the variation in conventional generation and load, the base TTC value for 1000 samples of load/generation is compared with all the 12 months' TTCs. The RES generation and load samples are varied to calculate the new TTCs. The new TTCs are then compared to the base TTC value. The tie line load flow limit is also checked. The TTC recorded for each case is (41*1000) 41,000 values, therefore, only the graphical representation in Fig. 2.5 is presented here for all the 12 months.

40 Sustainable developments by artificial intelligence & machine learning

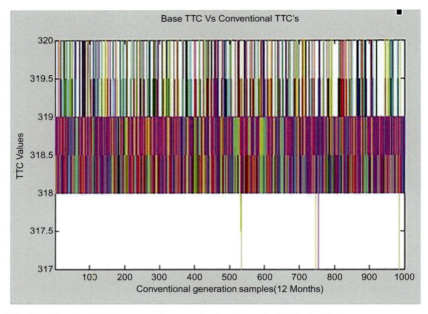

FIG. 2.5 Graph of TTC vs generation samples for 12 months. TTCs for 41 lines.

Tie line flow is recorded for the six lines. Table 2.16 represents the line flows for variation in conventional generation and load.

The above results are recorded for a total of 1000 generation and 1000 load samples for every month with variation as shown in Table 2.11. As sample values are very large for each parameter, final results are presented for each parameter without presenting the power flow run of the system.

TABLE 2.16 Tie line flow.

Tie line	Flow limit	Actual flow	Congestion
T-1	65	53.825	NO
T-2	32	30.757	NO
T-3	16	11.171	NO
T-4	16	20.000	YES
T-5	16	16.000	YES
T-6	65	66.106	YES

Tie line flow after inter area transaction.

2.5 Discussion and conclusion, future scope

2.5.1 Discussion

The results presented above show that the power flow pattern varied greatly due to uncertainty and location change of RES integration. The variation affects the power flow, which in turn changes the values of system parameters, leading the system towards congestion.

From Tables 2.17 and 2.18 it can be observed that, when RES uncertainty is introduced into the network with the change in its location, the sensitivity factors vary in responding to the change in the power flows. It can be clearly understood from the result that for wind generation, out of all the available locations, when the wind source is integrated with its intermittency on bus number 13 (case 5), the three sensitivity parameters show heavy variation, meaning that the system lines are highly stressed out for the respective location. Similarly, the solar source stresses the lines of the system heavily when integrated on bus number 2 (case 2).

The variation of GSDF value for line number 30 for wind and line 18 for solar are highly stressed. The variation of PTDF is highest for line numbers 14 and 15 for wind and solar scenarios, respectively. LODF reflects the contingency scenario during a line outage with uncertain RES generation. LODF resembles a different line, upsetting when there is a line outage and the RES output varies according to their distinctiveness. The LODF is analyzed to recognize the consequence of a contingency in tandem with the uncertainty and location impact.

TABLE 2.17 GSDF, PTDF, and LODF values for five cases of wind generation.

Parameter	Case I (Bus 23)		Case II (Bus 2)		Case III (Bus 22)	
	σ	Line	σ	Line	σ	Line
GSDF	0.019	1	0.012	30	0.016	1
PTDF	5.552	10	11.976	1	12.555	14
LODF	0.019	20	0.229	32	0.263	32
	Case IV (Bus 27)		Case V (Bus 13)			
	σ	Line	σ	Line		
GSDF	0.014	11	0.047	30		
PTDF	11.912	1	21.995	14		
LODF	0.266	32	1.011	20		

Highly stressed lines are represented.

TABLE 2.18 GSDF, PTDF, and LODF values for all five cases of solar generation.

Parameter	Case I (Bus 23)		Case II (Bus 2)		Case III (Bus 22)	
	σ	Line	σ	Line	σ	Line
GSDF	0.005	30	0.008	18	0.005	13
PTDF	3.231	14	5.855	15	4.658	12
LODF	0.022	41	0.052	30	0.043	30
	Case IV (Bus 27)		CaseV (Bus 13)			
	σ	Line	σ	Line		
GSDF	0.003	12	0.007	31		
PTDF	3.338	11	3.244	1		
LODF	0.014	13	0.021	13		

Highly stressed lined are represented.

From Tables 2.12 and 2.13, it is comprehensible that for whichever location of wind and solar farm in the system considered here, the alteration in LMP is beyond the accepted assortment, increasing the cost of energy due to congestion cost component, reflecting the occurrence of congestion. The variation in LMP is superior in case 5 (bus 13) for wind generation, which also has highly stressed lines from the observance of sensitivity factors. The wind generation introduces congestion in all the available locations in varying manners. For a solar uncertain generation, the congestion occurs only in case 2, the LMP variation is highest for case 2, representing the congestion scenario in the transmission system for uncertain solar generation. This pictures the worst-case scenario for solar generation. This means that to have minimum congestion with wind, the wind source can be integrated at the location with lower LMP values, whereas all other available locations can be deprived of solar other than location 2 without congestion. Tables 2.14 and 2.15 show the performance index along with the variance of LMP for all cases. The optimized location can be chosen based on these indices, according to the variation in RES. The graphical presentation in Figs. 2.3 and 2.4 resembles the variation of LMP for all five cases with respect to the base LMP of the system.

From Fig. 2.5, it is observed that, due to small variations in generation and load, the transaction of power from one area to another via transmission lines on the on tie lines may violate the transmission limits and induce congestion on the lines, which increases the congestion cost. The difference between maximum TTC and minimum TTC is the TRM value of the transmission line. From Table 2.16, it can be observed that when generation and load are varied and

Optimization of hybrid energy generation **Chapter | 2** **43**

power transaction is done between the areas, congestion occurs. The table shows that during the variation most of the tie lines undergo congestion. The power flow is recorded and limits are recorded for 1000 samples of generation and load for 12 months each. It can be observed that for each month the power flow limit is violated in the base case. That means there is congestion within the system during the interarea transaction. This will increase the LMP of the transaction, increasing the congestion cost component.

From the above results, it can be said that the RES source can be integrated efficiently at locations of bus numbers 22, 23, and 27 within the system. At these locations, the RES power output can be efficiently utilized without jeopardizing the system limits with minimum stressed lines, lower LMP values, and avoiding congestion in tie lines by maintaining the TTC and TRM of the lines, ensuring stable, reliable, and secure operation of the grid with RES integration. If the RES is integrated at location 13 or 2, it stresses the transmission lines, increases the LMP, and the stability limits are crossed, endangering the system's security. Hence, these two locations cannot be considered for optimization of hybrid generation in spite of higher availability.

The addition of RES into the grid certainly changes the power flow and the values of the parameters. These parameters reflect the congestion scenario post-integration. The vagaries are observed when RES is added to the grid.

2.5.2 Conclusion

This analysis provides a wider view and methodology for optimization of hybrid generation from the grid stability point of view in the presence of uncertainty in RES. Linear sensitivity factors, LMP, TTC, and TRM are the significant parameters providing an alert signal for transmission congestion and congestion cost. This analysis shows that these parameters are mainly dependent on the location of integration of RES irrespective of high availability and the power flow pattern. The higher the value of these parameters, the higher the probability of network congestion is. The comparison of these parameters with base values shows that adding RES at a suitable location is of utmost importance from an optimization point of view. Hence, it can also be concluded from the analysis that RES may add vagaries into the system, but choosing the optimized location from a transmission point of view can help to reduce the congestion therein.

This work also provides the methodology to calculate TTC and TRM values based on transmission power flow using DC-P-OPF. If real-time data is available, the study can be useful for system planners and operators. The congestion cost component in the LMP can be retrieved from the reliability parameter TTC. The TTC increases above the base case value indicating the system limit violation. This analysis helps to understand the variation in congestion cost components according to changes in generation and load. Congestion will increase the congestion cost within the system. This study establishes a methodology for any system that wishes to incorporate renewable energy sources into its system. In addition, it is evident that it is not easily predictable, without this kind of study, which case or line will be critically affected due to uncertainties mixed with a contingency. Hence, this study can be extended to any practical system if enough real-time data is available.

2.5.3 Future scope

This study can be extended to any practical system to integrate renewable sources into the existing grid. Optimal location can be found for integrating the new generation, which is extremely useful for system operators and planners. Reliability factors TTC and TRM can be used to calculate the available transfer capability (ATC), which is helpful in congestion management as well as deciding the role of markets for bidding. If sufficient data is available, the TTC and TRM can be predetermined using intelligent techniques such as ANN, PSOs, etc. Future work should consist of prior estimation of reliability parameters with wind uncertainty and managing congestion based on reliability parameters, finding out the optimal location of renewable source into the grid, and reducing congestion cost of the system, approaching towards economical congestion management.

Acknowledgment

The authors are thankful to IMD Pune, India, for providing the necessary data. The authors are also thankful to Poornima University, Jaipur, for providing research opportunities and suitable amenities. The authors extend their thanks to Dr. A.A. Dharme, Dr. V.V. Khatavkar, and Heramb Mayadeo for their guidance.

References

Ahmadi, H., & Ghasemi, H. (2011). Probabilistic optimal power flow incorporating wind power using point estimate methods. In *2011 10th International conference on environment and electrical engineering, EEEIC.EU 2011—Conference proceedings.* https://doi.org/10.1109/EEEIC.2011.5874815.

Albadi, M. H., El-Rayani, Y. M., El-Saadany, E. F., & Al-Riyami, H. A. (2017). Effect of large solar power plant on locational marginal prices in Oman. In *Canadian conference on electrical and computer engineering* Institute of Electrical and Electronics Engineers Inc. https://doi.org/10.1109/CCECE.2017.7946741.

Banerjee, B., Jayaweera, D., & Islam, S. M. (2012). Probabilistic optimisation of generation scheduling considering wind power output and stochastic line capacity. In *2012 22nd Australasian universities power engineering conference: \Green smart grid systems\, AUPEC 2012.*

Bhesdadiya, R. H., & Patel, R. M. (2014). Available transfer capability calculation methods: A review. *International Journal of Advanced Research in Electrical, Electronics and Instrumentation Engineering, 3.*

Bohn, R. E., Caramanis, M. C., & Schweppe, F. C. (1984). Optimal pricing in electrical networks over space and time. *RAND Journal of Economics*, *15*(3), 360–376. https://doi.org/10.2307/2555444.

Burke, D. J., & Malley, M. J. (2008). Optimal wind power location on transmission system—A probabilistic load flow approach. In *Proceedings of the 10th international conference on probabilistic methods applied to power systems*.

Celeska, M., & Najdenkoski, K. (2005). Estimation of Weibull parameters from wind measurement data by comparison of statistical methods. *IEEE Transaction*, *3*, 58–62.

Christie, R. D., Wollenberg, B. F., & Wangensteen, I. (2000). Transmission management in the deregulated environment. *Proceedings of the IEEE*, *88*(2), 170–195. https://doi.org/10.1109/5.823997.

Daneshi, H., & Srivastava, A. K. (2011). ERCOT electricity market: Transition from zonal to nodal market operation. In *IEEE power and energy society general meeting*. https://doi.org/10.1109/PES.2011.6039830.

Dhabai, P. B., & Tiwari, N. (2020a). Computation of locational marginal pricing in the presence of uncertainty of solar generation. In *2020 5th IEEE international conference on recent advances and innovations in engineering, ICRAIE 2020—Proceeding* Institute of Electrical and Electronics Engineers Inc. https://doi.org/10.1109/ICRAIE51050.2020.9358291.

Dhabai, P., & Tiwari, N. (2020b). Analysis of variation in power flows due to uncertain solar farm power output and its location in network. In *2020 IEEE international conference for innovation in technology, INOCON 2020* Institute of Electrical and Electronics Engineers Inc. https://doi.org/10.1109/INOCON50539.2020.9298308.

Dhabai, P., & Tiwari, N. (2020c). Effect of stochastic nature and location change of wind and solar generation on transmission lattice power flows. In *2020c International conference for emerging technology, INCET 2020* Institute of Electrical and Electronics Engineers Inc. https://doi.org/10.1109/INCET49848.2020.9154035.

Dharme, A., Khatavkar, V., Myadeo, H., & Dhabai, P. (2020). Computation of network rental in the presence of uncertainty of stochastic wind generation. In *2020 International conference on emerging technology* (pp. 1–7).

Dharmjit, & Tanti, D. K. (2012). Load flow analysis on IEEE 30 bus system. *International Journal of Scientific and Research Publications*, *2*.

Ela, E., Kirby, B., Lannoye, E., Milligan, M., Flynn, D., Zavadil, B., et al. (2010). Evolution of operating reserve determination in wind power integration studies. In *IEEE PES general meeting, PES 2010*. https://doi.org/10.1109/PES.2010.5589272.

Fangxing, L., & Bo, R. (2008). Dcopf-based lmp simulation: Algorithm, comparison with acopf, and sensitivity. In *2008 IEEE/PES transmission and distribution conference and exposition* (p. 1).

Jamil, I., Zhao, J., Zhang, L., Rafique, S. F., & Jamil, R. (2019). Uncertainty analysis of energy production for a $3 \times 50MW$ AC photovoltaic project based on solar resources. *International Journal of Photoenergy*, *2019*.

Kane, L., & Ault, G. (2014). A review and analysis of renewable energy curtailment schemes and principles of access: Transitioning towards business as usual. *Energy Policy*, *72*, 67–77. https://doi.org/10.1016/j.enpol.2014.04.010.

Karki, R., Hu, P., & Billinton, R. (2006). A simplified wind power generation model for reliability evaluation. *IEEE Transaction*, *21*, 173–182.

Khatavkar, V. V., Mayadeo, H., Dhabai, P., & Dharme, A. A. (2017). Impact of probabilistic nature and location of wind generation on transmission power flows. In *International conference on automatic control and dynamic optimization techniques, ICACDOT 2016* (pp. 189–193). Institute of Electrical and Electronics Engineers Inc. https://doi.org/10.1109/ICACDOT.2016.7877576.

46 Sustainable developments by artificial intelligence & machine learning

Kothari, D. P., & Dhillon, J. S. (2004). *Power system optimization* (pp. 172–175).

Li, Y., Li, W., Yan, W., Yu, J., & Zhao, X. (2014). Probabilistic optimal power flow considering correlations of wind speeds following different distributions. *IEEE Transactions on Power Systems, 29*(4), 1847–1854. https://doi.org/10.1109/TPWRS.2013.2296505.

Lowery, C., & O'Malley, M. (2014). Optimizing wind farm locations to reduce variability and increase generation. In *2014 International conference on probabilistic methods applied to power systems, PMAPS 2014—Conference proceedings* Institute of Electrical and Electronics Engineers Inc. https://doi.org/10.1109/PMAPS.2014.6960661.

Madrigal, M., Ponnambalam, K., & Quintana, V. H. (1998). Probabilistic optimal power flow. In *Vol. 1. Canadian conference on electrical and computer engineering* (pp. 385–388). IEEE.

Morales, J. M., Conejo, A. J., & Pérez-Ruiz, J. (2011). Simulating the impact of wind production on locational marginal prices. *IEEE Transactions on Power Systems, 26*(2), 820–828. https://doi.org/10.1109/TPWRS.2010.2052374.

Morgan, E. C. (2010). *Probability distribution for offshore wind.* Elsevier.

Morstyn, T., Teytelboym, A., Hepburn, C., & McCulloch, M. D. (2020). Integrating P2P energy trading with probabilistic distribution locational marginal pricing. *IEEE Transactions on Smart Grid, 11*(4), 3095–3106. https://doi.org/10.1109/TSG.2019.2963238.

Neuhoff, K., Boyd, R., & Grau, T. (2011). Renewable electric energy integration: Quantifying the value of design of markets for international transmission capacity. *IEEE Transactions, 2*, 107–114.

Nwulu, N. I. (2018). Modelling locational marginal prices using decision trees. In *Vol. 2017. 2017 international conference on information and communication technologies, ICICT 2017* (pp. 156–159). Institute of Electrical and Electronics Engineers Inc. https://doi.org/10.1109/ICICT.2017.8320181.

Omnisite. (2009). *Solar data sheet.* Omnisite.

Othman, M. M., & Musirin, I. (2011). A novel approach to determine transmission reliability margin using parametric bootstrap technique. *International Journal of Electrical Power & Energy Systems, 33*(10), 1666–1674. https://doi.org/10.1016/j.ijepes.2011.08.003.

Prabha, S. U., & Venkataseshaiah, C. (2009). Effect of uncertainties in the economic constrained available transfer capability in power systems. *Canadian Journal of Pure and Applied Sciences, 4*, 4522–4532.

Sarkar, V., & Khaparde, S. A. (2009). DCOPF-based marginal loss pricing with enhanced power flow accuracy by using matrix loss distribution. *IEEE Transactions on Power Systems, 24*(3), 1435–1445. https://doi.org/10.1109/TPWRS.2009.2021205.

Sathya, S. (2014). Modelling and control of hybrid systems solar wind battery usin three inputs DC-DC boost converters. *International Journal of Advanced Research in Electrical, Electronics and Instrumentation Engineering, 3.*

Shahidehpour, M., Yamin, H., & Li, Z. (2002). *Market operations in electric power systems.*

Shayesteh, E., Parsa Moghaddam, M., Haghifam, M. R., & Sheikh-EL-Eslami, M. K. (2009). Security-based congestion management by means of demand response programs. In *2009 IEEE Bucharest PowerTech: Innovative ideas toward the electrical grid of the future.* https://doi.org/10.1109/PTC.2009.5282069.

Shin, D. J., Lee, H. S., & Kim, J. O. (2003). Quantification method of TRM (transmission reliability margin) using probabilistic wad flow. In *IFAC power plant and power system controls* Elsevier.

Sjodin, E., Gayme, D. F., & Topcu, U. (2012). Risk-mitigated optimal power flow for wind powered grids. In *Proceedings of the American control conference* (pp. 4431–4437). Institute of Electrical and Electronics Engineers Inc. https://doi.org/10.1109/acc.2012.6315377.

Optimization of hybrid energy generation **Chapter | 2 47**

Thomas, A. S., & Padma, S. (2015). Study on the effect of renewable energy power generation on available transfer capability. *International Journal of Innovative Research in Science, Engineering and Technology, 4.*

Umale, G., & Warkad, S. B. (2016). Different models and properties on LMP calculations. *International Journal of Electrical, Electronics, 5*(1), 100–108.

Venkatesh, P., Ganangdas, R., & Narayan, P. P. (2004). *Assessment of available transfer capability in combined economic emission dispatch environment.*

Chapter 3

IoET-SG: Integrating internet of energy things with smart grid

M. Shahidul Islam[a], Md. Mehedi Islam[b], Sabbir Ahmed[a], Md. Sazzadur Rahman[a], Krishna Kumar[c], and M. Shamim Kaiser[a]

[a]Institute of Information Technology, Jahangirnagar University, Dhaka, Bangladesh, [b]Department of Electronics & Communication Engineering, Hajee Mohammad Danesh Science & Technology University, Dinajpur, Bangladesh, [c]Department of Hydro and Renewable Energy, Indian Institute of Technology Roorkee, Roorkee, Uttarakhand, India

3.1 Introduction

The traditional electricity grid is a system of transmission lines, substations, transformers, and other electrical components that supply electricity to our homes and businesses from the generating power station. On the other hand, the smart grid is a digital technology that enables bidirectional communication between the company and its customers and makes the grid smart by sensing along the transmission line (Ekpe & Umoh, 2019).

Since 2007, the industry has been preparing, designing, and studying a modern electrical grid framework known as smart grid (SG). Several countries have worked tirelessly to bring the technology to maturity. The Department of Energy and the Electric Power Research Institute in the United States are coordinating smart grid implementation through a project called "Intelligent Grid," which improves the communication mechanism between the power grid and a device to enhance power system efficiency and customer service (Fang, Misra, Xue, & Yang, 2012).

The current trend in energy grids and the emergence of the internet of things (IoT) makes energy systems accessible through the internet. The internet of energy things (IoET) enables all physical objects/things equipped with computing and communication capabilities (such as smart appliances, renewable energy sources, smart meters, and so on) to be conveniently integrated into the internet (Mahmud et al., 2018). A consumer may use a web-based interface and a smartphone application to monitor their appliances and other energy-related devices. The SG, on the other hand, is one of the most important infrastructures, combining conventional power grids, ICT networks, and renewable

---

Sustainable Developments by Artificial Intelligence and Machine Learning for Renewable Energies.
https://doi.org/10.1016/B978-0-323-91228-0.00013-6
Copyright © 2022 Elsevier Inc. All rights reserved.

50 Sustainable developments by artificial intelligence & machine learning

energy infrastructures. Smart devices, a variety of smart items, smart meters, and other smart devices will be integrated into the SG. As a result, combining IoET and SG will make grid equipment smart/intelligent and accessible anytime from anywhere (Asif-Ur-Rahman et al., 2019).

This chapter focuses on the challenges of conventional and smart grid systems, as well as how IoET-SG can help to overcome these barriers. Ongoing research on the implementation, strengths, and drawbacks of IoET-SG are discussed. Finally, upcoming research challenges, as well as effective solutions to these challenges, are outlined with the aim of providing the reader with specific research ideas and challenges (Ahmed Abdulkadir & Al-Turjman, 2021).

The organization of this chapter is as follows: Section 3.2 presents the basic advantages and disadvantages of a traditional grid system. Section 3.3 describes the features and advantages of the smart grid in relation to the traditional grid, Section 3.4 provides information about the IoET, Section 3.5 discusses the architecture and the main functions of the IoET-SG, and Section 3.6 lists research challenges and workable solutions to these problems. Finally, the chapter is concluded in Section 3.7.

3.2   Traditional grid

The traditional grid, the power grid of the last century, is a one-way transmission system. This implies that electricity flows unidirectionally from generators to substations, through transmission lines, and finally to buyers' outlets. The conventional grid is the interconnection of various power frameworks such as coordination machinery, power transformers, transmission lines, transmission substations, circulation lines, substations, and distinct types of loads. They lie at long distances from the electricity consumption zone and are transmitted through long transmission lines (Vandoorn, Meersman, De Kooning, & Vandevelde, 2012). The conventional grid has some advantages and disadvantages, and these are listed below.

(a) Advantages of Traditional Grid:

- Low deployment cost.
- Most conventional grids are linear, which means that the power flow is from generation centers to users, which makes the network implementation simple.
- In the conventional grid, the load predictability is easier due to its unidirectional flow. The electricity cannot be stored on a massive scale; thus, the load predictions are critical to keep supply and demand balance.
- The security arrangement is simpler. The current flows through the lines are known, thus if anything goes wrong, it can be detected.

(b) Limitation of Traditional Grid:

- The conventional grid is too rigid. It can only handle small renewable energy sources without adding to the protection systems complexity or losing predictability. This leads to energy efficiency issues.

- It is not self-healing, since if something goes wrong in generation, smaller areas will need to be fed from somewhere else. Though contingencies are considered, there are always blackouts.
- Limited processing capabilities and looping problem cause reductions in performance.
- It has a low security level.

3.3 Smart grid

The Smart Grid consists of millions of sensors, controllers, computers, and power lines that work together with emerging technology and equipment (Luthra, Kumar, Kharb, Ansari, & Shimmi, 2014). The SG is expected to provide the knowledge and resources for making informed decisions about our energy usage, not just about utilities and technologies. With the smart grid, electricity consumers will be able to take part at a never-before-seen pace.

The SG uses digital sensing technologies to monitor power consumption patterns of users and the state of the grid in real time and use this information to manage the grid and consumption. It has the potential to reduce energy demand and customer costs intelligently, so that electrical power supply agencies can use electricity efficiently and successfully satisfy the varying consumer demands (Paul, Rabbani, Kundu, & Zaman, 2014). The SG also incorporates features like bidirectional smart metering, remote access to devices, self-restoration, smart monitoring, improved security, reduced peak demand, reduced operations and management (O&M) cost, better integration to end consumer-owned micro power generation systems, and so on (Ma, Chen, Huang, & Meng, 2013). Fig. 3.1 shows the smart grid architecture where SG incorporates advanced metering infrastructure; smart controller; smart circuit breaker; load control switches; smart distribution boards; and smart appliances. Such a system includes centralized or consumer site renewable energy resources. In SG, the flow of electricity is bidirectional, that is, from the grid to the consumer and from the consumer to the grid.

The following are some of the advantages of using the smart grid:

- Transmission of electricity that is more powerful and reduces electricity losses such as distribution loss and transmission loss, and electricity adduction.
- There are no regulatory standards for SG technologies yet.
- There is a lack of official documentation on technology.

The SG has many features that are different from the conventional energy grid. Table 3.1 shows various characteristics of these two systems. Fig. 3.2 shows the main functionality of the SG.

3.4 Internet of energy things (IoET)

The IoT is the primary driving force for the advancement of our grid networks. In IoT, device-to-device connectivity allows users to communicate with the

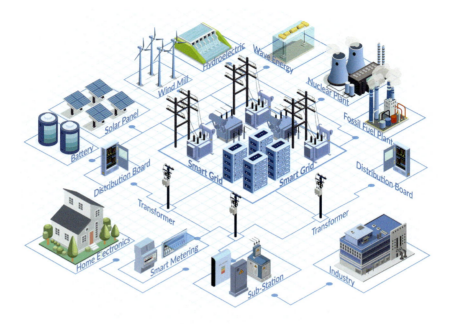

FIG. 3.1 Smart grid architecture.

TABLE 3.1 Various features of conventional and smart grid systems.

Features	Conventional grid	Smart grid
Technology	Electromechanical: The conventional electrical grid uses electromechanical equipment that is electrically powered. The systems do not communicate with each other using a network interface	Digital: The SG uses digital devices, which allows connectivity between devices and facilitates remote access and use of self-regulation
Generation	Centralized: With conventional energy infrastructure, power is transmitted from a centralized system. This removes the opportunity to integrate renewable energy sources into the grid	Distributed: Power from multiple plants and substations can be distributed through SG infrastructures to help balance the load, reduce peak times, and limit power outages
Distribution	One-Way Distribution: The infrastructure can only transmit power from the power plant, that is, the generating source, to the consumers	Two-Way Distribution: Though electricity is still delivered from the primary power plant, power can also be supplied by a secondary supplier in a SG system. Consumers with grid-tie renewable energy sources, such as solar panels, can send energy to the grid

TABLE 3.1 Various features of conventional and smart grid systems—cont'd

Features	Conventional grid	Smart grid
Sensors	Limited Sensor usage: The conventional grid uses limited sensors on the lines that may fail to identify the location of the fault. Thus, it requires high restoration time	Many sensors used: Multiple sensors are mounted on power lines of SG. This aids in pinpointing the location of the source and can aid in rerouting power to where it is needed while minimizing the areas affected by the outage
Monitoring	Manual: Energy distribution monitored manually	Automatic: Energy distribution monitored automatically
Restoration	Manual: Technicians must physically travel to the site of the failure for repair. The requirement for this will lengthen the time that outages persist	Self-Healing: After finding the fault on the line, SG heals the line automatically and sends the report to technicians at the monitoring center for problems related to infrastructure disruption
Equipment	Failure & Blackout: The conventional energy system is vulnerable to failures due to aging and limitations. Infrastructure failure can lead to blackouts, which means that the end-user does not receive power to their device, which causes downtime	Adaptive & Islanding: In SG, the power can be redirected to any problem area. This reduces the area affected by power outages, and this can be done on a per-home basis
Energy usage	Fewer: The conventional grid infrastructure is not prepared for consumers to choose how they want to receive energy	Many: Infrastructure can be shared using smart technology. This encourages more businesses and new sources of energy to come to the grid so that consumers can choose more from their energy supply

device through the internet (Aurna, Anika, Rubel, Kabir, & Kaiser, 2021). IoET is a system that allows different components of the grid network to communicate with one another. It also allows developers, researchers, policymakers, and users to connect with the grid. The role of IoET in the power system is to estimate resources, generate forecasts, reduce energy losses, reduce greenhouse gases, and maintain various parts of the intelligent grid system. An IoET-based energy management system can track energy usage in real-time and increase

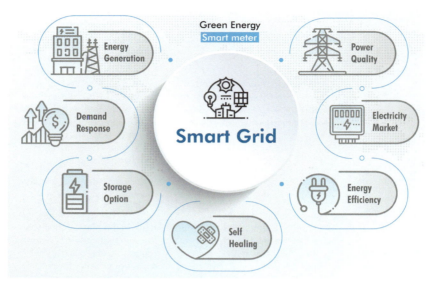

FIG. 3.2 The functionality of the smart grid.

understanding of energy efficiency at any stage of the supply chain. Fig. 3.3 shows the IoET layers, of which there are four: perception/physical layer, access layer, processing layer, and application layer (Asif-Ur-Rahman et al., 2019). The functionalities of these layers are given below.

(a) **Physical layer.**

This layer collects data from various nodes, including the generator subsystem, transmission and distribution subsystem, and consumption subsystem, using various sensors (Asif-Ur-Rahman et al., 2019).

(b) **Access layer.**

This layer consists of end users' routers, switches, and microcontrollers. These nodes collect the data from the physical layer and forward the data to the processing layer via high speed, such as 5G/6G access, networks (Kaiser et al., 2021). The microcontroller can be programmed to detect anomalies at the various sensing nodes using rule-based algorithms (Rahman, Ahmed, & Kaiser, 2016).

(c) **Processing layer.**

The processing layer is also called the cloud layer, where advanced machine learning algorithms are used to provide actionable insights through data analysis and also generate data visualization for the application layer.

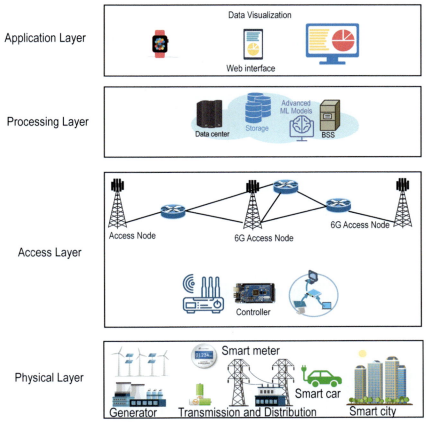

FIG. 3.3 The four IoET layers.

(d) Application layer.

This layer allows all the actors to interact with the systems and provides data visualization and insights.

3.5 IoET-SG system

Combining the IoET and SG will enable us to obtain energy data from the generator, transmission, distribution, and consumer levels (smart homes/buildings/infrastructure, internet of industry things, and so on). The key role of the IoET-SG system is in reducing energy losses and greenhouse gases, and the benefits of such IoET-SG systems will be to help the consumer to understand the usage pattern and policy-makers to understand the generation and demand pattern, reduce the efficient use of energy, reduce energy costs, and make the grid equipment smart (Persia, Carciofi, & Faccioli, 2017). Fig. 3.4 shows

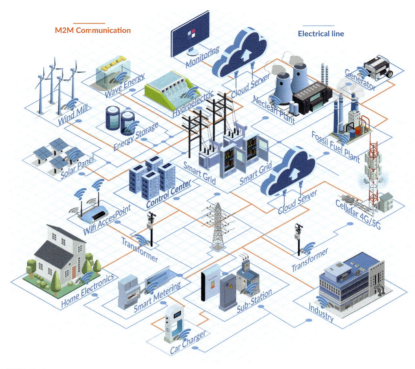

FIG. 3.4 IoET-SG architecture.

IoET-SG architecture, which collects energy data from the generator, transmission, distribution, and consumer levels (smart homes/buildings/infrastructure, internet of industry things, and so on). These data can then be analyzed using a big data platform and machine learning algorithm framework. The users, as well as the policy makers, can obtain actionable insights from these data.

The IoET-SG framework can be used to track, manage, and analyze the following items:

(a) Monitoring:

- Real-time monitoring of the power system (traditional and renewable)
- Distributed generation
- Transmission
- Line fault detection
- Power quality measurement substation

(b) Controlling:

- Power plant
- Smart transformer

- Smart appliance (load)
- Electric vehicle fleet
- Load balancing

(c) Management:

- Consumer (home/office) energy management
- Microgrid management
- Network management
- Battery energy management

(d) Regulation and Market:

- Energy democratization
- Virtual power plant

3.6 Research challenges and future guidelines

In this section, the main challenges and concerns for the implementation of IoET-SG and future research guidelines have been addressed. We have found the following research challenges, and the future research guidelines to these challenges are discussed.

(a) Load Balancing/Management.

The electrical grid must be capable of transmitting electricity in both directions (generating source to consumer and consumer to the grid). However, any fluctuation in the power generated at the end mile of transmission lines to the distribution level presents a challenge to researchers (Hossain, Kaiser, Ali, & Rizvi, 2015). Additionally, capacity for local, regional, and national power surplus storage is required. The output must become more adaptable in real-time using small plants such as heat cogeneration, small-scale run-of-river, and others that can be combined into dynamic, remote-controlled "virtual power plants." Load control/management based on artificial intelligence and machine learning, as well as demand-side management, are needed to balance generation and demand.

(b) Smart Metering.

A smart meter (SM) is a specialized measurement device that determines real-time electricity usage and stores the meter reader at predetermined intervals. The integration of SMs into power grids requires the use of a variety of techniques and applications. The design and interface of a system are determined by the service companies as well as the customer's needs (Diahovchenko, Kolcun, Čonka, Savkiv, & Mykhailyshyn, 2020). The advanced metering infrastructure, consisting of SMs, a communication network, hardware, and software, collects, stores, analyzes, and distributes electricity between the SM and the utility (or consumer). Advanced metering infrastructure transfers big energy data to a server via the communication interface. The choice of wireline

58 Sustainable developments by artificial intelligence & machine learning

or wireless communication interface and the interoperability are challenges for the researchers. Furthermore, the advanced metering infrastructure must cope with measuring uncertainties caused by poor power quality. As a result, an effective strategy must be developed to account for the impact of uncertainty due to poor power quality.

(c) Security and Privacy.

The IoET-SG system consists of millions of online heterogeneous IoET-SG nodes spread over a vast geographical area, and such a system is susceptible to major cyberattacks (Arnold, 2011; Kimani, Oduol, & Langat, 2019). Thus, cyber-security of the IoET-SG is a challenging issue, which attracts researchers. Internet-connected sensors, end devices (such as computers), and network devices are constant targets of online scanning, spying, ransom, theft, and destruction (Nurjahan, Nizam, Chaki, Al Mamun, & Kaiser, 2016). The most common cyber-attacks include device attacks, data attacks, privacy attacks, and network attacks. Therefore, a cyber-attack could result in catastrophic consequences as well as significant financial damage, as such an attack would halt the entire country. Thus, artificial intelligence, bio-inspired, and machine learning-based countermeasures need to be explored (Farhin, Kaiser, & Mahmud, 2021; Mahmud et al., 2018). Furthermore, distributed security measures (such as blockchain and holochain) can be used to address the privacy, trust, and security issues of the IoET-SG and SG (Zaman, Kaiser, Tasin Khan, & Mahmud, 2020).

(d) Big Energy Data.

Since the data collected from the IoET-SG system is big energy data. As a result, extracting actionable insights from it in real-time poses a challenge for researchers. However, parallel processing or stream processing techniques may be employed to extract the knowledge in real-time from the IoT data (Farhin et al., 2021).

(e) User Behavior Prediction.

The prediction of user behavior for energy management in IoT-SG is a challenge. A smart grid must tackle this task and coordinate decentralized power stations with the intelligent home's electricity. A statistical (or probabilistic) and artificial intelligence-based user behavior prediction model can be utilized.

(f) Energy Forecasting.

Energy forecasting plays a vital role in planning and controlling grid operations in the IoET-SG system. Advanced metering infrastructure in the IoET-SG transfers big energy data to a server via the communication interface. Thus, many data analytics applications, such as energy forecasting, have recently appeared. Such applications can be useful for scheduling generation, implementing demand response strategies, and ensuring financial benefits through optimum energy bidding. Many statistical and artificial intelligence/machine learning approaches may be used for forecasting energy generation, energy price, energy demand, and energy consumption.

(g) IoT Standard.

The IoT uses a range of technologies with various specifications for node-to-node linking. Standardization must be achieved for communication and networking, security and privacy protocols, and data fusion. The inconsistency of using various standards in IoT nodes is a new challenge.

(h) Self-healing.

Autonomous detection and recovery from faults is a way to reduce operational uncertainty and costs. In traditional systems, fault detection systems are used for both main grid and microgrid failure management. The key problem is the unidirectional flow of the traditional power systems, while the incorporation of diesel generators and grid-tie systems into the main grid and microgrid formation turns the flow of fault currents into a two-way stream. To solve this challenge, an adaptive system is proposed for the protection scheme with a monitoring function. However, such systems do not provide a self-healing feature, which is a research challenge. Thus, the statistical and machine-learning models can be used to heuristically classify a root cause of the fault and protect the system via isolation (Farhin et al., 2021). This approach can be enhanced by incorporating self-healing systems, which are able to recover the system fault automatically.

Fig. 3.5 shows the summarization of the research challenges, and the future research guidelines to these challenges.

FIG. 3.5 The research challenges and future trends of the IoET-SG system.

60 Sustainable developments by artificial intelligence & machine learning

3.7 Conclusion

The current trend in energy grids and the emergence of the internet of things (IoT) makes energy systems accessible through the internet. SG and IoET applications have many advantages, such as reducing energy cost, saving time, and making the grid equipment intelligent. However, the drawbacks of IoET-SG should not be overlooked. One of the most important limitations is data security and big energy data. In the IoET-SG, each connected device can be a port that transfers personal data to the infrastructure, thus, the concerns raised regarding security flaws, vulnerabilities, threats and data privacy, interoperability, and autonomous decision-making system. This chapter focused on the study, advantages, and challenges of IoT-SG, and discussed effective solutions to these challenges; in addition, we have outlined some primary challenges and general conclusions.

References

Ahmed Abdulkadir, A., & Al-Turjman, F. (2021). Smart-grid and solar energy harvesting in the IoT era: An overview. *Concurrency and Computation: Practice and Experience*, *33*(4). https://doi.org/10.1002/cpe.4896. John Wiley and Sons Ltd.

Arnold, G. W. (2011). Challenges and opportunities in smart grid: A position article. *Proceedings of the IEEE*, *99*(6), 922–927. Institute of Electrical and Electronics Engineers Inc https://doi.org/10.1109/JPROC.2011.2125930.

Asif-Ur-Rahman, M., Afsana, F., Mahmud, M., Shamim Kaiser, M., Ahmed, M. R., Kaiwartya, O., et al. (2019). Toward a heterogeneous mist, fog, and cloud-based framework for the internet of healthcare things. *IEEE Internet of Things Journal*, *6*(3), 4049–4062. https://doi.org/10.1109/JIOT.2018.2876088.

Aurna, N. F., Anika, F. S., Rubel, M. T. M., Kabir, K. H., & Kaiser, M. S. (2021). Predicting periodic energy saving pattern of continuous IoT based transmission data using machine learning model. In *2021 International conference on information and communication technology for sustainable development (ICICT4SD)* (pp. 428–433).

Diahovchenko, I., Kolcun, M., Čonka, Z., Savkiv, V., & Mykhailyshyn, R. (2020). Progress and challenges in smart grids: Distributed generation, smart metering, energy storage and smart loads. *Iranian Journal of Science and Technology - Transactions of Electrical Engineering*, *44*(4), 1319–1333. https://doi.org/10.1007/s40998-020-00322-8.

Ekpe, U. M., & Umoh, V. B. (2019). Comparative analysis of electrical power utilization in Nigeria: From conventional grid to renewable energy-based mini-grid systems. *American Journal of Electrical Power and Energy Systems*, *8*(5), 111–119.

Fang, X., Misra, S., Xue, G., & Yang, D. (2012). Smart grid—The new and improved power grid: A survey. *IEEE Communication Surveys and Tutorials*, 944–980. https://doi.org/10.1109/SURV.2011.101911.00087.

Farhin, F., Kaiser, M. S., & Mahmud, M. (2021). Secured smart healthcare system: Blockchain and bayesian inference based approach. In *Vol. 1309. Advances in intelligent systems and computing* (pp. 455–465). Springer Science and Business Media Deutschland GmbH. https://doi.org/10.1007/978-981-33-4673-4_36.

Hossain, M. R., Kaiser, M. S., Ali, F. I., & Rizvi, M. M. A. (2015). Network flow optimization by genetic algorithm and load flow analysis by Newton Raphson method in power system.

In *2nd international conference on electrical engineering and information and communication technology, iCEEiCT 2015* Institute of Electrical and Electronics Engineers Inc. https://doi.org/10.1109/ICEEICT.2015.7307388.

Kaiser, M. S., Zenia, N., Tabassum, F., Mamun, S. A., Rahman, M. A., Islam, M. S., et al. (2021). 6g access network for intelligent internet of healthcare things: Opportunity, challenges, and research directions. In *Vol. 1309. Advances in intelligent systems and computing* (pp. 317–328). Springer Science and Business Media Deutschland GmbH. https://doi.org/10.1007/978-981-33-4673-4_25.

Kimani, K., Oduol, V., & Langat, K. (2019). Cyber security challenges for IoT-based smart grid networks. *International Journal of Critical Infrastructure Protection*, *25*, 36–49. https://doi.org/10.1016/j.ijcip.2019.01.001.

Luthra, S., Kumar, S., Kharb, R., Ansari, M. F., & Shimmi, S. L. (2014). Adoption of smart grid technologies: An analysis of interactions among barriers. *Renewable and Sustainable Energy Reviews*, *33*, 554–565. https://doi.org/10.1016/j.rser.2014.02.030.

Ma, R., Chen, H. H., Huang, Y. R., & Meng, W. (2013). Smart grid communication: Its challenges and opportunities. *IEEE Transactions on Smart Grid*, *4*(1), 36–46. https://doi.org/10.1109/TSG.2012.2225851.

Mahmud, M., Kaiser, M. S., Rahman, M. M., Rahman, M. A., Shabut, A., Al-Mamun, S., et al. (2018). A brain-inspired trust management model to assure security in a cloud based IoT framework for neuroscience applications. *Cognitive Computation*, *10*(5), 864–873. https://doi.org/10.1007/s12559-018-9543-3.

Nurjahan, Nizam, F., Chaki, S., Al Mamun, S., & Kaiser, M. S. (2016). Attack detection and prevention in the Cyber Physical System. In *2016 International conference on computer communication and informatics, ICCCI 2016* Institute of Electrical and Electronics Engineers Inc. https://doi.org/10.1109/ICCCI.2016.7480022.

Paul, S., Rabbani, M. S., Kundu, R. K., & Zaman, S. M. R. (2014). A review of smart technology (smart grid) and its features. In *Proceedings of 2014 1st international conference on non conventional energy: Search for clean and safe energy, ICONCE 2014* (pp. 200–203). IEEE Computer Society. https://doi.org/10.1109/ICONCE.2014.6808719.

Persia, S., Carciofi, C., & Faccioli, M. (2017). NB-IoT and LoRA connectivity analysis for M2M/IoT smart grids applications. In *Vol. 2017. 2017 AEIT international annual conference: Infrastructures for energy and ICT: Opportunities for fostering innovation, AEIT 2017* (pp. 1–6). Institute of Electrical and Electronics Engineers Inc. https://doi.org/10.23919/AEIT.2017.8240558.

Rahman, S., Ahmed, M., & Kaiser, M. S. (2016). ANFIS based cyber physical attack detection system. In *2016 5th international conference on informatics, Electronics and vision, ICIEV 2016* (pp. 944–948). Institute of Electrical and Electronics Engineers Inc. https://doi.org/10.1109/ICIEV.2016.7760139.

Vandoorn, T. L., Meersman, B., De Kooning, J. D., & Vandevelde, L. (2012). Analogy between conventional grid control and islanded microgrid control based on a global DC-link voltage droop. *IEEE Transactions on Power Delivery*, *27*(3), 1405–1414.

Zaman, S., Kaiser, M. S., Tasin Khan, R., & Mahmud, M. (2020). Towards SDN and Blockchain based IoT countermeasures: A survey. In *2020 2nd international conference on sustainable technologies for industry 4.0, STI 2020* Institute of Electrical and Electronics Engineers Inc. https://doi.org/10.1109/STI50764.2020.9350392.

Chapter 4

Evolution of high efficiency passivated emitter and rear contact (PERC) solar cells

Sourav Sadhukhan[a], Shiladitya Acharya[a], Tamalika Panda[a], Nabin Chandra Mandal[a], Sukanta Bose[a], Anupam Nandi[a], Gourab Das[a], Santanu Maity[a], Susanta Chakraborty[b], Partha Chaudhuri[a], and Hiranmay Saha[a]

[a]*School of Advanced Materials, Green Energy and Sensor Systems (SAMGESS), Howrah, India,* [b]*Department of Computer Science and Technology (CST), Indian Institute of Engineering Science and Technology (IIEST) Shibpur, Howrah, India*

4.1 Introduction

Passivated emitter and rear contact solar cell (PERC) on mono silicon substrate has become a mainstream research area in solar cells because of its promising industrial mass production (Fig. 4.1). Among the many c-Si-based solar cell technologies, back surface field (BSF) cells, which currently have 30% market share, will disappear by 2024, and PERC/PERL/PERT/TOPCon solar cells will dominate the market for the next few years (Fig. 4.1) (International Technology Roadmap for Photovoltaic, 2020).

Aluminum back surface field (BSF) provides a substandard quality of passivation on the rear side of screen printed Al BSF solar cells (De Rose, Magnone, Zanuccoli, Sangiorgi, & Fiegna, 2013; Gatz, Dullweber, & Brendel, 2011; Verhoef & Sinke, 1990; Wijekoon et al., 2013). There is also less optical generation due to a significant portion of infrared light reaching the rear aluminum contact not being reflected back into the silicon substrate; rather, it will absorb the IR part of the solar spectrum, leading to an increase in the cell temperature (Blakers, Wang, Milne, Zhao, & Green, 1989; Lorenz, John, Vermang, & Poortmans, 2010; Wijekoon et al., 2013; Xiao & Shuyan, 2014). These optical and electrical losses can be minimized by PERC architecture (Blakers, 2019; Blakers et al., 1989; Gassenbauer et al., 2012; Gatz, Hannebauer, et al., 2011; Green, 2015) as shown in Fig. 4.2. PERC (passivated emitter and rear cell) structure was initially developed in 1989 by the University of New South Wales in lab scale (Blakers et al., 1989) and redeveloped in 2002 by Fraunhofer

---

Sustainable Developments by Artificial Intelligence and Machine Learning for Renewable Energies.
https://doi.org/10.1016/B978-0-323-91228-0.00007-0
Copyright © 2022 Elsevier Inc. All rights reserved.

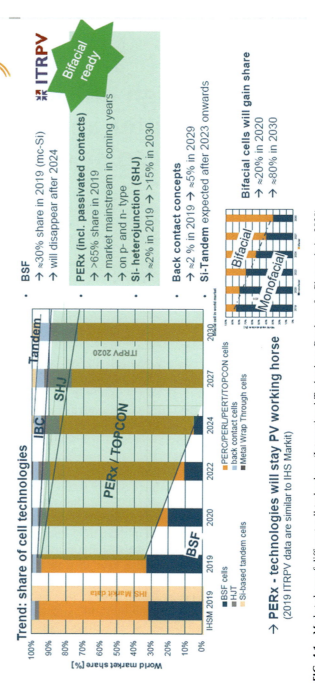

FIG. 4.1 Market share of different cell technology (International Technology Roadmap for Photovoltaic, 2020).

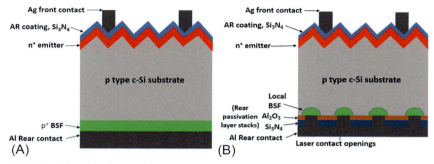

FIG. 4.2 Schematic diagram of (A) p type Al-BSF and (B) p type PERC solar cells.

ISE by using the pilot-line laser fired contact process. Realization of PERC solar cell needs local contact formation by laser ablation (Engelhart et al., 2007; Huang et al., 2017; Schneiderlöchner, Preu, Lüdemann, & Glunz, 2002; Ye et al., 2016) through dielectric layer stacks. Although SiO_2 (silicon oxide) was originally used as rear passivating layer for both p-type and n-type PERC solar cells (Blakers et al., 1989; Xiao & Shuyan, 2014), nowadays Al_2O_3 (aluminum oxide), hafnium oxide (HfO_2), aluminum nitride (AlN), and gallium oxide (Ga_2O_3) are the popular materials being used as passivation layer for p-type substrate, as these provide excellent chemical passivation and field effect passivation too, with boron doped silicon surface, due to the presence of negative fixed charges (Angarita, Palacio, Trujillo, & Arroyave, 2017; Balaji et al., 2015; Cuevas, Allen, Bullock, Wan, & Zhang, 2015; Dingemans & Kessels, 2012; Dingemans, Terlinden, Verheijen, Van de Sanden, & Kessels, 2011; Fan, Sun, Wilkes, & Gupta, 2019; Hsu et al., 2019; Kelly, Abu-Zeid, Arnell, & Tong, 1996; Kim et al., 2019; Koski, Hölsä, & Juliet, 1999; Pawlik et al., 2014; Preu, Lohmüller, Lohmüller, Saint-Cast, & Greulich, 2020; Schmidt et al., 2012; Sharma, 2013). Silicon nitride (Si_3N_4) is generally used as capping layer on the rear side. The Si_3N_4 capping layer on the rear side protects passivating layers from being damaged by Al paste during contact formation and reflects light internally, resulting in better optical generation (Blakers, 2019; Gassenbauer et al., 2012; Gatz, Hannebauer, et al., 2011; Xiao & Shuyan, 2014). The Al_2O_3/Si_3N_4 dielectric layer stack plays three important roles: (i) it provides good chemical passivation; (ii) it provides field effect passivation; and (iii) it provides better reflection of light on the rear side (Preu et al., 2020). From an industrial point of view, PERC solar cells can be easily manufactured by performing only a few add-ons to the existing setup for Al-BSF solar cells. As a result, the PERC cell is gradually becoming the most cost-efficient choice for mass production of crystalline silicon solar cells.

4.2 Photon absorption and optical generation

The absorption of sunlight and creation of electron-hole pairs is an important phenomenon in the operation of solar cells. The excitation of an electron directly from the valence band (which leaves a hole behind) to the conduction band by absorption of light is called fundamental absorption. The momentum and total energy of all particles involved in the absorption process is conserved.

The conduction band minimum occurs at same crystal momentum from that of the valence band maximum in direct bandgap semiconductors. But conduction band minimum occurs at a different crystal momentum from that of valence band maximum in indirect bandgap semiconductors such as Si and Ge. Therefore, to conserve the electron momentum, phonons must be involved in the photon absorption process in indirect bandgap semiconductors. Phonons are low-energy particles with relatively high momentum and represent the lattice vibrations in the semiconductor. This is illustrated in Fig. 4.3. Note that light absorption takes place by either phonon absorption or phonon emission. The absorption coefficient is represented as:

$$\alpha_a(h\nu) = \frac{A(h\nu - E_G + E_{ph})^2}{\left(e^{E_{ph}/kT} - 1\right)} \quad (4.1)$$

$$\alpha_e(h\nu) = \frac{A(h\nu - E_G - E_{ph})^2}{\left(1 - e^{-E_{ph}/kT}\right)} \quad (4.2)$$

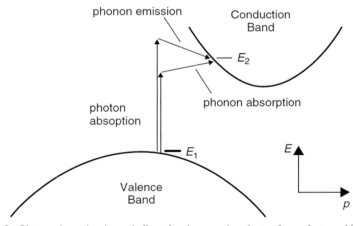

FIG. 4.3 Photon absorption in an indirect bandgap semiconductor for a photon with energy $h\nu < E_2 - E_1$ and a photon with energy $h\nu > E_2 - E_1$. Energy and momentum in each case are conserved by the absorption and emission of a phonon, respectively.

when a phonon is absorbed (α_a) and emitted (α_e), respectively. Since both processes are possible, therefore:

$$\alpha(h\nu) = \alpha_a(h\nu) + \alpha_e(h\nu) \tag{4.3}$$

Since both a phonon and an electron are needed to make the indirect gap absorption process possible, the absorption coefficient depends not only on the density of full initial electron states and empty final electron states. But also on the availability of phonons (both emitted and absorbed) with the required momentum. Thus, compared with direct transitions, the absorption coefficient for indirect transitions is relatively small. As a result, light penetrates more deeply into indirect bandgap semiconductors than direct bandgap semiconductors. This is illustrated in Fig. 4.4 for Si, an indirect bandgap semiconductor, and GaAs, a direct bandgap semiconductor.

The absorption length is a useful quantity for solar materials. It gives the depth where the intensity of light has dropped by a factor of e^{-1} or ~36%.

The optical generation rate, i.e., rate of creation of electron-hole pairs per cm^3 per second as a function of position within a solar cell, can be written as:

$$G(x) = (1-s) \int_\lambda (1 - r(\lambda)) f(\lambda) \alpha(\lambda) e^{-\alpha x} d\lambda \tag{4.4}$$

where s is the grid-shadowing factor, $r(\lambda)$ is the reflectance, $\alpha(\lambda)$ is the absorption coefficient, and $f(\lambda)$ is the incident photon flux (number of photons incident per unit area per second per wavelength).

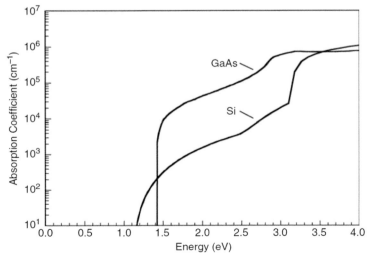

FIG. 4.4 Absorption coefficient as a function of photon energy for Si (indirect bandgap) and GaAs (direct bandgap) at 300 K. Their bandgaps are 1.12 and 1.42 eV, respectively.

68 Sustainable developments by artificial intelligence & machine learning

The sunlight is assumed to be incident at $x=0$. Here, the absorption coefficient has been shown in terms of the light's wavelength through the relationship $h\nu = hc/\lambda$. The photon flux, $f(\lambda)$ is obtained by dividing the incident power density at each wavelength by the photon energy.

The optical generation resulting from an illumination spectrum is determined as the product of the absorbed photon density and the quantum yield (Synopsys, 2014). The quantum yield model describes how many of the absorbed photons are converted to generated electron-hole pairs. Complex refractive index for all materials is the parameter generally used to calculate absorbed photon density in all materials of a solar cell. The complex refractive index can be written as:

$$\tilde{n} = n + i \cdot k \tag{4.5}$$

With:

$$n = n_0 + \Delta n_\lambda + \Delta n_{carrier} + \nabla n_T \tag{4.6}$$

$$k = k_0 + \Delta k_\lambda + \Delta k_{carrier} \tag{4.7}$$

The real part, n, of the complex refractive index is composed of the base refractive index n_0 and correction terms. The correction terms are due to the dependency on wavelength, carrier density, and temperature.

The imaginary part, k, is composed of the base extinction coefficient k_0 and the correction terms. The correction terms are due to the dependency on wavelength and carrier density. To make the results more practical, empirical formulae are used for correction terms.

The wavelength dependency of the complex refractive index can be written as:

$$\Delta n_\lambda = C_{n,\lambda} \cdot \lambda + D_{n,\lambda} \cdot \lambda^2 \tag{4.8}$$

$$\Delta k_\lambda = C_{k,\lambda} \cdot \lambda + D_{k,\lambda} \cdot \lambda^2 \tag{4.9}$$

where $C_{n,\lambda}$, $D_{n,\lambda}$, $C_{k,\lambda}$, and $D_{k,\lambda}$ are constants and linear or spline interpolation is considered if values of n and k for different wavelengths is provided in tabulated format.

The temperature dependency of the complex refractive index can be written as:

$$\Delta n_T = n_0 \cdot C_{n,T} \cdot (T - T_{par}) \tag{4.10}$$

where $C_{n,T}$ is a constant and T_{par} is the temperature at which parameters are considered. The real part of the complex refractive index depends on the absorption by free carriers as:

$$\Delta n_{carr} = -C_{n,carr} \cdot \frac{q^2 \lambda^2}{8\pi^2 c^2 \varepsilon_0 n_0} \left(\frac{n}{m_n} + \frac{p}{m_p} \right) \tag{4.11}$$

where $C_{n,carr}$ is constant, m_n and m_p are effective masses, and n and p are carrier concentrations for electron and hole, respectively; ε_0 is the permittivity, c is speed of light, q is the elementary charge, and λ is the wavelength.

Change in extinction coefficient due to free carrier absorption can be written as:

$$\Delta k_{carr} = \frac{10^{-4}}{4\pi}\left[\left(\frac{\lambda}{\mu m}\right)^{\Gamma_{k,carr,n}}\frac{C_{k,carr,n}}{cm^2}\cdot\frac{n}{cm^{-3}} + \left(\frac{\lambda}{\mu m}\right)^{\Gamma_{k,carr,p}}\frac{C_{k,carr,p}}{cm^2}\cdot\frac{p}{cm^{-3}}\right]$$

(4.12)

Where $C_{k,carr,n}$, $\Gamma_{k,carr,n}$, $C_{k,carr,p}$, and $\Gamma_{k,carr,p}$ are fitting parameters (Synopsys, 2014).

A ray tracing model (Rose et al., 2012; Synopsys, 2014; Xu et al., 2013) is used to calculate absorbed photon and optical generation, as a material absorption occurs for the imaginary component (extinction coefficient, k) of the complex refractive index. The absorption coefficient can be written as:

$$\alpha_\lambda = \frac{4\pi k}{\lambda}$$

(4.13)

The photon absorption rate in an element can be written as:

$$G^{opt}(x, y, z, t) = I(x, y, z)\left[1 - e^{-\alpha L}\right]$$

(4.14)

where, $I(x,y,z)$ is the rate intensity (units of s^{-1}) and L is the length of the light ray in the element.

The absorption rate is calculated in $cm^{-3}\,s^{-1}$. Quantum yield determines the fraction of this value that is to be added in the continuity equation as optical generation. Quantum yield accounts for the thermalization by interband absorption and intraband absorption (Synopsys, 2014).

4.3 Loss mechanisms in PERC solar cells

The maximum theoretical efficiency of a solar cell using a single p-n junction was first calculated by William Shockley and Hans-Joachim Queisser at Shockley Semiconductor in 1961. In this first calculation, the 6000 K black-body spectrum was used as an approximation to the solar spectrum (Shockley & Queisser, 1961). In 2016, Sven Rühle calculated that the theoretical maximum achievable efficiency for a single-junction crystalline silicon solar cell with bandgap of 1.1 eV is 32.23%, with $V_{oc} = 842$ mV, $J_{sc} = 44.23$ mA/cm^2, $FF = 0.86$ at photon flux of AM 1.5G, and with power 1000 W/m^2 (Rühle, 2016). Considering more realistic absorption conditions, mainly free carrier absorption (FCA) and near bandgap photon light trapping, yielded a maximum efficiency of 29.8% (Tiedje, Yablonovitch, Cody, & Brooks, 1984). Richter et al. gave a recent update for crystalline silicon considering the development of knowledge on the efficiency-limiting material constants, which yielded a current maximum efficiency of 29.4% for a spectral photon flux $\Phi_{AM1.5}$ according to the IEC reference spectrum AM1.5 for photovoltaic devices (International Electrotechnical Commission, 2019; Richter, Hermle, & Glunz, 2013). But in practical cases, the commercially available cell efficiency lies far below that due to various losses in solar cells. Since the solar cell is an optoelectronic device, it suffers from mainly optical losses and electrical losses due to various reasons, which are described in the following section.

4.3.1 Optical losses

1. Loss due to reflection
2. Incomplete absorption
3. Shadowing

4.3.2 Electrical losses

1. Resistive losses
2. Recombination losses

4.3.2.1 Loss due to reflection

When sunlight falls on the surface of a solar cell, a part of the incident light beam is reflected. This is known as loss of reflection or reflection loss of the solar cell. The reflection of a bare silicon surface is over 30% due to its high refractive index. This loss due to reflection can be minimized by texturing the front surface and applying antireflection coatings (ARC) (Fig. 4.5). In 1974, Haynos reported the first evidence of reflection reduction in Si solar cells using Si surface texturing (Haynos, Allison, Arndt, & Meulenberg, 1974). A textured surface reduces the reflection by increasing the number of reflections

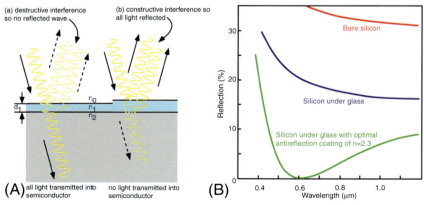

FIG. 4.5 (A) Application of antireflection coating (ARC) to minimize the surface reflection; (B) Comparison of surface reflection from a silicon solar cell.

of a single ray on the surface before being reflected back to the surroundings. Generally, textured surfaces are pyramidal in structure. The pyramids on the surface are random, with a maximum base of 5–6 μm and height of 4–5 μm. Front surface reflectance strongly depends on the size and shape of the texturing (Faust, 1960; Lee, 1969; Price, 1973).

Antireflection coating on a solar cell is a thin layer of dielectric material with a specially chosen thickness so that interference effects in the coating cause the wave reflected from the antireflection coating top surface to be out of phase with the wave reflected from the semiconductor surface. The wavelength in the dielectric material is one quarter the wavelength of the incoming wave. For a quarter wavelength antireflection coating of a transparent material with a refractive index n_1 and light incident on the coating with a free-space wavelength λ_0, the thickness, d_1, which causes minimum reflection is calculated by:

$$d_1 = \frac{\lambda_0}{4n_1} \tag{4.15}$$

Reflection is further minimized if the refractive index of the antireflection coating is the geometric mean of that of the materials on either side; that is, glass or air and the semiconductor. This is expressed by:

$$n_1 = \sqrt{n_0 n_2} \tag{4.16}$$

Mostly, plasma-enhanced chemical vapor deposition (PECVD) coated silicon nitride (SiN_x) is used as an ARC in commercially available solar cells (Gray, 2011; PVEducation, n.d.; Sze & Ng, 2007; Würfel & Würfel, 2016). A typically used SiN_x layer for the front side of PERC devices has a refractive index $n_1 = 2.00$ at $\lambda = 633$ nm and an extinction coefficient $k = 0.0139$ at $\lambda = 350$ nm (Duttagupta, Ma, Hoex, Mueller, & Aberle, 2012). The SiN_x layer thickness, d_1, is in the range of 70 nm $\leq d_1 \leq 80$ nm (Duttagupta et al., 2012; Lohmüller et al., 2020; Lv et al., 2020). Some researchers have also developed double layer antireflection coating (DLARC) to reduce the front surface reflection to a greater extent (Goetzberger, Knobloch, & Voss, 1998). But the development and implementation of DLARC is complex and expensive. In a recent study it is shown that gradient-index SiO_xN_y/Si_3N_4 double-layer antireflection coating (DLARC) can significantly reduce the reflectivity for short wavelengths and the short-circuit current density of the cell can be increased by 0.32 mA/cm^2. Concave pyramid-like textures and SiO_xN_y/Si_3N_4 DLARC can improve the photoelectric conversion efficiency of the PERC solar cell by 0.20% (Zhou et al., 2020).

4.3.2.2 Incomplete absorption

The energy of a photon having wavelength λ can be written as:

$$E = \frac{hc}{\lambda} \tag{4.17}$$

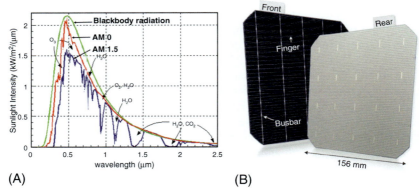

FIG. 4.6 (A) A comparison of solar radiation outside the Earth's atmosphere with the amount of solar radiation reaching the Earth [ASTM] (Sekuler & Blake, 1985) (B) Photograph of front and rear sides of a monofacial p-type Czochralski-grown silicon (Cz-Si) PERC solar cell with five busbar contacts (Werner et al., 2017).

where λ is the wavelength (in m), h is Planck's constant ($6.626 \cdot 10^{-34}$ J s), and c is the speed of light in vacuum ($2.998 \cdot 10^8$ m/s).

Using the values of λ and c in Eq. (4.17), we find that E (in eV) $= 1.24/\lambda$ (in μm). Since Si has a bandgap of 1.12 eV, it can absorb photons having energy ≥ 1.12 eV. Therefore, all the photons of wavelengths between 300 and 1120 nm of solar radiation AM 1.5G (ASTM G173-03) can be absorbed. Fig. 4.6A shows the solar irradiance AM 1.5 along with AM 0 and blackbody radiation (Sekuler & Blake, 1985).

Interband absorption

When a photon is absorbed across the bandgap in a semiconductor, it is absorbed to create an electron-hole pair. The excess energy (photon energy minus the bandgap) of the new electron-hole pair is assumed to thermalize, resulting eventually in lattice heating.

Intraband absorption

In the case of the photon energy being smaller than the bandgap, the photon can be absorbed to increase the energy of a carrier. The excess energy relaxes eventually, contributing to lattice heating.

Free-carrier absorption

In this type of absorption, electrons in the conduction band absorb the energy of a photon and move to an empty state higher in the conduction band (correspondingly for holes in the valence band); this is typically only significant for photons with $E < E_G$ since the free-carrier absorption coefficient increases with increasing wavelength:

$$\alpha_{fc} \propto \lambda^\gamma \tag{4.18}$$

where $1.5 < \gamma < 3.5$.

Thus, in single-junction solar cells, it does not affect the creation of electron-hole pairs and can be ignored (although free-carrier absorption can be exploited to probe the excess carrier concentrations in solar cells for the purpose of determining recombination parameters). However, free-carrier absorption is a consideration in tandem solar cell systems in which a wide bandgap (E_{G1}) solar cell is stacked on top of a solar cell of smaller bandgap ($E_{G2} < E_{G1}$) (Gray, 2011; Preu et al., 2020; PVEducation, n.d.; Sze & Ng, 2007; Würfel & Würfel, 2016).

Single junction solar cells having thicker substrates absorb more sunlight than thinner substrate solar cells. The standard thickness of silicon substrate is 180 μm, and 200 μm and 250 μm cells are also being fabricated. Nowadays, for fabrication of PERC solar cell generally 180–200 μm substrates are being used. It is seen that photons having wavelengths >950 nm of solar irradiance (AM 1.5G) are less absorbed in the substrate and escape from the substrate. These escaped photons may be absorbed in the substrate if they reflect back from the rear side of the cell. The full area Al on the rear side of an Al BSF solar cell can reflect only 60%–70% of escaped photons. To increase the absorption probability of escaped photons, dielectric layer stacks with a lower refractive index than Si are used on the rear side of PERC solar cells (Angarita et al., 2017; Balaji et al., 2015; Cuevas et al., 2015; Dingemans & Kessels, 2012; Dingemans, Terlinden, et al., 2011; Fan et al., 2019; Hsu et al., 2019; Kelly et al., 1996; Kim et al., 2019; Koski et al., 1999; Pawlik et al., 2014; Preu et al., 2020; Schmidt et al., 2012; Sharma, 2013).

4.3.2.3 Shadowing

To generate electricity from solar cells, it is important to collect the photo-generated carriers. For this purpose, metallization is performed on the front and backside of the solar cell. Metal fingers are used on the front side of commercially available solar cells, along with busbars to make the flow of current from cells to the external load. Fig. 4.6B shows the metallization pattern on front and rear sides of a p-type PERC solar cell. But front metal contacts for mono-facial and both side metal contacts for bi-facial solar cells drastically reduce the illumination area of the cells. On the other hand, if the spacing and size of the metal contacts are not optimized then electrical losses may be higher. Therefore, the sizes, shapes, and spacing of fingers and busbars should be well optimized to achieve electrical conductivity as well as higher optical transparency (Horzel et al., 1995; Mette, 2007; Preu, Kleiss, Reiche, & Bucher, 1995).

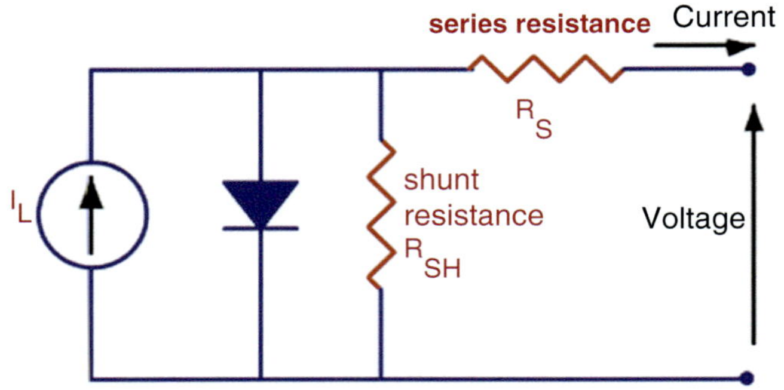

74 Sustainable developments by artificial intelligence & machine learning

FIG. 4.7 Solar cell equivalent circuit with parasitic resistances.

4.3.2.4 Resistive losses

Electrical losses include resistive losses and losses due to the recombination of the charge carriers. Resistive losses come from the combined effect of series resistance (R_s) and shunt resistance (R_{sh}) (Fig. 4.7). Series resistance is the algebraic summation of resistances of all the components that come in the path of the current flow. This includes the resistance of the base, the resistance of the emitter, and the resistance of metal semiconductor contacts at both the front and backside of the solar cell, along with the resistance of metal contacts. Series resistance does not affect the solar cell at open-circuit voltage since the overall current flow through the solar cell is zero. The main effect of series resistance is on the fill factor of the solar cell, and excessively high series resistance reduces the short-circuit current. On the other hand, shunt resistance is the combined effect of cracking, material defects, and improper edge isolation, which provide a path of flow for leakage current inside the solar cell. Fill factor and open circuit voltage are affected by shunt resistance. Short-circuit current density and *FF* are enormously affected due to the resistive losses. The lower value of series resistance is necessary for commercial crystalline silicon solar cells to have better *FF* and higher power conversion efficiency. In contrast, to receive higher short-circuit current density and better *FF*, the magnitude of the shunt resistance should be as high as possible.

The dependence of *FF* on series and shunt resistances can be written as:

$$FF_s = FF_0(1 - r_s) \quad \text{and} \quad FF_{sh} = FF_0\left(1 - \frac{1}{r_{sh}}\right) \tag{4.19}$$

where r_s and r_{sh} are called normalized series and shunt resistances; the fill factor, which is not affected by series and shunt resistances, is denoted by FF_0 (Gray, 2011; PVEducation, n.d.; Sze & Ng, 2007; Würfel & Würfel, 2016).

4.3.2.5 Recombination losses

In a solar cell, recombination is the process by which light-generated excess carriers get recombined. The recombination of the charge carriers in materials is a natural phenomenon. It is the opposite process to that of the process of generation of the charge carriers. Under illumination, electron-hole pairs are generated inside the solar cell due to the absorption of photon energy. But when this illumination is removed, the photo-generated carriers recombine to return to thermal equilibrium. Furthermore, under illumination, not all the photo-generated carriers are collected by the metal contacts in the solar cell. So, recombination also takes place under illumination. There are mainly four types of recombination mechanisms, which take place simultaneously inside the solar cell:

1. Radiative recombination
2. Auger recombination
3. Shockley-Read-Hall (SRH) recombination
4. Surface recombination

Fill factor (FF), open-circuit voltage (V_{oc}), short-circuit density (J_{sc}), and overall efficiency (η) of a solar cell are strongly influenced by all these four recombination processes. Auger recombination results when kinetic energy is imparted to a third carrier. This is also a nonradiative process.

Radiative recombination

This is also known as band-to-band recombination, as both valance and conduction bands are involved in this recombination process. It is exactly the reverse phenomenon of the photo-generation process. In this process, an excited electron falls back directly from the conduction band to the valence band and thus recombines with the hole present in the valence band (Fig. 4.8A). During this transition, the excited electron releases its energy in terms of photons, i.e., just opposite to the generation of the electron-hole pair by absorbing the photon energy. The expression for band-to-band radiative recombination for a nondegenerate semiconductor under nonequilibrium conditions is given as:

$$U_{rad} = B_r \left(np - n_i^2\right) \quad (4.20)$$

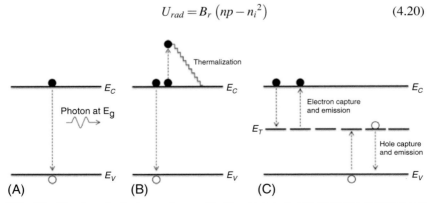

FIG. 4.8 The three basic types of recombination processes: (A) Radiative band-to-band; (B) nonradiative Auger; and (C) nonradiative recombination centers (traps).

where B_r is called the radiative recombination coefficient, and depends on a material's bandgap as well as the strength of the absorption coefficient.

Values range from 10^{-9} to 10^{-11} cm^3/s for direct bandgap materials and from 10^{-13} to 10^{-15} cm^3/s for indirect bandgap materials.

$$\tau_{n,rad} = \frac{1}{B_r N_A} \tag{4.21}$$

where $\tau_{n,rad}$ is the electron minority carrier radiative lifetime, i.e., the lifetime of the electrons in p-type materials. Since B_r is very small for an indirect bandgap semiconductor, radiative processes are very slow and much less probable in indirect bandgap materials (Fischer, 2003; Rahman, 2012).

Auger recombination

Auger recombination is a three-particle process that is more significant at the high level of carrier concentration. In this process, an electron in the conduction band recombines with a hole in the valence band but the excited electron does not release its energy in terms of photon energy. The excited electron transfers its energy to another electron present in the conduction band. After receiving the excess energy, the second electron, now having higher kinetic energy, goes into higher energy levels, as shown in Fig. 4.8B. After a certain time, the second electron again comes back to the conduction band edge from the higher energy levels. The same phenomenon may also happen in the valence band. In this case, two holes and one electron are required, exactly opposite to the previous case, where two electrons and one hole are involved. For doped materials under low injection conditions, Auger lifetime ($\tau_{n,p}(s)$) can be written as:

$$\tau_{n,Auger} = \frac{1}{B_{Auger,n} N_A^2} \tag{4.22}$$

for electrons in p-type material and:

$$\tau_{p,Auger} = \frac{1}{B_{Auger,p} N_B^2} \tag{4.23}$$

for holes in n-type materials.

Values of B_{Auger} for Si and GaAs are both in the order of 10^{-31} cm^6/s. In indirect bandgap materials, since the Auger processes are also able to conserve momentum, these processes are the dominant recombination pathway, and thus are the efficiency-limiting loss mechanism for high purity Si or Ge solar cells (Fischer, 2003; Rahman, 2012; Tyagi & Van Overstraeten, 1983).

Shockley-Read-Hall (SRH) recombination

Trap-assisted nonradiative recombination is modeled by Shockley-Read-Hall (Shockley & Read, 1952) and is often the practical limiting loss mechanism in most solar cells. The presence of energy levels in between the valence band

and conduction band in a semiconductor material is responsible for this SRH recombination mechanism. This type of recombination is also known as trap-assisted recombination as the presence of energy levels in between the valence band and the conduction band acts as a trap center of the carriers (Fig. 4.8C). Structural defects and impurities in a material give rise to the creation of these mid gap energy levels. As the number of defects increases, the SRH recombination rate also increases.

The electron occupation probability, f^n, of a trap can be written as (Synopsys, 2014):

$$f^n = \frac{\sum c_i^n}{\sum (c_i^n + e_i^n)} \tag{4.24}$$

where c_i^n denotes an electron capture rate for an empty trap and e_i^n denotes an electron emission rate for a full trap.

For a single defect level, SRH recombination rate can be written as:

$$U_{SRH} = \frac{np - n_i^2}{\tau_{p0}(n + n_1) + \tau_{n0}(p + p_1)} \tag{4.25}$$

where:

$$n = (n_0 + \triangle n) \text{ and } p = (p_0 + \triangle p) \tag{4.26}$$

n_0 and p_0 are equilibrium carrier concentrations, $\triangle n$ and $\triangle p$ are excess carrier concentrations for electron and hole, respectively, and τ_{p0} and τ_{n0} are the fundamental hole and electron lifetimes, which are related to the thermal velocity of charge carriers v_{th}, the density of recombination defects N_t, and the capture cross-sections σ_n and σ_p, for the specific defect, and written as:

$$\tau_{p0} = \frac{1}{\sigma_p v_{th} N_t} \quad \text{and} \quad \tau_{n0} = \frac{1}{\sigma_n v_{th} N_t} \tag{4.27}$$

n_1 and p_1 are statistical factors defined as:

$$n_1 = n_i e^{\left(E_{trap} - E_i\right)/kT} \quad \text{and} \quad p_1 = n_i e^{\left(E_i - E_{trap}\right)/kT} \tag{4.28}$$

where n_i is intrinsic carrier concentration, E_i is intrinsic Fermi level energy, E_{trap} is energy of trap level, k is Boltzmann constant 1.3806×10^{-23} J/K, and T is absolute temperature (Black, 2016; Fischer, 2003; Rahman, 2012; Sharma, 2013; Tyagi & Van Overstraeten, 1983).

For doped semiconductors under low-level injection conditions, we can simplify the SRH recombination for p type doping such that (Black, 2016; Fischer, 2003; Rahman, 2012; Tyagi & Van Overstraeten, 1983):

$$U_{SRH} \approx \frac{\triangle n}{\tau_{n0}} \tag{4.29}$$

Surface recombination

In the previous recombination mechanism, we have observed that defect states and impurities inside the material acting as trap centers of the charge carriers. Similarly, the surface of crystalline silicon is also not defect-free or impurity-free. Discontinuity in the atomic arrangement leads to the growth of unfinished bonds at the surface of silicon. These unfinished bonds result in the presence of energy levels in the bandgap of the material. Therefore, these mid gap energy levels can trap the carriers effectively. In general, there are a large number of defects at the surface of silicon. This large number of defects is sufficient to produce higher surface recombination velocities of the carriers (Fig. 4.9).

For a single defect at the surface, the rate of surface recombination, U_{Surf}, is written as:

$$U_{SRH, surf} = \frac{n_s p_s - n_i^2}{\frac{1}{S_{p0}}(n_s + n_1) + \frac{1}{S_{n0}}(p_s + p_1)} \quad (4.30)$$

where $n_s = (n_{s0} + \Delta n_s)$ and $p_s = (p_{s0} + \Delta p_s)$, n_{s0}, p_{s0} are equilibrium carrier concentrations and Δn_s, Δp_s are excess carrier concentrations for electron and hole at surface, and S_{n0} and S_{p0} are related to the density of surface states per unit area, N_{ts}, and the capture cross-sections, σ_n and σ_p, for the specific defect:

$$s_{p0} = \sigma_p N_{ts} \vartheta_{th} \quad \text{and} \quad s_{n0} = \sigma_n N_{ts} \vartheta_{th} \quad (4.31)$$

For p-type materials near the surface, the SRH recombination expression simplifies to:

$$U_{SRH, surf} \approx S_n \Delta n_s \quad (4.32)$$

So the surface recombination velocity can be related to the fundamental properties of the surface defects through Eq. (4.30) (Black, 2016; Rahman, 2012; Sharma, 2013).

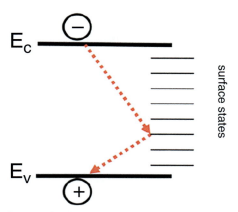

FIG. 4.9 Schematic diagram of surface recombination mechanism.

As has already been stated at the starting of this section, these four recombination mechanisms can take place simultaneously inside the semiconductor material. The effective carrier lifetime, τ_{eff}, is given by:

$$\frac{1}{\tau_{eff}} = \frac{1}{\tau_{Bulk}} + \frac{1}{\tau_{surface}} \tag{4.33}$$

Considering radiative, Shockley-Read-Hall (SRH), and Auger recombination, the bulk lifetime can be expressed as:

$$\frac{1}{\tau_{Bulk}} = \frac{1}{\tau_{Radiative}} + \frac{1}{\tau_{Auger}} + \frac{1}{\tau_{SRH}} \tag{4.34}$$

4.4 Carrier transport equations

The total hole and electron currents (vector quantities) are the sum of their drift and diffusion components (Boukortt, Patanè, & Hadri, 2019; Gray, 2011; Sze & Ng, 2007; Würfel & Würfel, 2016):

$$\vec{J}_p = \vec{J}_p^{drift} + \vec{J}_p^{diff} = q\mu_p p \vec{E} - qD_p \nabla p = -q\mu_p p \nabla\varnothing - qD_p \nabla p \tag{4.35}$$

$$\vec{J}_n = \vec{J}_n^{drift} + \vec{J}_n^{diff} = q\mu_n n \vec{E} + qD_n \nabla n = -q\mu_n n \nabla\varnothing + qD_n \nabla n \tag{4.36}$$

Under nonequilibrium conditions, we can replace the Fermi energy in our carrier density formulae with the quasi-Fermi energy, such that (Gray, 2011; Sze & Ng, 2007; Würfel & Würfel, 2016):

$$\begin{aligned} n &= N_i e^{(E_{Fn} - E_i)/kT} \\ p &= N_i e^{(E_i - E_{Fp})/kT} \end{aligned} \tag{4.37}$$

Current flow, recombination and generation can be related to one another though the continuity equations. These equations are simply a statement of the conservation of charge for each carrier type. The continuity equations for electrons and holes are (Gray, 2011; Sze & Ng, 2007; Würfel & Würfel, 2016):

$$\begin{aligned} \frac{\partial n}{\partial x} &= (G_n - U_n) + \frac{1}{q}\vec{\nabla}\cdot\vec{J}_n \\ \frac{\partial p}{\partial x} &= (G_p - U_p) - \frac{1}{q}\vec{\nabla}\cdot\vec{J}_p \end{aligned} \tag{4.38}$$

In order to apply the continuity equations to the analysis of a pn junction, we make a set of simplifying assumptions to arrive at a useful set of transport equations that can be solved in closed form:

- Analysis will be restricted to the quasineutral regions of the diode such that $E \approx 0$. Also, we assume any disturbance from equilibrium is small (low level injection) so that the thermal equilibrium values are basically unchanged (e.g., $n \cong n_0$ and $\nabla p \ll n_0$ for n-type material).

80 Sustainable developments by artificial intelligence & machine learning

- The thermal carrier concentrations (n_0, p_0) are assumed to be independent of position. Therefore, we are only concerned with derivatives with respect to excess electron and hole populations.
- For solar cells, we are typically only concerned with behavior at a fixed bias voltage and with constant illumination from the sun. Thus, operation in the steady state leads to $\frac{\partial p}{\partial t} = \frac{\partial n}{\partial t} = 0$.
- Under the low-level injection condition, we can replace the net recombination term with the recombination mechanisms by:

$$U_n = \nabla n \left(\frac{1}{\tau_{Radiative}} + \frac{1}{\tau_{Auger}} + \frac{1}{\tau_{SRH}} \right) = \frac{\nabla n}{\tau_n} \qquad (4.39)$$

where τ_n is the net minority carrier lifetime due to radiative, Auger, and SRH recombination.

- The only source of generation is photogeneration, so that electrons and holes are produced in pairs and $G_n = G_p = G_L$.

Applying these assumptions, the transport equations for minority carriers become:

$$D_n \frac{\partial^2 \nabla n}{\partial x^2} - \frac{\nabla n}{\tau_n} + G_L = 0 \qquad (4.40)$$

for minority carrier electrons in a p-type material, and:

$$D_p \frac{\partial^2 \nabla p}{\partial x^2} - \frac{\nabla p}{\tau_n} + G_L = 0 \qquad (4.41)$$

for minority carrier holes in an n-type material.

Eqs. (4.40) and (4.41) are known as the minority carrier diffusion equations. These represent the working relationships that, in combination with Poisson's equation, are used to model the carrier densities and currents in a minority carrier device, with respect to applied voltage and illumination (Belghachi, 2013; Boukortt et al., 2019; Gray, 2011; PVEducation, n.d.; Sze & Ng, 2007; Würfel & Würfel, 2016).

4.4.1 Solar cell parameters

Four basic parameters characterize a solar cell. The parameters are briefly discussed below:

(a) **Short-Circuit Current (I_{sc}):** The maximum amount of current following through the solar cell when its terminals are made short, i.e., the voltage across the solar cell terminals becomes zero. In this case, the short-circuit current will be equal to the photogenerated current. Theoretically, the maximum possible current density in a single junction c-Si solar with AM 1.5G and input power 1000 W/m^2 is 46.3 mA/cm^2 (IEC, 2008). In a solar cell with perfectly passivated surface and uniform generation, the equation for the short-circuit current can be approximated as:

$$J_{sc} = qG(L_n + L_p) \qquad (4.42)$$

where G is the generation rate and L_n and L_p are the electron and hole diffusion lengths, respectively.

Although this equation makes several assumptions that are not true for the conditions encountered in most solar cells, the above equation nevertheless indicates that the short-circuit current depends strongly on the generation rate and the diffusion length.

(b) **Open Circuit Voltage (V_{oc}):** The maximum voltage a solar cell can generate when its terminals are made open. During this time, the cell current will automatically become zero. Mathematically, the equation for open circuit voltage of a solar cell under illumination is (Boukortt et al., 2019; Gray, 2011; PVEducation, n.d.; Sharma, 2013; Würfel & Würfel, 2016):

$$V_{oc} = \frac{nKT}{q} \ln\left(\frac{J_{sc}}{J_0} + 1\right) \qquad (4.43)$$

where J_0 = recombination current density or dark saturation current density and n is the ideality factor.

Now, recombination current density can be written as (Gray, 2011; PVEducation, n.d.; Würfel & Würfel, 2016):

$$J_0 = J_{01} + J_{02} \qquad (4.44)$$

where J_{01} = recombination current density due to recombination in the quasi-neutral regions and J_{02} = recombination current density due to recombination in the space-charge region.

For a p-type BSF solar cell, J_{01} can be written as (Gray, 2011):

$$J_{01} = J_{01,p} + J_{01,n} \qquad (4.45)$$

For an n-type emitter region:

$$J_{01,p} = q\frac{n_i^2 D_p}{N_D L_p} \left\{ \frac{D_p/L_p \sinh\left[(W_n - x_n)/L_p\right] + S_{F,eff} \cosh\left[(W_n - x_n)/L_p\right]}{D_p/L_p \cosh\left[(W_n - x_n)/L_p\right] + S_{F,eff} \sinh\left[(W_n - x_n)/L_p\right]} \right\}$$

$$(4.46)$$

And for a p-type base region:

$$J_{01,n} = q\frac{n_i^2 D_n}{N_A L_n} \left\{ \frac{D_n/L_n \sinh\left[(W_p - x_p)/L_n\right] + S_{BSF} \cosh\left[(W_p - x_p)/L_n\right]}{D_n/L_n \cosh\left[(W_p - x_p)/L_n\right] + S_{BSF} \sinh\left[(W_p - x_p)/L_n\right]} \right\}$$

$$(4.47)$$

82 Sustainable developments by artificial intelligence & machine learning

where D_p, L_p, $S_{F, eff}$ are diffusivity, diffusion length, and effective surface recombination velocity, respectively, for minority holes in an n-type emitter, and D_n, L_n, and S_{BSF} are the same for minority electrons in a p-type base region; Wp and Wn are the thicknesses of emitter and base regions, respectively, xn and xp are width of depletion region in emitter and base regions, respectively, and J_{02} can be written as (Gray, 2011):

$$J_{02} = q\frac{n_i W_D}{\tau_D} \tag{4.48}$$

From Eq. (4.48), it is seen that J_{02} is bias-dependent, since the depletion width, W_D, is a function of the applied voltage. Recombination current due to recombination in the depletion region (J_{02}) can be ignored, which is a reasonable and common assumption for a good solar cell, especially at larger forward biases.

From Eq. (4.43) it can be understood that open-circuit voltage (V_{oc}) of a solar cell is dependent on short-circuit current density (J_{sc}) and reverse saturation current density (J_0). But chances of variation in reverse saturation current are higher than the short-circuit current. Again, reverse saturation current is proportional to the net recombination in the solar cell. Therefore, to obtain higher V_{oc}, lower reverse saturation current is required, i.e., recombination of the photo-generated carriers should be as low as possible.

The V_{oc} can also be determined from the carrier concentration as (Gray, 2011; Hsu et al., 2019; PVEducation, n.d.; Sze & Ng, 2007; Würfel & Würfel, 2016):

$$V_{oc} = \frac{kT}{q} \ln\left(\frac{(N_A + \nabla n)\nabla n}{n_i^2}\right) \tag{4.49}$$

where kT/q is the thermal voltage, N_A is the acceptor type doping concentration, Δn is the excess carrier concentration, and n_i is the intrinsic carrier concentration.

The determination of V_{oc} from the carrier concentration is also termed "implied V_{oc}."

(c) **Fill Factor (FF):** Fig. 4.10 represents the I-V curve of a solar cell. FF is defined as the ratio of the maximum power output of the solar cell ($V_{mpp}I_{mpp}$) to the product of open-circuit voltage (V_{oc}) and short-circuit current (I_{sc}). FF is given as:

$$FF = \frac{V_{mpp}I_{mpp}}{V_{oc}I_{sc}} \tag{4.50}$$

The effect of resistive losses, including recombination losses, limits the fill factor of a solar cell. To obtain a high value of FF, electrical losses must be lower.

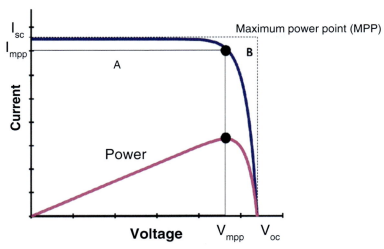

FIG. 4.10 *I-V* curve of a solar cell.

(d) **Efficiency (η):** The efficiency of a solar cell is determined as the fraction of incident power that is converted to electricity and is defined as (Gray, 2011; PVEducation, n.d.; Würfel & Würfel, 2016):

$$\eta = \frac{V_{oc}I_{sc}FF}{P_{in}} \quad (4.51)$$

The efficiency of a solar cell is measured under STC (standard test condition), which corresponds to $P_{in} = 100$ mW/cm², AM1.5G, and temperature = 25°C.

4.5 PERC technology

In 1989, at UNSW (the University of New South Wales), A.W. Blakers and his team were the first to try to develop PERC solar cells (Blakers et al., 1989). In 2002, researchers at the Fraunhofer ISE institute reported 22.1% of conversion efficiency of PERC solar cells by applying a laser fire contact (LFC) process (Schneiderlöchner et al., 2002). Since then, scientists, researchers, and engineers around the world have seriously attempted PERC solar cells in the PV industry. As discussed in the introduction, different cost-efficient techniques/processes like surface passivation applying dielectric materials through low temperature deposition techniques, screen-printed Al LBSF, and laser ablation have been utilized to enhance the performance of the cell. Credit for the industrialization of PERC cells goes to Trina Solar who, in 2015, demonstrated cell efficiency of 22.13% using mono-crystalline p-Si wafers and 21.25% on multi-crystalline p-Si wafers in the production line (Energytrend, 2015). Since then,

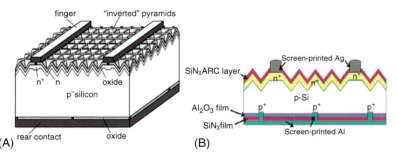

FIG. 4.11 (A) Schematic drawing of the lab PERC cell (Benick, Bateman, & Hermle, 2010); (B) Schematic drawing of the industrial PERC cell (Huang et al., 2017).

different PV companies have been associated with the production of PERC cells, and it has become the most cost-efficient option for mass production of c-Si solar cells. At the start of 2020, the efficiency of PERC solar cells was above 22% in industrial production lines.

The schematics of lab PERC and industrial PERC are represented in Fig. 4.11A and B, respectively. In general, the following differences can be observed:

1. Cost-efficient industrial processes are used to replace high-cost lab processes:
   (a) CZ c-Si wafers are used to replace high-cost float zone (FZ) c-Si wafers;
   (b) Inverted pyramids are replaced with random pyramids;
   (c) Antireflective coating (ARC) of ZnS/MgF_2 deposited through physical vapor deposition method (PVD) is replaced by using plasma-enhanced chemical vapor deposition (PECVD) SiN_x;
   (d) Rear dielectric passivation is done using low-temperature dielectric deposition processes instead of high-temperature thermal oxidation;
   (e) Laser ablation technique is used for dielectric opening in place of photolithography;
   (f) Metallization using PVD (evaporation) (front side: Ti/Pd/Ag, rear side: Al) is replaced with screen-printing (front side: Ag paste, rear side: Al paste).
2. It is observed that the rear contact in lab PERC cells made by PVD Al with ~400°C postmetallization anneal (PMA) cannot provide p^+-BSF, but both p^+-LBSF and rear local contacts can be facilitated by screen-printed and RTP (rapid thermal processing) fired Al (real peak firing temperature ~720–750°C, time: 2–4 s) due to the properties of Al (dopants and metallization) after screen-printing and firing for industrial PERC cells.

Alternatively, the difference in the cell structure of the industrial PERC to lab PERC brings significant impacts on cell device design, as discussed in the following. Although the lab PERC (PVD Al with a PMA of ~400°C) shows lower

SRV as compared to industrial PERC (screen-printed and RTP fired Al) due to superior purity of material, PVD Al with a PMA does not provide BSF, due to which recombination of minority carriers in the cell rear cannot be suppressed. Photo-generated minority carriers recombine in the rear local Al contact regions as these contacts act like sinks to suck those photo-generated minority carriers. Hence, rear side proper passivation using dielectric material is absolutely necessary in order to accomplish a low rear (back) SRV (BSRV), which also makes the design window of cell-rear parameters narrower. Further, the rear contact pitch needs to be well spaced out so that the minority carriers have to pass through a long distance, longer than the $L_{D,\ bulk}$, to arrive at the rear contact regions. On the other hand, the collection of the photo-generated majority carriers can be hampered owing to a long distance of the rear contact regions. This, in turn, increases lateral transportation resistance. Therefore, a trade-off is required to solve this contradiction. One thing to note is that, because of a much smaller dimension than the contact pitch and bulk minority carrier diffusion length ($L_{D,\ bulk}$), the size of the rear contact window is not that significant for determining the total recombination in lab PERC cells.

For the industrial PERC cell, due to the multifunctions of LBSF and rear local contacts using screen-printed and fired Al, the design tolerance of rear local contact pattern (pitch and fraction, especially the former) is broad.

4.5.1 PERC process flow

Fig. 4.12 shows the process flow diagram for fabrication of p-type PERC solar cell.

4.5.2 Surface passivation

Reduction in surface recombination velocity is indispensable for solar cells. The mechanisms of surface passivation can be classified into two categories, namely field-effect and chemical passivation (Fig. 4.13).

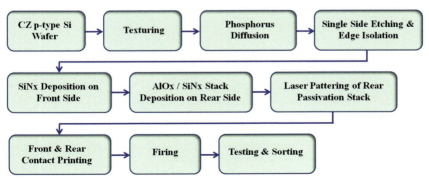

FIG. 4.12 Process flow diagram of industrial p-type PERC solar cell fabrication.

86 Sustainable developments by artificial intelligence & machine learning

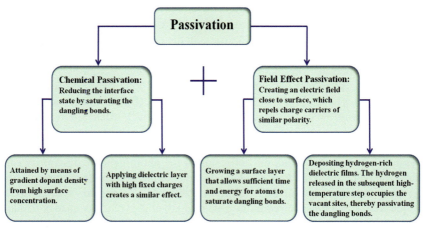

FIG. 4.13 Different types of passivation.

Passivation of the surface of the solar cell for the reduction of the SRV using dielectric materials comprises both of the abovementioned mechanisms. Well-known dielectric materials like silicon oxide (SiO_2), silicon nitride (SiN_x), aluminum oxide (Al_2O_3), hafnium oxide (HfO_2), aluminum nitride (AlN), and gallium oxide (Ga_2O_3) are used worldwide for dielectric surface passivation. As shown in Fig. 4.14, different dielectrics have different fixed charge

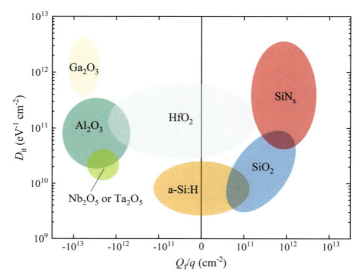

FIG. 4.14 Interface defect density (D_{it}) and fix charge density (Q_f) of common silicon surface passivation layers (Cuevas et al., 2015; Preu et al., 2020).

density (Q_f) and interface defect density (D_{it}), so the share of field-effect or chemical passivation in the total surface passivation is also different between various dielectrics.

Great care is required when applying the dielectric passivation method in the practical solar cell fabrication process, as dielectric passivation is done in the middle of the processing steps in cell fabrications; consequently, the passivation quality can be deteriorated due to the effect of the other processing steps, as those steps could be responsible for damage to the dielectric layer. The most important benchmark to achieve significant surface passivation of solar cells is the "final passivation," which is the passivation measured after completion of the co-firing step with screen-printed contacts (front and back), rather than a good quality preliminary or intermediary passivation. So special attention is required in monitoring the intermediary passivation quality after each processing step (e.g., RTP firing, laser ablation, and screen printing) to maintain or even enhance "final passivation" quality.

4.5.2.1 Passivation by SiO_2

Generally, bare silicon possesses surface recombination velocity (SRV) higher than 10^5 cm/s. This high value of SRV is not suitable for high-efficiency solar cells. Having an extremely low D_{it} (10^9–10^{10} eV^{-1} cm^{-2}) and moderate positive Q_f ($\sim 10^{10}$–10^{11} cm^{-2}), thermally grown silicon dioxide (SiO_2) represents one of the most suitable surface passivation dielectric material for both the n-Si and p-Si substrates (Aberle, 1999; Aberle, Glunz, & Warta, 1992; Eades & Swanson, 1985; Füssel, Schmidt, Angermann, Mende, & Flietner, 1996; Glunz, Biro, Rein, & Warta, 2012; Jin, Weber, Dang, & Jellett, 2007; King, 1990; Reed & Plummer, 1988; Wang, 1992). By applying thermally grown SiO_2 as a surface passivation layer, a very low surface recombination velocity of about 41 cm/s can be obtained for 0.7 Ω-cm FZ p-type Si wafers (King, Sinton, & Swanson, 1987). Again, it was reported that evaporated Al on the top of the oxide and subsequent annealing at a higher temperature in a forming gas atmosphere can reduce the surface recombination velocity effectively (Kerr & Cuevas, 2002). Furthermore, SiO_2 capped with Al can act as an excellent back reflector, which helps to reflect the unabsorbed light into the absorbing layer of the solar cells. This results in a better red response of the solar cells. However, the requirement of a high-temperature oxidation process for the development of high-quality thermal oxide is a major drawback for achieving high throughput. It is also observed that the passivation process with high temperature causes significant degradation of the bulk lifetime, especially in the case of lower cost defective multicrystalline silicon wafers. Here, PECVD SiO_x comes out as a potential candidate for the solar PV industry owing to a higher deposition rate and lower processing temperature ($<450°C$). It was reported that in the as-deposited state, a single-layer PECVD SiO_x possesses a D_{it} of

88 Sustainable developments by artificial intelligence & machine learning

$\sim 10^{11} \, eV^{-1} \, cm^{-2}$ and a positive Q_f of $\sim 10^{11} \, cm^{-2}$ (Dingemans, van de Sanden, & Kessels, 2011; Duttagupta, Ma, Hoex, & Aberle, 2014; Leguijt et al., 1996). A subsequent postanneal treatment in a forming gas atmosphere can effectively enhance its passivation feature by reducing the D_{it} ($\sim 10^{11} \, eV^{-1} \, cm^{-2}$) at the Si-SiO$_2$ interface and regulating the Q_f ($\sim 10^{11} \, cm^{-2}$) in the interface (Dingemans, Mandoc, Bordihn, Van De Sanden, & Kessels, 2011). During the postannealing treatment, the H atom in the forming gas can diffuse into the interface of SiO$_x$/Si and Si bulk, hence passivating both the dangling bonds at the surface and defects in Si bulk.

4.5.2.2 Passivation by SiN$_x$

PECVD coated silicon nitride (SiN$_x$) has been used as an antireflection coating (ARC) over the years in commercial solar cells. It has excellent surface passivation properties of n$^+$-emitter (Aberle, 1999; Kerr, 2002) along with effective antireflection properties. It possesses a fairly strong positive Q_f of about 10^{11}–$10^{12} \, cm^{-2}$ (Dauwe, Mittelstadt, Metz, & Hezel, 2002; Sharma, 2013) along with a medium level of D_{it} ($\sim 10^{11} \, eV^{-1} cm^{-2}$) (Aberle, 1999; Kerr, 2002) and hence it can effectively passivate the phosphorous doped emitter (a wide range of doping level) on p-type silicon. Surface recombination velocity of as low as 10 cm/s has been reported by using SiN$_x$ as a surface passivation layer on p-type Si (Lauinger, Schmidt, Aberle, & Hezel, 1996). However, due to its high positive charge density, it is not a suitable material to use as a back passivation layer in p-PERC (PERC solar cells using p-type wafer as substrate) solar cells as electrons from the back are attracted by this highly dense positive charge. This causes parasitic shunting and hence reduces the short-circuit current density (J_{sc}). However, Chen et al. (2007) reported that passivation of p$^+$-emitter is also possible if the PECVD SiN$_x$ layer possesses proper Si—N bond density and also adequate thermal treatment after deposition. Although SiN$_x$ is generally not used for backside passivation, an appropriate thin thermal oxide/SiN$_x$ stack can provide excellent surface passivation quality along with good back reflective properties without parasitic shunting (Schultz, Hofmann, Glunz, & Willeke, 2005).

4.5.2.3 Passivation by Al$_2$O$_3$

An Al$_2$O$_3$ layer can provide excellent field-effect passivation due to its high negative fixed charge density ($Q_f = 10^{12}$–$10^{13} \, cm^{-2}$) along with good chemical passivation ($D_{it} \approx 10^{10}$–$10^{11} \, eV^{-1} cm^{-2}$) on the p-Si substrate, as a result of which it can provide very low surface recombination velocity (Angarita et al., 2017; Balaji et al., 2015; Dingemans & Kessels, 2012; Dingemans, Terlinden, et al., 2011; Fan et al., 2019; Kelly et al., 1996; Kim et al., 2019; Koski et al., 1999; Pawlik et al., 2014; Schmidt et al., 2012; Sharma, 2013). Apart from this, different characteristics like antiultraviolet radiation, good low-light performance, and resistance to humidity make Al$_2$O$_3$ the most suitable candidate to bring benefits to module performance in different environments (Dingemans & Kessels, 2012). In 2015, Trina Solar was able to produce industrial PERC using Al_2O_3 passivation (Energytrend, 2015).

In recent years, Al_2O_3 passivation has become the conventional technology for surface passivation of p-Si and is globally applied in PERC cell processing in both lab and industry. Different techniques like plasma-enhanced chemical vapor deposition (PECVD), atomic layer deposition (ALD), atmospheric pressure chemical vapor deposition (APCVD), and physical vapor deposition (PVD) can be used to synthesize Al_2O_3 (Dingemans & Kessels, 2012; Foroughi-Abari & Cadien, 2012; Granneman, Vermont, Kuznetsov, & Coolen, 2010; McIntosh, Provancha, & Black, 2011; Preu et al., 2009; Schmidt et al., 2008; Veith et al., 2012; Werner et al., 2010). Among these methods, ALD is considered to be one of the most suitable deposition techniques for industry usage owing to some exceptional advantages like pinhole-free coating, mono-layer film growth control, and low substrate temperature process (Dingemans & Kessels, 2012). The conventional chemical vapor deposition (CVD) method is modified to design the ALD system. In an ALD process for deposition of Al_2O_3, the two precursors trimethylaluminum ($Al(CH_3)_3$) and water (H_2O) or ozone (O_3) are alternatively chemically reacting into the vacuum chamber to react at the substrate surface, thus attaining a mono-layer growth controlled (able to control precisely at the atomic scale) and self-limited reaction at the surface. The basic features are presented in Fig. 4.15. In an ideal ALD process, the reaction happens only on the substrate surface (Chunduri & Schmela, 2019). Generally, ALD systems are of two types: batch ALD and spatial ALD. Batch ALD are further categorized into thermal ALD and plasma-assisted ALD.

The conventional ALD systems are generally batch type and time-based, having very slow reactors. To overcome this constraint, innovative product platforms have been developed by equipment manufacturers. Keeping the same operating principle, i.e., alternate dosing of the precursor and oxidant is done along with purging being carried out in the identical reactor chamber, products have been adapted to larger batch sizes. The deposition of the Al_2O_3 layer on the wafer rims along with the peripherals of the front side, known as the wraparound deposition, is one of the inherent limitations that these systems suffer from. As a workaround, these systems are promoted by the manufacturers for both side deposition of the wafer. If the aluminum oxide is deposited after the deposition of silicon nitride on the emitter side, then the deposited Al_2O_3 layer (having fixed negative charge) will not able to get in touch with the emitter surface of the silicon; instead, it has the contact with the silicon nitride. Thus, the formation of the inversion layer is avoided. This type of ALD batch tool is supplied by companies like Korea's NCD and China's Leadmicro. Some manufacturers have taken ALD technology to the spatial domain, where instead of using one chamber for the alternate pumping of the reactants, the reactor is divided into separate zones for sub-ALD-steps in which the wafer is exposed to the

90 Sustainable developments by artificial intelligence & machine learning

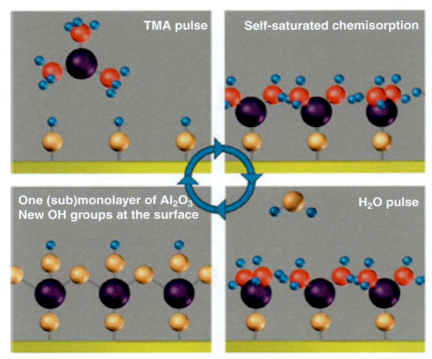

FIG. 4.15 A typical cycle of the ALD Al$_2$O$_3$ process. The expose of Al(CH$_3$)$_3$ and H$_2$O (or O$_3$) on the substrate is separated by a purge step, where the inner gas (N$_2$) is used to empty the byproducts and unreacted precursors (Glunz et al., 2012).

precursor. PECVD (Roth & Rau or Meyer Burger) (Preu et al., 2009; Veith et al., 2012), Batch ALD (Beneq and ASM), inline spatially ALD (Levi-tech, Solaytec, and Ideal Energy) (Richter, Benick, Hermle, & Glunz, 2011; Schmidt, Veith, & Brendel, 2009) are a few examples of industrial Al$_2$O$_3$ deposition tools that are showing excellent industrial potential. Companies like Levi-tech, SoLay-Tec, and Ideal Energy are producing Spatial ALD systems having max throughput of 4800, 4800, and 7200 wafers/h, respectively, whereas NCD and Leadmicro demonstrated max throughput of 4500 and 10,000 wafers/h using a time-based ALD system (Chunduri & Schmela, 2019). On the other hand, PECVD-based Al$_2$O$_3$ systems are not far behind. Meyer Burger, Centrotherm, Fullshare, and SC New Energy are providing max throughput of 6300, 6000, 3400, and 3000 wafers/h, respectively, using a PECVD system with either remote plasma or direct plasma (Chunduri & Schmela, 2019). Continuous improvement of the industrial Al$_2$O$_3$ tools is boosted by the competition between different leading manufacturing companies, resulting in reductions in the process costs.

4.5.2.4 Dielectric stack passivation

One of the major issues in using Al_2O_3 or SiO_2 as a single passivating layer for PERC cells is that it suffers from firing during contact formation (Dingemans & Kessels, 2012; Kho, Black, & McIntosh, 2009; Richter et al., 2011; Schmidt et al., 2009; Sze & Ng, 2007; Talló & Mcintosh, 2009) due to which high intermediate passivation can be obtained during cell processes but final passivation deteriorates. It has been observed from earlier studies that the poor firing stability of thin Al_2O_3 (<20 nm) as the passivation layer is mainly attributed to the increment of D_{it} (Dingemans & Kessels, 2012; Schmidt et al., 2009), which is due to the dissociation of interfacial Si—H bonds at elevated temperatures (Dingemans et al., 2009). Researchers have tried to overcome this issue with thicker Al_2O_3 (e.g., >30 nm). Although this thick layer showed firing stability to some extent as compared to thin layers, it suffers from blistering (Li, Repo, Von Gastrow, Bao, & Savin, 2013; Vermang et al., 2011) and is also not cost-effective from an industrial production perspective.

In place of a single dielectric passivation layer, if dielectric stacks, e.g., SiO_2/SiN_x or Al_2O_3/SiN_x, were used, then it was observed that passivation quality remains unaltered or even improves, which is mainly due to hydrogen passivation provided by the stacks (Chen et al., 2011; Dingemans et al., 2009; Dingemans & Kessels, 2012; Dingemans, Mandoc, et al., 2011; Hofmann et al., 2008; Schmidt et al., 2009). The idea of dielectric stacks passivation is extensively used in different structures of c-Si solar cells (Schmidt et al., 2008). The role of the first layer (Al_2O_3 or SiO_2) is to provide surface passivation through a superior amalgamation of field-effect and chemical passivation. The second layer (SiN_x), which is often called the capping layer, is responsible for the improvement in final passivation either by improving the chemical passivation or by upgrading field-effect passivation. Moreover, dielectric stacks passivation at the rear side also enhances the optical performance of the solar cell through its capability of trapping long-wavelength light, thus decreasing parasitic absorption.

In industrial PERC solar cells, Al_2O_3/SiN_x is normally applied for rear surface passivation. Here, the objective of the presence of the SiN_x is to enhance the protection of the Al_2O_3 layer from penetration and damage created by the screen-printed and fired Al paste. Additionally, the PECVD SiN_x procedure can facilitate activation of the Al_2O_3 passivation. Furthermore, this SiN_x capping layer is responsible for improving the firing stability of the Al_2O_3 layer due to H passivation (Dingemans et al., 2009; Hoex, Heil, Langereis, van de Sanden, & Kessels, 2006).

Recent studies showed that the SiO_2/Al_2O_3 stack also has very promising firing stability. Here, upon changing the SiO_2 interlayer thickness (d_{SiO_2}), tuning of the effective charge density (Q_{eff}) of the whole stacks can be done, which in turn regulates the space-charge field (Aboaf, Kerr, & Bassous, 1973; Dingemans, Terlinden, et al., 2011; Mack et al., 2011; Terlinden, Dingemans, Vandalon, Bosch, & Kessels, 2014; van de Loo et al., 2015).

Additionally, excellent chemical passivation of the Si interface due to H passivation by the SiO_2/Al_2O_3 stack is observed by many researchers (Dingemans, Beyer, van de Sanden, & Kessels, 2010; Dingemans, Einsele, Beyer, van de Sanden, & Kessels, 2012; Dingemans, van de Sanden, & Kessels, 2010). Researchers have found the SiO_2/Al_2O_3 stack can effectively passivate both p-Si (Lin et al., 2012; Mack et al., 2011; van de Loo et al., 2015) and n-Si (Bordihn, Dingemans, Mertens, & Kessels, 2013; Pasanen, Vähänissi, Theut, & Savin, 2017; van de Loo et al., 2015, 2017) due to its excellent chemical passivation property along with a weak field-effect passivation quality.

4.5.3 LBSF and rear local contact

Principally, PERC solar cells can be considered as being composed of two kinds of subcells in parallel, where the first one is the cell whose rear end is passivated with a low-SRV dielectric material and the other is a cell of rear local contacts (for industrial PERC: also LBSF). The lab PERC does not contain the function of p^+-LBSF due to the physical vapor deposited Al (with $\sim$400°C PMA), which only acts as rear local contacts. On the other hand, both p^+-LBSF and rear local contacts were formed in the industrial PERC cell due to the properties of Al dopants and Al metallization processed using Al paste through screen printing and RTP firing. Due to the formation of LBSF in industrial PERC through the "screen printed Al with RTP firing" process, the probability of suppression of the minority carriers from recombining at the cell rear is enhanced, as compared to lab PERC cell. Optimally designed and processed rear metalization can make ohmic contact results in the reduction of rear surface recombination along with series resistance (R_s). Concerning the design point of view of rear local-contact arrays, point- or segment-arrays normally are advantageous compared to line arrays due to having low R_s with a lower contact fraction, along with a lower BSRV. But due to wider process tolerance for both laser ablation and Al screen-printing, line- or segment-designs are more suitable in mass production.

A narrow dimension (e.g., $\sim$20–30 µm) opening of the rear dielectric stack is the crucial requirement for the formation of the rear local contact pattern. To achieve this, laser ablation has become the most suitable choice due to its precision and cost-effectiveness. Laser ablation of dielectric (Al_2O_3, SiN_x, or SiO_2) layers from the Si surface is based on different absorption coefficients of laser radiation in Si and dielectric layers in addition to the lift-off of these layers because of thermal expansion of the molten Si underneath. Due to lower absorption coefficients of the dielectrics (Al_2O_3/SiN_x or SiO_2/SiN_x), as compared to the underlying Si, a major challenge is to selectively remove dielectrics without creating a considerable level of crystal damage in the Si. Therefore, precise adjustment of laser energy through the optimization of laser process parameters is vital for the rear contact opening. Generally, laser systems having ultrashort pulse duration in the picosecond range are used to accomplish this objective due to less thermal-induced heating, melting, and debris around the area

(Gall et al., 2010; Gatz, Bothe, Müller, Dullweber, & Brendel, 2011; Heinrich, Bähr, Stolberg, & Wütherich, 2011; Knorz, Aleman, Grohe, Preu, & Glunz, 2009; Löffler, Wipliez, de Keijzer, Bosman, & Soppe, 2009; Thorsten et al., 2012). Recent research results pointed out that nanosecond lasers can effectively attain this goal as well through process optimization (Ferré et al., 2014; Gall et al., 2010; Heinrich et al., 2011; Knorz et al., 2009; Lossen, Wald, & Bahr, 2007).

4.5.4 Rear polishing

It has already been discussed that rear side passivation and ablating the dielectric stack for contact formation are two key prerequisites for PERC cell realization, while rear surface polishing is another essential requirement. Rear side polishing is done by etching away the random pyramid textured surface, as it is well known that a polished surface exhibits a lower degree of recombination compared to a textured one due to lower surface area and also the dangling bonds. It was also observed that coverage of the passivation layer becomes better on a planar surface when PECVD is used as the deposition tool. There is no harm in removing the textured surface as the rear side of the cell as it does not actively participate in light absorption and also is not engaged in direct light trapping. And there is an even more elementary reason: using traditional diffusion, the wafer is doped on either side unless single side diffusion is realized in a belt furnace or the rear surface is masked explicitly. If the phosphorus is present on the rear side, it must be removed.

To achieve rear polishing, wet-benches used for edge isolation have also been designed to etch the rear surface. The process chemistry needs to be adjusted to attain the required surface roughness. The technology has evolved to be robust and leading wet-bench suppliers such as RENA, RCT, Singulus, Schmid, and SC New Energy are offering rear polishing tools mostly integrated into PSG etch and edge isolation systems (Chunduri & Schmela, 2018).

4.5.5 PERC performance

One of the major reasons for the widespread implementation of PERC technology is the advancement it has demonstrated in terms of efficiency improvements. The technology makes noteworthy improvements, as can be noticed from the graph of record efficiencies shown in Fig. 4.16. For multicrystalline PERC cells, SCHOTT Solar was the first to declare 18.7% efficiency in 2010. Jinko Solar announced 21.63% efficiency in September 2016 and in October 2017 they declared cell efficiency of 22.04%, which is still the record efficiency for a multicrystalline PERC cell.

However, in today's market environment, PERC is largely driven by monocrystalline silicon cell technology (Fig. 4.17). In 2012, Suntech announced 20.3% cell efficiency, which was the record efficiency at that point in time.

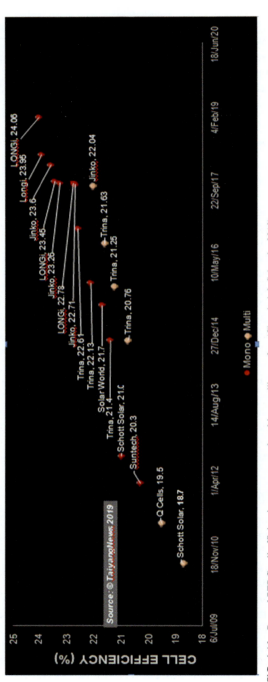

FIG. 4.16 Record PERC cell efficiencies on mono and multi crystalline wafers (Chunduri & Schmela, 2019).

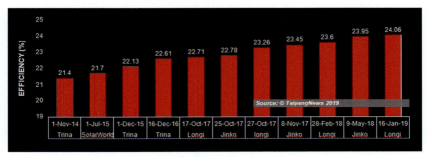

FIG. 4.17 Graphical representation of increasing peak monocrystalline PERC efficiencies (Chunduri & Schmela, 2019).

Recently, companies have approached 24% efficiency level, but those cells are not pure PERC cells anymore. Jinko Solar's 23.95% efficiency cell is a passivated contact technology adapted to the PERC platform, whereas LONGi achieved 23.6% cell efficiency using improved fine line double printing, an optimized doping profile, and grid design. Applying further modifications LONGi has been successful in enhancing the efficiency further, up to 24.06% (National Center of Supervision and Inspection on Solar Photovoltaic Product Quality in China (CPVT) certified), which is still the record efficiency for a monocrystalline PERC cell (Chunduri & Schmela, 2019).

Not only the record efficiency but also the average production efficiency for PERC has increased significantly. The average production efficiency of some top cell makers, such as Jinko Solar, LONGi, and Tongwei, is about 22.2% to 22.4%, which is around a 0.8% gain in average production efficiency in 1 year, higher than for any other technology.

4.5.6 Improvements of PERC solar cells

The PERC technology is advancing at a faster pace than expected to replace Al-BSF technology in the global field of the solar PV market. A lot of research is going on, including wafer processing, passivating materials, contact formation technology concerning cells, and also of different aspects of modules, but still, there are several unexplored prospective areas to improve the performance of "standard" PERC cells. In 2015, researchers of ISFH published a roadmap on how to perk up PERC efficiencies beyond 24%, which is presented in Fig. 4.18.

The key proposals of this roadmap are still valid:

- Advanced emitter structures such as selective emitters;
- Boron-added Al paste for the rear;
- Wafers with 1 ms lifetime;
- Multiwires instead of busbars;
- 10 μm narrow fingers with a high aspect ratio.

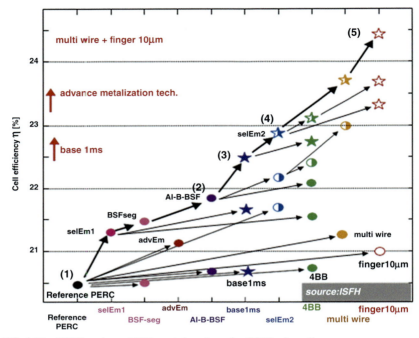

FIG. 4.18 PERC roadmap: ISFH-created roadmap for PERC solar cells.

In summary, the majority of PERC advancements have generally followed the abovementioned guidelines. However, improving the emitter doping profile is key, which brings selective emitters back into the picture. While the first attempt to implement selective emitters in solar cell production was not very successful, it is definitely a low hanging fruit to quickly gain on efficiency by up to 0.3% absolute in a reasonably simple manner. Nearly every leading PERC maker is using selective emitters today (Chunduri & Schmela, 2019).

Front surface passivation is another promising route to improve PERC performance. The usual practice is to apply the classical silicon nitride that acts as both the antireflection layer as well as the emitter passivation layer. Reports suggest that stacking silicon oxide and silicon nitride or depositing a gradient film can improve the front surface passivation, which is already applied by nearly every top cell maker. There has also been notable progress with reducing finger-width to levels as low as 35–40 μm (Chunduri & Schmela, 2019).

4.5.7 Further improvements

Considering PERC as the basis, further improvement can be made by changing its design to passivated emitter rear locally-diffused (PERL) and passivated emitter rear totally-diffused (PERT) cells, as shown in Fig. 4.19. In PERT

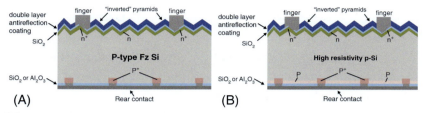

FIG. 4.19 Schematic diagram of (A) p-type PERL solar cell, and (B) p-type PERT solar cell.

structure, the rear side is totally diffused with boron for p-type substrates. But in PERL structure, the rear side is locally diffused (Preu et al., 2020; Xiao & Shuyan, 2014). From a crucial perspective, the industrial PERC cell is similar to the lab PERL (passivated emitter and rear locally-diffused (Wang, 1992; Zhao, Wang, & Green, 1999)) cell in terms of cell structure. For the lab PERL, the p$^+$-LBSF is formed by boron doping; while in the case of industrial PERC, the p$^+$-LBSF is achieved by Al doping. J. Benick and B. Hoex are the first who were able to apply Al$_2$O$_3$ successfully into B-emitter passivation and n-PERT cell and reported 23.2% cell efficiency in the year 2008 (Benick et al., 2008). In 2010, Fraunhofer ISE was able to improve the efficiency to 23.9% by using Al$_2$O$_3$ passivated B-emitter in n-PERL structure (Glunz et al., 2010).

4.5.8 Bifacial PERC

Above all, bifacial is the most alluring aspect of PERC technology. Unlike the other sophisticated cell architectures, PERC is not fundamentally bifacial. That's because the configuration of the PERC cell contains a solid local-BSF forming layer of Al paste. Nonetheless, the requirements to turn PERC into a bifacial solar cell are relatively small—it is only to apply an aluminum grid as a substitute to spreading paste over the full area.

PERC's bifacial potential is perhaps the foremost attribute that will make the technology spread wider and sustain longer than many predicted. The PERC approach is the easiest and most cost-effective bifacial structure. Its unquestionable advantage is cost, which is more or less equivalent to the monofacial PERC cell process. PERC almost avoids the usage of Ag on the rear side, while other bifacial configurations require much more of the expensive metal. Still, as with standard solar cells, ribbons cannot be soldered to Al, thus the rear busbars are still Ag-based.

One of the research papers presented by the ISFH group at EUPVSEC 2015 showed the pathway to realize the bifacial PERC technology for the very first time to the PV community (Dullweber et al., 2015). Their study suggests optimization of the rear Al finger grid in place of the conventional full-area Al rear layer while using the identical PERC manufacturing sequence. In this particular

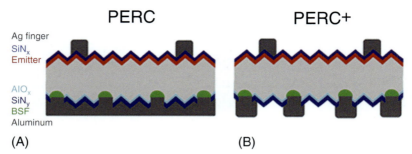

FIG. 4.20 Schematic drawing of (A) an industry typical monofacial PERC solar cell with full-area Al rear layer, and (B) a bifacial PERC+ solar cell with rear Al finger grid. The drawings are not to scale.

case, Al paste consumption of the PERC+ cells is drastically reduced to 0.15 g as compared to 1.6 g for the conventional PERC cells (Fig. 4.20).

From then on, intensive R&D activities have been carried out by leading cell manufacturers with the collaboration of the reputed research institutes/universities. Presently, bifacial PERC is a quite well-established technology and it is predicted in International Technology Roadmap for Photovoltaic (2020) that in the near future it will be the front runner technology in the field of solar PV. Increasing the bifaciality to today's level of 70%–75% by reducing the aluminum fingers will happen eventually with further developments in pastes and complementing module technologies such as multibusbars (Chunduri & Schmela, 2019).

4.6 Fabrication of PERC solar cells

Fig. 4.21 shows the process flow diagram for fabrication of PERC type c-Si solar cell. The process flow for the fabrication of PERC type c-Si solar cell starts

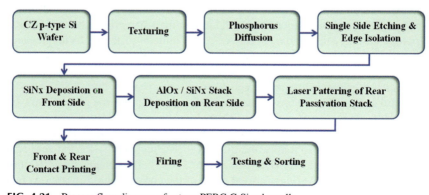

FIG. 4.21 Process flow diagram of p-type PERC C-Si solar cells.

with a 200 μm thick, 1–3 Ω cm resistivity and ⟨100⟩ oriented p-type Czochralski c-Si (Cz-Si) wafer. The wafers usually have surface damage in μm sizes, especially damage created by saw cutting of wafers. After saw damage removal, the surface of the wafer is textured chemically to reduce the optical reflectivity from the wafer surface. Then an n-type emitter layer is grown by thermal diffusion using a diffusion furnace followed by PSG removal. A SiO_2 passivation layer is formed by thermal oxidation followed by antireflection (AR) coating on the emitter layer for emitter passivation as well as light management. Subsequently, edge isolation and rear side polishing is carried out using reactive ion etching to eliminate the electrical shorting between top and bottom surfaces and improve light reflection near IR region from the rear side. After that, Al_2O_3 is deposited by ALD to passivate the rear surface followed by SiN_x deposition as capping layer. Further, green laser is used to selectively remove passivation layer from the rear side for the formation of local BSF during metallization. Finally, printing and firing is done for metallization with Ag paste at the front side and Al paste at the rear side for electrical contacts. The process equipment and methodology are discussed in the following sections.

4.6.1 Saw damage removal, texturization, and cleaning

Texture etching is a well-established process step for crystalline silicon solar cell fabrication. Generally, 156 mm × 156 mm (pseudo-square) ⟨100⟩ oriented p-type monocrystalline silicon wafers with resistivity 1–3 Ω cm and thickness 200 μm are used as the starting material for texturization. The process of saw damage removal (SDR) and subsequent texturization is carried out in the custom made automatic wet-chemical bath (Fig. 4.22A). The wafers are arranged in the wafer carrier as shown in Fig. 4.22B for transportation through the system. The baths with chemicals and DI water are arranged in such a way that the saw damage removal, texturization, and cleaning process is completed in a

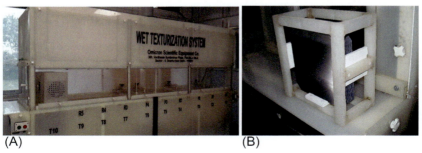

FIG. 4.22 (A) Photograph of a wet chemical bath manufactured by Omicron Scientific Equipment Co., India. (B) Photograph of wafer carrier to carry the silicon wafers in the chemical bath system (Laboratory of DST-IIEST solar PV Hub, IIEST, Shibpur).

FIG. 4.23 Process flow for saw damage removal, texturization, and wafer cleaning of c-Si wafers.

single run. The capacity of each bath is 20 L. Nitrogen purging and heating facilities are available in the SDR and texturization baths. A wafer drier with heating mechanism is attached at the end of the system. The process flow diagram is shown in Fig. 4.23.

4.6.1.1 Saw damage removal

The wafers are first treated with a highly concentrated (10 wt%) KOH solution with deionized water (DI-W) at 80–85°C for 2–3 min for saw damage removal. The surface damage on the silicon wafers is removed by isotropic etching of silicon with high concentrated KOH. In DI-W, KOH breaks into K^+ and OH^- ions. The chemical reaction of silicon in the presence of KOH is as follows (Moynihan et al., 2010):

$$Si + H_2O + 2KOH \rightarrow K_2SiO_3 + 2H_2O \qquad (4.52)$$

During SDR, about 10 μm of silicon etches out from both surfaces of the wafers. Then the wafers are rinsed with DI-W for 4–5 min.

4.6.1.2 Texturization

The SDR silicon wafers are then treated with a low concentration KOH (1.8 wt%) solution along with isopropyl alcohol (IPA) as surfactant and DI-W at 75–80°C for 35–40 min followed by rinsing with DI-W. The anisotropic etching of silicon wafers gives rise to a textured surface. The etch rate of the silicon $\langle 110 \rangle$ plane is faster than the $\langle 100 \rangle$ plane, while the $\langle 111 \rangle$ plane etches at a slower rate than the $\langle 100 \rangle$ plane. Different etching rates of different silicon crystal planes leads to formation of pyramidal structures on the $\langle 100 \rangle$ plane (Singh, Kumar, Lal, Singh, & Das, 2001). Pyramid sizes over the silicon surface depend on the concentration of KOH and IPA, temperature, and reaction time.

4.6.1.3 Wafer cleaning

After the texturization process, the textured wafers are treated with a RCA1 and RCA2 wet chemical cleaning process followed by diluted HF (1%–2%) treatment. Finally, the wafers are rinsed with DI-W and dried with a hot air drier.

Passivated emitter and rear contact solar cells **Chapter | 4** **101**

4.6.2 Diffusion and oxidation

4.6.2.1 Phosphorus diffusion

The p-n junction formation is essential to fabricate crystalline silicon solar cells as the function of a solar cell depends on its junction characteristics like junction depth, doping concentration, and uniformity over the surface area. The thermal diffusion of phosphorus is widely used to form an n-type emitter layer in a p-type silicon wafer. The concentration of emitter dopants (phosphorus) is much higher than the boron concentration of p-type substrate, which leads to formation of n^+-type emitter on the p-type silicon substrate. The emitter thickness range (0.2–$0.5\,\mu m$) depends on diffusion process parameters.

Process steps for phosphorus diffusion

The diffusion is carried out in two steps, predeposition and drive-in. In the predeposition step, liquid $POCl_3$ is vaporized by N_2 gas bubbling in to the $POCl_3$ bubbler. At high temperature—about $820°C$—the $POCl_3$ vapor is deposited on the surface of the silicon wafers and reacts with O_2, and phosphosilicate glass (PSG) is formed, which is a hard material of phosphorus doped silicon dioxide; as a result, impurity atoms get deposited on the surface of the silicon wafer.

After that, in the drive-in step, the wafers are heated at $825°C$ in the presence of nitrogen to initiate the diffusion of impurity atoms (phosphorous) into the silicon wafers and, as a result, an n^+-type emitter layer is formed. Then the temperature of the diffusion furnace is reduced to $750°C$ for low temperature oxidation (LTO) with O_2 gas; as a result, hard PSG becomes softened. Finally, furnace temperature is reduced to $600°C$ for unloading the diffused wafer. The process flow for thermal diffusion of phosphorus is shown in Fig. 4.24.

PSG removal

After the drive-in step, a hard PSG layer is formed which must be removed. The diffused wafers are then treated with low temperature oxidation (LTO) to soften the PSG layer and removed using dilute HF ($5\,wt\%$) solution. Then the wafers are rinsed with DI-W for $5\,min$ and dried with N_2 flow.

4.6.2.2 Thermal oxidation

Thermal oxidation is a way to produce a thin layer of oxide on the surface of a material. This technique forces an oxidizing agent to diffuse into the material at high temperature and react with it. Thermal oxidation can be applied to different types of materials. In this work, thermal oxidation is used for silicon substrates to produce a thin layer of silicon dioxide (SiO_2) for passivation of the silicon wafer surface. There are two types of thermal oxidation of silicon, dry oxidation and wet oxidation, which are described by the following equations:

$$Si + O_2 \rightarrow SiO_2 \tag{4.53}$$

$$Si + 2H_2O \rightarrow SiO_2 + 2H_2 \tag{4.54}$$

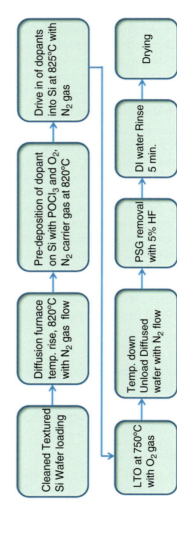

FIG. 4.24 Process flow diagram of thermal diffusion.

Silicon dioxide growth occurs 54% above and 46% below the original surface of silicon as silicon is consumed. The wet oxidation rate is faster than the dry oxidation process. Hence, the dry oxidation process is suitable for thin oxide layer formation to passivate the silicon surface.

Cleaned phosphorous diffused silicon wafers are placed inside the quartz tube of an oxidation furnace with pure (UHP) dry nitrogen (N_2) flow; as a result, nitrogen prevents oxidation from occurring until the furnace reaches the required temperature. After the specified temperature in the oxidation tube is reached, the N_2 gas flow is shut off and pure (UHP) O_2 gas flow is started to the tube. After the oxidation is complete, N_2 is reintroduced into the tube to prevent further oxidation from occurring during the temperature decreasing in the furnace. Then the wafers are removed from the chamber (Fig. 4.25).

4.6.3 Reactive ion etching

Reactive ion etching (RIE) is an etching technology used in the fabrication of semiconductor devices for microfabrication. It uses chemically reactive plasma to remove semiconductor substrate or material deposited on wafers. The plasma

(A) (B)

FIG. 4.25 Photograph of SVCS (SVFUR-AH3) diffusion and oxidation furnace (Laboratory of DST-IIEST solar PV Hub, IIEST, Shibpur).

is generated under low pressure by an electromagnetic field. From neutral gases and glow discharge utilization, chemically reactive high energy species are generated to react with materials being etched to form volatile byproducts. Etch gases are thus selected on product volatility considerations.

Reactive ion etching (RIE) is widely used for etching thin layers of silicon, silicon dioxide, silicon nitride, and metals. It is a dry etching system using both ions (physical) and radicals (chemical) from generated plasma and used extensively for semiconductor device fabrication. Compared to a conventional wet etching process, RIE has higher anisotropy, better uniformity and control, and better etch selectivity. In this work, the system is used for controlled etching of thin silicon layers for edge isolation of diffused silicon wafers. It is required to prevent electrical shorting between top and bottom phosphorus doped n^+ layers across the edges as it is formed over the exposed areas including edges of wafers during the diffusion process. Sometimes this etching system is used for rear side etching purposes for passivated emitter and rear cells (PERC) fabrication.

A typical RIE system consists of two parallel plates inside a cylindrical vacuum chamber. Source gases enter the reaction chamber from the top plate through shower-type holes and the vacuum pump system sucking through the bottom. Depending upon the requirement of the etching process, the types and amount of gases used can be varied. Very low chamber pressure is maintained, in the order of millitorr, during the etching process. Sulfur hexafluoride (SF_6) etchant gas is commonly used for silicon etching as F is the most widely used etching species (Williams & Muller, 1996).

A strong RF (radio frequency) electromagnetic field is applied to the bottom plate for plasma generation in the system. Typically, a few hundred watts RF power is used with 13.56 MHz frequency. In each cycle of the field electrically accelerated, electrons are moved up and downwards, striking both upper and lower plates in the chamber. However, the heavier ions move very slowly with respect to electrons. A schematic sketch of a reactive ion etching system is shown in Fig. 4.26A.

Accumulated electrons on the bottom plate build up charge. This built up charge generates a large negative voltage on the bottom plate while the plasma itself develops a slightly positive charge due to the higher concentration of positive ions with respect to free electrons. Due to the large voltage difference, the positive ions tend to drift toward the bottom plate and collide with the samples to be etched.

4.6.4 Plasma-enhanced chemical vapor deposition (PECVD)

The plasma-enhanced chemical vapor deposition (PECVD) method is an excellent alternative to thin film deposition. Low temperature deposition, good film quality, uniformity, and reproducibility are several advantages of the PECVD method, which is why it is suitable for largescale commercial production.

Passivated emitter and rear contact solar cells **Chapter | 4** **105**

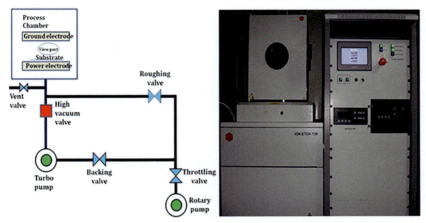

FIG. 4.26 (A) Schematic of reactive ion etching system; (B) Photograph of reactive ion etching system (Laboratory of DST-IIEST solar PV Hub, IIEST, Shibpur).

FIG. 4.27 Photograph of PECVD system (Laboratory of DST-IIEST solar PV Hub, IIEST, Shibpur).

Recently, this technique has been used by industries for SiO_x, SiN_x, a-Si:H, and μc-Si:H deposition for solar cell production (Luft & Tsuo, 1993; Plummer, Deal, & Griffin, 2009). Fig. 4.27 shows a PECVD syatem.

PECVD uses a DC, RF, or microwave excitation source to generate a glow discharge plasma within the source gases in which the energy is transferred into the gas mixture in an isolated chamber (Schlüter & Shivarova, 1998). The energetic particles (electrons and ions) within the plasma collide with the source gas molecules and dissociate them into reactive radicals, ions, neutral atoms and molecules, and other highly excited species. The secondary reactions take place within the plasma between the charged and neutral particles and interact with the substrate surface, which determines the structure and composition of the deposited material. In the case of DC glow discharge, plasma is generated in a low pressure chamber between two metal plates when one is connected to a DC power supply and the other one grounded. In the case of RF glow discharge, plasma is created in a similar way using an RF power supply.

106 Sustainable developments by artificial intelligence & machine learning

The RF-PECVD glow discharges are further subdivided into two categories: capacitively coupled discharges and inductively coupled discharges, where plasma is created between two parallel metal plates and by the RF-coils outside the discharge, respectively. Microwave discharge (2.45 GHz) can also produce glow discharge plasma for deposition purposes. ECR (electron cyclotron resonance) microwave discharge is a special variant of this technique in which energetic ions are created by microwaves in combination with a magnetic field. In this case, plasma is generated in one zone of the plasma chamber, which is used to dissociate the source gas molecules located in a different zone within the plasma chamber (Schlüter & Shivarova, 1998). Therefore these techniques are termed remote plasma techniques.

PECVD is a physical deposition technique where reactive species are produced by an electrical discharge leading to plasma. The key process steps of PECVD are source gas diffusion, electron impact dissociation, gas-phase chemical reaction, radical diffusion, and deposition. In a glow discharge deposition process when silane (SiH_4) is used as a source gas, the electron impact process is transformed into reactive neutral species, like SiH, SiH_2, SiH_3, Si_2H_6, H, and H_2, and ionized species, like SiH^+, SiH_2^+, SiH_3^+, and so on. After the introduction of the precursor gases, a low temperature glow discharge is ignited and sustained by an electric field between the two parallel electrodes. The electrons are accelerated by the electric field and gain adequate energy to decompose the gas molecules into neutral radicals and ions. In the bulk plasma, a complex gas phase reaction happens between the radicals, ions, and molecules. Such secondary reactions are so critical that they mostly control the electronic and structural properties of the deposited films.

The properties of the deposited film are determined by whether its structure is amorphous or microcrystalline, its thickness, chemical composition, which strongly depends on the deposition parameters like power density, the ratio of silane, hydrogen, and other precursors such as gas flow, chamber pressure, excitation frequency, type of plasma source, etc. The doped films are prepared by introducing diborane or phosphine in the gas mixture.

4.6.5 Atomic layer deposition (ALD)

Atomic layer deposition (ALD) is one kind of thin-film deposition technique derived from the sequential gas phase chemical process. During this process, a film is grown on a substrate when it is exposed to alternating gaseous species known as precursors. In this technique, the precursors are never present at the same time; they are inserted in the deposition chamber as a nonoverlapping series of sequential pulses. The precursor molecules react with the surface in each of these pulses in a self-limiting way. Therefore, the maximum amount of material consumed on the surface after a single exposure of all the precursors that are deposited on the surface is known as the ALD cycle. It is determined by the nature of the precursor-surface interaction (Oviroh, Akbarzadeh, Pan, Coetzee, & Jen, 2019). It can be possible to determine the uniformity of the grown material by varying the number of cycles. The ALD method is suitable for very thin layer deposition with conformal growth of thickness and composition at the atomic level. Now, ALD is an active field of research in industries and research laboratories (Puurunen, 2005).

There are various types of ALD processes available, such as thermal ALD, plasma ALD, photo-assisted ALD, metal ALD, etc. The thermal ALD deposition technique is described for the deposition of a field effect passivation layer of Al_2O_3 on the rear side of passivated emitter and rear contact (PERC) monocrystalline silicon solar cells. The system consisted of different units, including a main system with precursors, along with ALD valves, vacuum pump, compressor, N_2 generator, and software system. The main system is a bench-top unit consisting of different precursors, in which Al_2O_3 and H_2O in the form of precursor has been used with Al_2O_3 as source material. Chamber pressure was kept within 100–120 millitorr when the precursor valves were off. The pump should be rated 12cfm using NW-25(1″) or NW-40 tubing (1.5″) in addition to a reducer. A compressed dry air source was used near the rear side of the chamber. An N_2 generator was used to constantly provide N_2 gas. All the gas lines were made of stainless steel. N_2 should be ultrahigh purity and it should be regulated to 20 ± 3 psi. Trimethyl aluminum precursor (TMA) from Sigma Aldrich kept at room temperature and deionized water is used as source oxidant for Al_2O_3 layer deposition. The total deposition process was carried out in inert N_2 ambient with 10 sccm flow rate. After loading the samples, the chamber pressure was kept within 30 millitorr by pumping out the gases. The process temperature was maintained at 175°C. The entire deposition process consists of a number of cycles depending on the thickness of the layer. Each cycle provides thickness within 9 Å and, as the cycle proceeds, thickness is enhanced for each cycle (Fig. 4.28).

4.6.6 Laser ablation

Laser scribing is a laser technology used to scribe materials, typically used for research purposes and industrial manufacturing applications. The system is designed based on the principle of laser light—generally pulsed laser—which is focused through microscope optics. It can selectively remove materials from the surface of semiconductor devices without damaging it. The laser system is successfully used to remove passivation layer from the rear side of passivated emitter and rear contact (PERC) silicon solar cells for the formation of a local back surface field (BSF) during metallization.

One of the cost-effective and easy techniques to manufacture PERC solar cells for production lines is the selective ablation of a dielectric passivation layer from the rear side of PERC solar cells using lasers (Kim, Park, & Kim, 2013)

108 Sustainable developments by artificial intelligence & machine learning

FIG. 4.28 Photograph of thermal ALD (GEMSTAR XT Bench-top ALD) system (Laboratory of DST-IIEST solar PV hub partner, Meghnad Saha Institute of Technology (MSIT), Kolkata).

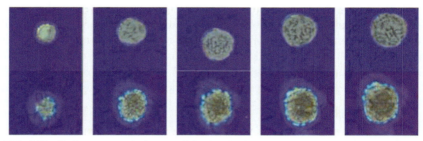

FIG. 4.29 Surface topography images of single pulse irradiated samples with green (above) and UV (below) laser sources at different fluences: From left to right, 0.09 J/cm^2, 0.16 J/cm^2, 0.24 J/cm^2, 0.30 J/cm^2, and 0.40 J/cm^2 for green laser; 0.29 J/cm^2, 0.80 J/cm^2, 1.53 J/cm^2, 1.88 J/cm^2, and 2.55 J/cm^2 for UV laser (Bounaas et al., 2013).

(Fig. 4.29). Many research and development groups and companies have been using laser systems for the selective ablation of dielectric layers from the rear side of PERC solar cells in order to form a uniform local Al-BSF, so that recombination under the metal contacts would be minimized (Jaffrennou, 2011). For this purpose, in the green (visible) wavelengths region, the frequency doubled Nd:YAG laser is commonly used at a frequency of 532 nm with a pulsed power output.

4.6.7 Metallization

Two types of metallization techniques are used for solar cell metallization: metallization by screen printing and the firing process used for fabrication of crystalline silicon Al-BSF and PERC solar cells.

Screen printing and firing is a well-established metallization process for solar cell manufacturing industries. Screen printing and firing of metal pastes on both front SiN_x antireflection coated (ARC) emitter and rear sides of solar cells is done for metal electrodes formation to get electrical contacts with the external circuits as it is a less expensive and simple technology, is compatible with industrial processes, along with easy automation and high throughput. In the screen printing metallization process, first the aluminum (Al) is screen printed through a stainless steel mesh patterned screen on the rear side of solar cells using a screen, followed by a drying process through a belt furnace at 250–275°C for ~2 min. After the rear side Al paste printing and drying, the front contact electrode is formed by screen printing of silver (Ag) paste a through stainless steel mesh patterned screen with fingers and busbars using a separate screen printer to avoid contamination between Ag and Al pastes. After front side Ag paste printing, the wafers are again dried through a separate belt furnace at 250–275°C for ~2 min to avoid contamination between Ag and Al pastes. The printing quality depends on various parameters such as squeeze pressure, squeeze speed, fiducial alignment accuracy, gap between screen and wafer, and solvent content of paste. Finally, the wafers are subject to co-firing through a three zone belt furnace (firing furnace) at a very high temperature—greater than 800°C—for a few seconds. Initially, a burn-out process takes place at temperature zone 1 in the firing furnace where the temperature is set between 350°C and 510°C. In this zone, organic binders from the pastes are removed. Al-Si alloy formation takes place in temperature zone 2, where the temperature is set between 750°C and 850°C. Formation of Al-BSF during the firing of the wafers with the rear Al contacts (Urrejola, Peter, Plagwitz, & Schubert, 2011) occurs in temperature zone 3, where the temperature is set between 880°C and 940°C; at the same time, the front Ag paste forms a contact with the n^+ emitter layer by penetrating through the SiN_x ARC with the help of glass frit particles of Ag paste. Then the wafers are passed through a cooling zone to cool down. The actual temperatures obtained by the wafers are less than the set temperatures at different zones in the firing furnace due to fast movement of the belt. Optimization of the entire firing process is crucial as it is responsible for ohmic contact formation on both back and front sides of the solar cells at the same time. The series and shunt resistances of solar cells and, hence, the fill factor and efficiency are largely dependent on the temperature profile of the firing process. The schematic of process flow for the entire printing and firing process is shown in Fig. 4.30A.

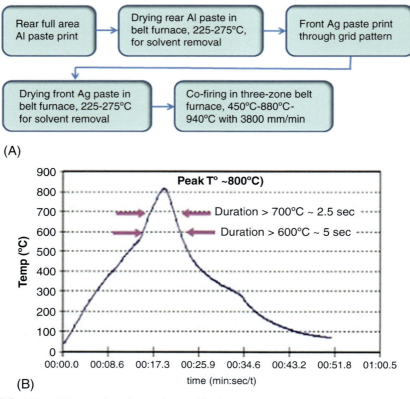

FIG. 4.30 (A) Process flow diagram for metallization through screen printing; (B) Recommended firing profile for Dupont front Ag paste PV17B (Dupont, n.d.).

Usually the temperature profile used for firing of pastes is that provided by the manufacturers of the paste. The temperature profile for front Ag paste PV17B made by Dupont is shown in Fig. 4.30B. The screen printers used for printing on the back with Al and front with Ag pastes are shown in Fig. 4.31A. The drying and firing belt furnaces are shown in Fig. 4.31B.

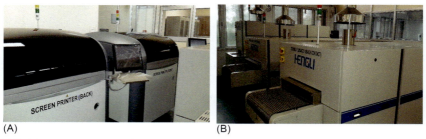

FIG. 4.31 (A) Photograph of screen printers DEK (Horizon 03ix); (B) Photograph of belt furnaces (Hengli) for drying and firing (Laboratory of DST-IIEST solar PV Hub, IIEST, Shibpur).

4.7 Characterization equipment

4.7.1 Scanning electron microscopy (SEM)

The scanning electron microscope (SEM) is a special type of electron micro-scope that produces images of a sample by scanning the surface with a focused beam of electrons. It is widely used in materials research laboratories and semi-conductor industries. An FE-SEM is used to study the surface morphology of SDR and textured silicon wafers. This system has nine stages to insert nine sam-ples at a time. One nanometer resolution is achievable using SEM with the help of a field emission (FE) electron gun. Magnification depends on the scanning system rather than the lenses, so that a surface in focus can be imaged with a wide range of magnifications from $3\times$ to $150,000\times$. A photograph of the FE-SEM system is shown in Fig. 4.32.

4.7.2 Four point probe measurement

A four point probe is a very simple but important instrument for measurement of sheet resistivity of semiconductor wafers. A four point probe instrument, as shown in Fig. 4.33B, is used for measurement of sheet resistivity of the dif-fused n^+ emitter layer and resistivity of crystalline silicon wafers. Generally, the sheet resistivity value represents the lateral conductivity of a thin film. Although sheet resistance can be measured by a simple two point probe method, this suffers from the effect of parasitic resistances such as probe con-tact resistance, probe resistance, and spreading resistance (Smits, 1958). In the four point method, the position of probes is maintained at the same distance from each other. Here, the current is passed through the two outer probes while the voltage is measured between the two inner probes, as shown in Fig. 4.33A. The semiconductor sheet resistance is calculated using the current and voltage readings from the probes with the following relation (Shaffner & Schroder, 2000):

$$R_{sh} = \frac{\pi}{\ln 2} \frac{V}{I} \approx 4.53 \frac{V}{I} \qquad (4.55)$$

where, R_{sh} is the sheet resistance, V is the voltage measured from the inner two voltage probes, I is the current passed through the two outer current probes, and 4.53 is the geometry correction factor of the four point probe measurement.

The doping profile of the emitter of a PERC solar cell can be measured by performing electrochemical capacitance-voltage (ECV) measurements (Saraei, Eshraghi, Tajabadi, & Massoudi, 2018).

4.7.3 Thickness profilometer

A thickness profilometer is a typical measuring instrument used to measure sur-face roughness, critical step dimensions, and curvatures (Fig. 4.34). In a profil-ometer, a diamond stylus is moved across the sample laterally and also

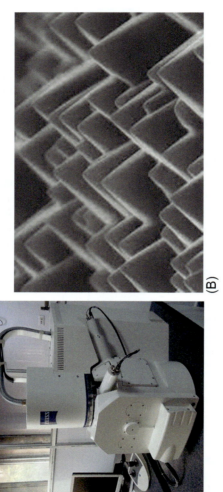

FIG. 4.32 (A) Photograph of FE-SEM (Carl Zeiss) system (Laboratory of DST-IIEST solar PV Hub, IIEST, Shibpur); (B) Textured surface (PVEducation, n.d.).

Passivated emitter and rear contact solar cells **Chapter | 4** 113

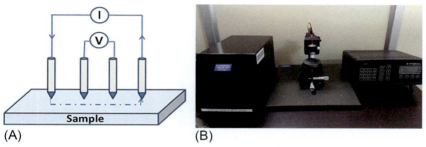

FIG. 4.33 (A) Schematic diagram of a four point probe measurement method; (B) Photograph of Jandel four point probe measurement system (Laboratory of DST-IIEST solar PV Hub, IIEST, Shibpur).

FIG. 4.34 Photograph of Bruker thickness profilometer (Dektak XT) (Laboratory of DST-IIEST solar PV Hub, IIEST, Shibpur).

vertically. The vertical movement of the stylus on the sample measures the surface roughness. It can measure vertical stylus displacement for small surface variations as a function of position. It can measure small vertical features ranging from 10 nm to 1 mm. During measurement, the vertical position of the diamond stylus creates an analog signal that is converted into a digital signal, which is stored, analyzed, and then displayed on the monitor. The horizontal resolution can be controlled by the scan sped as well as the data signal sampling rate.

Metallization is done for electrical contact formation of Al-BSF and PERC type solar cells by a screen printing technique using Ag and Al pastes. The aspect ratio of the Ag fingers must be high in order to obtain low series resistance on the top and rear of the solar cell (Preu et al., 2020) Therefore, measurement of height and width of the front Ag fingers as well as rear Al contacts is needed for optimization of electrical properties of the solar cell.

114 Sustainable developments by artificial intelligence & machine learning

FIG. 4.35 (A) Photograph of Keithley *C-V* measurement system; (B) Photograph of Agilent *I-V* measurement system (Laboratory of DST-IIEST solar PV Hub, IIEST, Shibpur).

4.7.4 *I-V* and *C-V* measurement

The *I-V* and *C-V* measurement system is a very useful tool for characterization of semiconductor devices, components, and materials (Fig. 4.35). This measurement tool helps the semiconductor manufacturer and researcher to monitor critical processing steps for identifying the contamination and material quality. This test equipment is widely used by almost every semiconductor manufacturer and research laboratory for measurements including capacitance vs voltage (*C-V*), current vs voltage (*I-V*), conductance vs voltage (*G-V*), gate oxide integrity (GOI), dielectric constant, and interface trap density (ITD).

In the capacitance vs voltage measurement, the quality of the diode formed by MIS structure can be determined. The *C-V* measurement is a reliable guide for evaluating the quality of ultrathin SiO_x (1.5–2 nm) (Mandal, Biswas, et al., 2020). *C-V* plots for MIS devices are analyzed to get the interface state densities (Nss).

4.7.5 X-ray photo electron spectroscopy (XPS)

As the energy of an X-ray with particular wavelength is known and because the emitted electrons' kinetic energies are measured, the electron binding energy of each of the emitted electrons can be determined by using an equation that is based on the work of Rutherford (1914):

$$E_{binding} = E_{photon} - (E_{kinetic} - \varphi) \tag{4.56}$$

where $E_{binding}$ is the binding energy (BE) of the electron, E_{photon} is the energy of the X-ray photons being used, $E_{kinetic}$ is the kinetic energy of the electron as measured by the instrument, and φ is the work function of the spectrometer (not the material).

This equation is essentially a conservation of energy equation. The work function term φ is an adjustable instrumental correction factor that accounts for the few eV of kinetic energy given up by the photoelectron as it becomes absorbed by the instrument's detector. It is a constant that rarely needs to be adjusted in practice. A typical XPS spectrum is a plot of the number of electrons detected (sometimes per unit time) (Y-axis, ordinate) versus the binding energy of the electrons detected (X-axis, abscissa).

X-ray photoelectron spectroscopy (XPS) is used to study different silicon oxide passivations prepared by thermal (TO), chemical (CO), and plasma (PO) techniques.

4.7.6 Lifetime and Suns-V_{oc} measurement

The photoconductance-based contactless lifetime measurement method is one of the standard measurement methods for determining the effective minority carrier lifetime of semiconductor samples. The powerful measurement and analysis technique of this instrument is based on an eddy-current method. In this method, sheet conductivity of the sample is measured using an RF sensor, and a light sensor is used to measure the intensity of the flash light (filtered xenon flash lamp) to which the sample is exposed. These measured values are then used to calculate the effective carrier lifetime of the sample (Sinton, Cuevas, & Stuckings, 1996).

Measurements can be done by two methods: quasisteady-state photoconductance (QSSPC) method and transient photoconductance decay (transient PCD) method. This instrument is very useful for the estimation of material quality, passivation quality, and diffusion of dopants in c-Si solar cells.

Suns-V_{oc} measurement technique plays an important role in optimization of the metallization process for solar cell fabrication. Measurement of open circuit voltage as a function of light intensity is the main principle of the Suns-V_{oc} technique (Kerr, Cuevas, & Sinton, 2002). The Suns-V_{oc} measurement can be performed with a P-N junction diode when contacts are made on both sides of the junction. This measurement provides the solar cell I-V characteristic without the effects of series resistance and hence can be used to characterize shunting (Sinton & Cuevas, 2000). Sinton WCT-120 is a Suns-V_{oc} measurement instrument shown in Fig. 4.36. It monitors the illumination intensity using a reference solar cell. The Suns-V_{oc} is estimated from decay characteristics of the photo generated charge carriers. This system uses a flash lamp with a slow decay. This system is also adopted for different dimensions of solar cell structures. It is also possible to determine the reverse saturation current densities and the shunt resistance values using a two diode analysis method. Fitting a Suns-V_{oc} curve is simpler than fitting a J_{sc}-V_{oc} curve. Optimization of the co-firing process is done during metallization using a Sinton WCT-120 instrument.

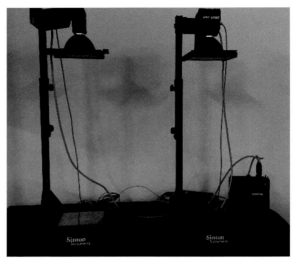

FIG. 4.36 Photograph of Sinton WCT-120 lifetime test instrument (right) and Suns-V_{oc} measurement system (left) (Laboratory of DST-IIEST solar PV Hub, IIEST, Shibpur).

4.7.7 Reflectance and external quantum efficiency (EQE) measurement

Reflectance, transmittance, absorbance, external quantum efficiency (EQE), and internal quantum efficiency (IQE) measurements play an important role for characterization of solar cells and are helpful to optimize the fabrication processes of solar cells. A typical optical spectrometer such as the Bentham PVE300 is widely used in various research laboratories and solar industries for this purpose.

4.7.7.1 Reflectance measurement

Light management is a big issue for solar cells fabrication. Texturization and antireflection (AR) coating are essential to reduce reflectance of incident solar lights from the top surface of solar cells. Reflectance of textured monocrystalline silicon wafers and silicon nitride (SiN_x) AR coated textured wafers are measured using a Bentham PVE300 (Bentham, n.d.), as shown in Fig. 4.37A. It includes a single monochromator, a monochromatic probe based on a Bentham PVE300, and a Xenon/Quartz halogen light source with a spectral range of 300–2500 nm. A DTR6 integrating sphere is included with this instrument for total reflectance as well as transmittance measurement. The entire internal surface of the integrated sphere is coated with barium sulfate (Ba_2SO_4). For reflectance measurement, first, calibration of equipment in reflection mode is done using a barium sulfate (Ba_2SO_4) sample, which is kept at the reflectance

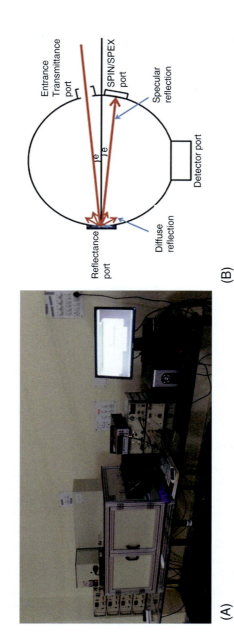

FIG. 4.37 (A) Photograph of Bentham PVE300 for reflectance measurement (Laboratory of DST-IIEST solar PV Hub, IIEST, Shibpur); (B) Ray diagram for reflectance measurement.

118 Sustainable developments by artificial intelligence & machine learning

port of the integrated sphere and reflectance is measured within a spectral range of 300–1100 nm and saved as a reference data file. Then, the Ba_2SO_4 sample is replaced by the actual sample to be measured at the reflectance port and the reflectance measured using the saved reference data file as reference. The ray diagram for this measurement using the integrated sphere is shown in Fig. 4.37B. For transmission measurement, the actual sample to be measured is placed at the transmittance port, keeping the Ba_2SO_4 sample at the reflectance port.

4.7.7.2 External quantum efficiency (EQE) measurement

The external quantum efficiency (EQE) is defined as the number of free electrons (which are produced by the incident photons) collected from the device to the external circuit of the device per photon incident on it (Lee & Price, 2017). It is directly obtained from the spectral response measurement using a Bentham PVE300, as shown in Fig. 4.37. Spectral response can provide detailed optical behavior of the device under monochromatic light illumination.

The device short-circuit current density (J_{sc}) can be calculated from the spectral response measured data. It may be used to predict expected J_{sc} value under standard testing conditions. This is simply calculated by the following equation over the spectral response range of the device under test:

$$J_{sc} = \int S_t(\lambda) \cdot E_0(\lambda) \cdot d\lambda \qquad (4.57)$$

where J_{sc} is integrated short-circuit current density in $A\,m^{-2}$, $S_t(\lambda)$ is the spectral response of the device in $A\,W^{-1}\,nm^{-1}$, and $E_0(\lambda)$ is the AM1.5 reference spectrum in $W\,m^{-2}\,nm^{-1}$.

4.7.8 Current-voltage (I-V) measurement

The most important characteristic of a complete solar cell is the current-voltage (*I-V*) measurement. Solar cell *I-V* is measured under a light source of air mass (AM) 1.5G standard spectra with irradiance $1000\,W/m^2$ (1 Sun). During measurement, the temperature is maintained at 25°C. A solar cell tester is an integrated system including a solar simulator and *I-V* measurement system. A Photo Emission Tech., Inc. (PET CT150AAA) solar cell tester is used to perform *I-V* measurement, as shown in Fig. 4.38.

A PET solar cell tester includes a steady state light source 1000 W Xenon Short Arc lamp with maximum illuminated area 156 mm × 156 mm and a built-in lamp power supply along with standard and advanced *I-V* measurement software. In this system, the lamp is surrounded with an ellipsoidal reflector, which collects most of the lamp output, and a uniform diverging beam produced using an optical integrator. After that, the beam is directed onto a collimating lens with the help of a mirror. A special filter is placed between the collimating lens and

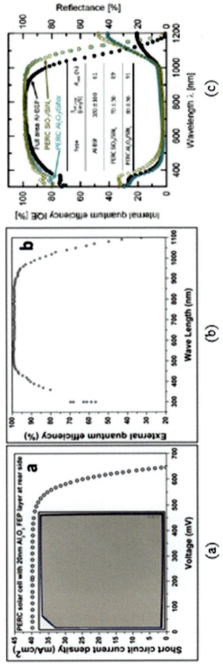

FIG. 4.38 Photograph of PET CT150AAA solar cell tester measurement (Laboratory of DST-IIEST solar PV Hub, IIEST, Shibpur).

120 Sustainable developments by artificial intelligence & machine learning

the mirror, as a result, the shape of the radiation spectra is matched with air mass 1.5 (AM 1.5G) (Photoemission, n.d.). The system also includes a computer controlled high quality solar simulator, which has an intensity measurement facility with feedback control for long-term stability, a flexible cell test fixture with four-probe cell contacting technique, multiple cell contacting probes for larger cells, and temperature controlled chuck with vacuum hold. This fixture can accommodate from small to up to 165 mm × 165 mm cells. Adjustable cell stops are attached in the X and Y-axes, which consistently locate the cells for testing. Multiple spring loaded busbars are attached for cell contacting from topside. Fig. 4.39A shows I-V characteristics of PERC solar cell with 20 nm Al_2O_3 FEP layer and 80 nm SiN_x capping layer at the rear side of the solar cell.

PET cell tester is capable of measuring solar cell parameters with a diverse range such as I_{sc}, V_{oc}, I_{max}, V_{max}, P_{max}, FF, R_{sh}, or R_s, complete light and dark I-V curves, along with cell conversion efficiency (η).

FIG. 4.39 (A) Current-voltage characteristic; (B) EQE of PERC solar cell with 20 nm Al_2O_3 FEP layer and 80 nm SiN_x capping layer at the rear side of the solar cell (Mandal, Acharya, et al., 2020); (C) Comparison of internal quantum efficiency (IQE) and reflectance between PERC solar cells with SiO_2/SiN_x and Al_2O_3/SiN_x passivtion stacks at the rear and a full-area Al BSF reference cell (Xiao & Shuyan, 2014).

4.8 Conclusion

From this study, we can conclude that the use of passivating layer stacks on the rear side of PERC solar cells improves optical generation as well as decreasing carrier loss due to recombination, compared to a conventional Al-BSF solar cell. Therefore efficiency of the cell increases. As total rear contact area decreases, it increases total series resistance in the case of PERC solar cells, which in turn decreases the fill factor of the cell. Chemical passivation strongly depends on interface defects present at different interfaces of the PERC solar cell. The field effect passivation strongly depends on the negative fixed charge present in the passivating layer. With increase in fixed charge recombination current decreases, hence passivation quality improves. By including ALD, single side etcher, and green laser in existing Al BSF processes, flow line PERC solar cells can easily be fabricated. For that reason, industrial fabrication of PERC solar cells will be mainstream in the coming decades.

References

Aberle, A. G. (1999). *Crystalline silicon solar cell—Advanced surface passivation and analysis.* Sydney: Centre for Photovoltaic Engineering, Univ. of New South Wales.

Aberle, A. G., Glunz, S., & Warta, W. (1992). Impact of illumination level and oxide parameters on Shockley–Read–Hall recombination at the Si-SiO2 interface. *Journal of Applied Physics, 71,* 4422–4431.

Aboaf, J. A., Kerr, D. R., & Bassous, E. (1973). Charge in SiO2-Al2O3 double layers on silicon. *Journal of Electrochemical Society Solid-State Science and Technology, 120,* 1103.

Angarita, G., Palacio, C., Trujillo, M., & Arroyave, M. (2017). Synthesis of alumina thin films using reactive magnetron sputtering method. *Journal of Physics: Conference Series, 850*(1), 012022. IOP Publishing.

Balaji, N., Park, C., Raja, J., Ju, M., Venkatesan, M. R., Lee, H., et al. (2015). Low surface recombination velocity on p-type Cz–Si surface by sol–gel deposition of Al2O3 films for solar cell applications. *Journal of Nanoscience and Nanotechnology, 15*(7), 5123–5128.

Belghachi, A. (2013). Detailed analysis of surface recombination in crystalline silicon solar cells. In *2013 international renewable and sustainable energy conference (IRSEC)* (pp. 161–166). IEEE.

Benick, J., Bateman, N., & Hermle, M. (2010). Very low emitter saturation current densities on ion implanted boron emitters. In *Proceedings of 25th European photovoltaic solar energy conference and exhibition, Valencia, Spain* (pp. 1169–1173).

Benick, J., Hoex, B., van de Sanden, M. C. M., Kessels, W. M. M., Schultz, O., & Glunz, S. W. (2008). High efficiency n-type Si solar cells on Al2O3-passivated boron emitters. *Applied Physics Letters, 92,* 253504.

Bentham. http://www.bentham.co.uk/solcell.htm.

Black, L. E. (2016). *New perspectives on surface passivation: Understanding the Si-Al2O3 interface.* Springer.

Blakers, A. (2019). Development of the PERC solar cell. *IEEE Journal of Photovoltaics, 9*(3), 629–635.

Blakers, A. W., Wang, A., Milne, A. M., Zhao, J., & Green, M. A. (1989). 22.8% efficient silicon solar cell. *Applied Physics Letters, 55*(13), 1363–1365.

122 Sustainable developments by artificial intelligence & machine learning

Bordihn, S., Dingemans, G., Mertens, V., & Kessels, W. M. M. (2013). Passivation of n+-type Si surfaces by low temperature processed SiO2/Al2O3 stacks. *IEEE Journal of Photovoltaics, 3*, 925–929.

Boukortt, N., Patanè, S., & Hadri, B. (2019). Development of high-efficiency PERC solar cells using Atlas Silvaco. *Silicon, 11*(1), 145–152.

Bounaas, L., Auriac, N., Grange, B., Monna, R., Pirot, M., De Vecchi, S., et al. (2013). Laser ablation of dielectric layers and formation of local Al-BSF in dielectric back passivated solar cells. *Energy Procedia, 38*, 670–676.

Chen, J., Cornagliotti, E., Loozen, X., Simoen, E., Vanhellemont, J., Lauwaert, J., et al. (2011). Impact of firing on surface passivation of p-Si by SiO2/Al and SiO2/SiNx/Al stacks. *Journal of Applied Physics, 110*, 126101.

Chen, F., Romijn, I., Weeber, A., Tan, J., Hallam, B., & Cotter, J. (2007). Relationship between PECVD silicon nitride film composition and surface and edge passivation. In *22nd European photovoltaic solar energy conference and exhibition, Milan* (pp. 1053–1060).

Chunduri, S. K., & Schmela, M. (2018). *PERC solar cell technology*. 2018 Edition Taiyang News. www.taiyangnews.info.

Chunduri, S. K., & Schmela, M. (2019). *High efficiency cell technologies*. 2019 Edition Taiyang News. www.taiyangnews.info.

Cuevas, A., Allen, T., Bullock, J., Wan, Y., & Zhang, X. (Eds.). (2015). *Skin care for healthy silicon solar cells 2015 IEEE 42nd photovoltaic specialist conference (PVSC)* (pp. 1–6). IEEE.

Dauwe, S., Mittelstadt, L., Metz, A., & Hezel, R. (2002). Experimental evidence of parasitic shunting in silicon nitride rear surface passivated solar cells. *Progress in Photovoltaics, 10*, 271–278.

De Rose, R., Magnone, P., Zanuccoli, M., Sangiorgi, E., & Fiegna, C. (2013). Loss analysis of silicon solar cells by means of numerical device simulation. In *2013 14th international conference on ultimate integration on silicon (ULIS)* (pp. 205–208). IEEE.

Dingemans, G., Beyer, W., van de Sanden, M. C. M., & Kessels, W. M. M. (2010). Hydrogen induced passivation of Si interfaces by Al2O3 films and SiO2/Al2O3 stacks. *Applied Physics Letters, 97*, 152106.

Dingemans, G., Einsele, F., Beyer, W., van de Sanden, M. C. M., & Kessels, W. M. M. (2012). Influence of annealing and Al2O3 properties on the hydrogen-induced passivation of the Si/SiO2 interface. *Journal of Applied Physics, 111*, 093713.

Dingemans, G., Engelhart, P., Seguin, R., Einsele, F., Hoex, B., van de Sanden, M. C. M., et al. (2009). Stability of Al2O3 and Al2O3/a-SiNx:H stacks for surface passivation of crystalline silicon. *Journal of Applied Physics, 106*, 114907.

Dingemans, G., & Kessels, W. M. M. (2012). Status and prospects of Al2O3-based surface passivation schemes for silicon solar cells. *Journal of Vacuum Science & Technology A: Vacuum, Surfaces, and Films, 30*(4), 040802.

Dingemans, G., Mandoc, M. M., Bordihn, S., Van De Sanden, M. C. M., & Kessels, W. M. M. (2011). Effective passivation of Si surfaces by plasma deposited SiOx/a-SiNx:H stacks. *Applied Physics Letters, 98*, 222102.

Dingemans, G., Terlinden, N. M., Verheijen, M. A., Van de Sanden, M. C. M., & Kessels, W. M. M. (2011). Controlling the fixed charge and passivation properties of Si (100)/Al2O3 interfaces using ultrathin SiO2 interlayers synthesized by atomic layer deposition. *Journal of Applied Physics, 110*(9), 093715.

Dingemans, G., van de Sanden, M. C. M., & Kessels, W. M. M. (2010). Influence of the deposition temperature on the c-Si surface passivation by Al2O3 films synthesized by ALD and PECVD. *Electrochemical and Solid-State Letters, 13*, H76.

Dingemans, G., van de Sanden, M. C. M., & Kessels, W. M. M. (2011). Excellent Si surface passivation by low temperature SiO2 using an ultrathin Al2O3 capping film. *Physica Status Solidi Rapid Research Letters, 5*, 22–24.

Dullweber, T., Kranz, C., Peibst, R., Baumann, U., Hannebauer, H., Fülle, A., et al. (2015). The PERC+ cell: A 21%-efficient industrial bifacial PERC solar cell. In *31st European photovoltaic solar energy conference and exhibition* (pp. 341–350). https://doi.org/10.4229/EUPVSEC2015 2015-2BO.4.3.

Dupont. http://www.dupont.com/content/dam/dupont/products-and-services/solar-photovoltaic-materials/solar-photovoltaic-materials-anding/documents/solamet_%20PV19B_datasheet.pdf.

Duttagupta, S., Ma, F. J., Hoex, B., & Aberle, A. G. (2014). Excellent surface passivation of heavily doped p+ silicon by low-temperature plasma-deposited SiOx/SiNy dielectric stacks with optimised antireflective performance for solar cell application. *Solar Energy Materials & Solar Cells, 120*, 204–208.

Duttagupta, S., Ma, F., Hoex, B., Mueller, T., & Aberle, A. G. (2012). Optimised antireflection coatings using silicon nitride on textured silicon surfaces based on measurements and multidimensional modelling. *Energy Procedia, 15*, 78.

Eades, W. D., & Swanson, R. M. (1985). Calculation of surface generation and recombination velocities at the Si-SiO2 interface. *Journal of Applied Physics, 58*, 4267–4276.

Energytrend. (2015). https://www.energytrend.com/news/Trina_Solar_Mono_si_PERC_PV_Cell_Hits_Efficiency_Record.html.

Engelhart, P., Hermann, S., Neubert, T., Plagwitz, H., Grischke, R., Meyer, R., et al. (2007). Laser ablation of SiO2 for locally contacted Si solar cells with ultra-short pulses. *Progress in Photovoltaics: Research and Applications, 15*(6), 521–527.

Fan, P., Sun, Z., Wilkes, G. C., & Gupta, M. C. (2019). Low-temperature laser generated ultrathin aluminum oxide layers for effective c-Si surface passivation. *Applied Surface Science, 480*, 35–42.

Faust, J. W., Jr. (1960). In H. C. Gatos (Ed.), *The surface chemistry of metals and semiconductors* (p. 151). New York: John Wiley & Sons.

Ferré, R., Florian, S., Beier, B., Dullweber, T., Jahn, C., & Gatz, S. (2014). 20.5% industrial PERC cells with PECVD Al2O3/SiNx rear-passivation and flat-top nanosecond laser contact opening. In *29th European photovoltaic solar energy conference and exhibition, Amsterdam* (pp. 617–620).

Fischer, B. (2003). *Loss analysis of crystalline silicon solar cells using photoconductance and quantum efficiency measurements.* Cuvillier Verlag.

Foroughi-Abari, A., & Cadien, K. (2012). Atomic layer deposition for nanotechnology. In *Nanofabrication* (pp. 143–161). Vienna: Springer.

Füssel, W., Schmidt, M., Angermann, H., Mende, G., & Flietner, H. (1996). Defects at the Si/SiO2 interface: Their nature and behaviour in technological processes and stress. *Nuclear Instruments and Methods in Physics Research Section A: Accelerators, Spectrometers, Detectors and Associated Equipment, 377*, 177–183.

Gall, S., Brune, J., Moorhouse, C., Manuel, S., Pirot, M., Monna, R., et al. (2010). Comparison of nanosecond 248 nm and picosecond 355nm laser processes for selective ablation of SiNx material. In *25th European photovoltaic solar energy conference and exhibition and 5th world conference on photovoltaic energy conversion, Valencia* (pp. 1845–1848).

Gassenbauer, Y., Ramspeck, K., Bethmann, B., Dressler, K., Moschner, J. D., Fiedler, M., et al. (2012). Rear-surface passivation technology for crystalline silicon solar cells: A versatile process for mass production. *IEEE Journal of Photovoltaics, 3*(1), 125–130.

124 Sustainable developments by artificial intelligence & machine learning

Gatz, S., Bothe, K., Müller, J., Dullweber, T., & Brendel, R. (2011). Analysis of local Al doped back surface fields for high efficiency screen-printed solar cells. *Energy Procedia, 8*, 318–323.

Gatz, S., Dullweber, T., & Brendel, R. (2011). Evaluation of series resistance losses in screen-printed solar cells with local rear contacts. *IEEE Journal of Photovoltaics, 1*(1), 37–42.

Gatz, S., Hannebauer, H., Hesse, R., Werner, F., Schmidt, A., Dullweber, T., et al. (2011). 19.4%-efficient large-area fully screen-printed silicon solar cells. *Physica Status Solidi RRL: Rapid Research Letters, 5*(4), 147–149.

Glunz, S. W., Benick, J., Biro, D., Bivour, M., Hermle, M., Pysch, D., et al. (2010). Preu, n-type silicon-enabling efficiencies > 20% in industrial production. In *35th IEEE photovoltaic specialists conference, Honolulu* (pp. 050–056).

Glunz, S. W., Biro, D., Rein, S., & Warta, W. (2012). Field-effect passivation of the SiO2-Si interface. *Journal of Applied Physics, 683*, 683–691.

Goetzberger, A., Knobloch, J., & Voss, B. (1998). *Crystalline silicon solar cells*. Wiley.

Granneman, K. V. E. H. A., Vermont, P., Kuznetsov, V., & Coolen, M. (2010). High-throughput, in-line ALD Al2O3 system. In *25th European photovoltaic solar energy conference and exhibition, Valencia* (pp. 1640–1644).

Gray, J. L. (2011). *Handbook of photovoltaic science and engineering* (2nd ed.). John Wiley & Sons, Ltd., ISBN:978-0-470-72169-8.

Green, M. A. (2015). The passivated emitter and rear cell (PERC): From conception to mass production. *Solar Energy Materials and Solar Cells, 143*, 190–197.

Haynos, J., Allison, J., Arndt, R., & Meulenberg, A. (1974). The COMSAT nonreflective silicon solar cell: a second generation improved cell. In *International conference on photovoltaic power generation* (p. 18).

Heinrich, G., Bähr, M., Stolberg, K., & Wütherich, T. (2011). Investigation of ablation mechanisms for selective laser ablation of silicon nitride layers. *Energy Procedia, 8*, 592–597.

Hoex, B., Heil, S. B. S., Langereis, E., van de Sanden, M. C. M., & Kessels, W. M. M. (2006). Ultra-low surface recombination of c-Si substrates passivated by plasma assisted atomic layer deposited Al2O3. *Applied Physics Letters, 89*, 042112.

Hofmann, M., Kambor, S., Schmidt, C., Grambole, D., Rentsch, J., Glunz, S. W., et al. (2008). PECVD-ONO: A new deposited firing stable rear surface passivation layer system for crystalline silicon solar cells. *Advances in Optoelectronics, 2008*, 485467.

Horzel, J., de Clercq, K., Evrard, O., Szlufcik, J., Frisson, L., Duerinckx, F., et al. (1995). Advantages of a new metallisation structure for the front side of solar cells. In *Vol. 2. 13th European photovoltaic solar energy conference* (p. 1368).

Hsu, C.-H., Huang, C.-W., Cho, Y.-S., Wan-Yu, W., Wuu, D.-S., Zhang, X.-Y., et al. (2019). Efficiency improvement of PERC solar cell using an aluminum oxide passivation layer prepared via spatial atomic layer deposition and post-annealing. *Surface and Coatings Technology, 358*, 968–975.

Huang, H., Lv, J., Bao, Y., Xuan, R., Sun, S., Sneck, S., et al. (2017). 20.8% industrial PERC solar cell: ALD Al2O3 rear surface passivation, efficiency loss mechanisms analysis and roadmap to 24%. *Solar Energy Materials and Solar Cells, 161*, 14–30.

IEC. (2008). *Photovoltaic devices—Part 3. Measurement principles for terrestrial photovoltaic (PV) solar devices with reference spectral irradiance data* (2nd ed.). International Electrotechnical Commission.

International Electrotechnical Commission (Ed.). (2019). IEC 60904-3: Photovoltaic devices—Part 3: Measurement principles for terrestrial photovoltaic (PV). In *Solar devices with reference spectral irradiance data* (2nd ed.). International Electrotechnical Commission.

International Technology Roadmap for Photovoltaic. (2020). https://itrpv.vdma.org/.

Jaffrennou, P. (2011). Laser ablation of SiO2/SiNx and AlOx/SiNx back side passivation stacks for advanced cell architectures. In *Proceedings of the 26th world conference and exhibition on photovoltaic solar energy conversion.*

Jin, H., Weber, K. J., Dang, N. C., & Jellett, W. E. (2007). Defect generation at the Si–SiO2 interface following corona charging. *Applied Physics Letters, 90,* 262109.

Kelly, P. J., Abu-Zeid, O. A., Arnell, R. D., & Tong, J. (1996). The deposition of aluminium oxide coatings by reactive unbalanced magnetron sputtering. *Surface and Coatings Technology, 86,* 28–32.

Kerr, M. J. (2002). *Surface, emitter and bulk recombination in silicon and development of silicon nitride passivated solar cells.* Australian National University.

Kerr, M. J., & Cuevas, A. (2002). Very low bulk and surface recombination in oxidized silicon wafers. *Semiconductor Science and Technology, 17,* 35–38.

Kerr, M. J., Cuevas, A., & Sinton, R. A. (2002). Generalized analysis of quasi-steady-state and transient decay open circuit voltage measurements. *Journal of Applied Physics, 9,* 399.

Kho, T. C., Black, L. E., & McIntosh, K. R. (2009). Degradation of Si-SiO2 interfaces during rapid thermal annealing. In *24th European photovoltaic solar energy conference and exhibition* (pp. 21–25). Hamburg: WIP.

Kim, K., Borojevic, N., Winderbaum, S., Duttagupta, S., Zhang, X., Park, J., et al. (2019). Investigation of industrial PECVD AlOx films with very low surface recombination. *Solar Energy, 186,* 94–105.

Kim, M., Park, S., & Kim, D. (2013). Highly efficient PERC cells fabricated using the low cost laser ablation process. *Solar Energy Materials & Solar Cells, 117,* 126–131.

King, R. R. (1990). *Studies of oxide-passivated emitters in silicon and applications to solar cells.* Stanford University.

King, R. R., Sinton, R. A., & Swanson, R. M. (1987). Low surface recombination velocities on doped silicon and their implications for point contact solar cells. In *19th IEEE photovoltaic specialists conference, New Orleans* (pp. 1168–1173).

Knorz, A., Aleman, M., Grohe, A., Preu, R., & Glunz, S. W. (2009). Laser ablation of antireflection coatings for plated contacts yielding solar cell efficiencies above 20%. In *24th European photovoltaic solar energy conference, Hamburg* (pp. 1002–1005).

Koski, K., Hölsä, J., & Juliet, P. (1999). Properties of aluminium oxide thin films deposited by reactive magnetron sputtering. *Thin Solid Films, 339*(1–2), 240–248.

Lauinger, T., Schmidt, J., Aberle, A. G., & Hezel, R. (1996). Record low surface recombination velocities on 1 Ω cm p-silicon using remote plasma silicon nitride passivation. *Applied Physics Letters, 68,* 1232.

Lee, D. B. (1969). Anisotropic etching of silicon. *Journal of Applied Physics, 40,* 4569.

Lee, S., & Price, K. J. (2017). Spectral responses in quantum efficiency of emerging kesterite thin-film solar cells. In *Optoelectronics—Advanced device structures.* https://doi.org/10.5772/68058.

Leguijt, C., Lölgen, P., Eikelboom, J. A., Weeber, A. W., Schuurmans, F. M., Sinke, W. C., et al. (1996). Low temperature surface passivation for silicon solar cells. *Fuel and Energy Abstracts, 40,* 297–345.

Li, S., Repo, P., Von Gastrow, G., Bao, Y., & Savin, H. (2013). Effect of ALD reactants on blistering of aluminum oxide films on crystalline silicon. In *39th IEEE photovoltaic specialists conference, Tampa* (pp. 1265–1267).

Lin, F., Duttagupta, S., Shetty, K. D., Boreland, M., Aberle, A. G., & Hoex, B. (2012). Excellent passivation of p+ silicon surfaces by inline plasma enhanced chemical vapour deposited SiOx/AlOx stacks. *Japanese Journal of Applied Physics, 51,* 10NA17.

Löffler, J., Wipliez, L. A., de Keijzer, M. A., Bosman, J., & Soppe, W. J. (2009). Depth-selective laser ablation for monolithic series interconnection of flexible thin-film silicon solar cells. In *24th European photovoltaic solar energy conference, Hamburg* (pp. 2704–2706).

Lohmüller, E., Greulich, J., Saint-Cast, P., Lohmüller, S., Schmidt, S., Belledin, U., et al. (2020). Front side optimization on boron- and gallium-doped Cz-Si PERC solar cells exceeding 22% conversion efficiency. In *37th European photovoltaic solar energy conference*.

Lorenz, A., John, J., Vermang, B., & Poortmans, J. (2010). Influence of surface conditioning and passivation schemes on the internal rear reflectance of bulk silicon solar cells. In *Proceedings of the 25th European photovoltaic solar energy conference and exhibition-EPVSEC* (pp. 2059–2061).

Lossen, J., Wald, M., & Bahr, M. (2007). Selective laser ablation of dielectric layers. In *22nd European photovoltaic solar energy conference, Milan* (pp. 1061–1067).

Luft, W., & Tsuo, Y. S. (1993). *Hydrogenated amorphous silicon alloy deposition processes*. Marcel Dekker, Inc.

Lv, Y., Zhuang, Y. F., Wang, W. J., Wei, W. W., Sheng, J., Zhang, S., et al. (2020). Towards high-efficiency industrial p-type mono-like Si PERC solar cells. *Solar Energy Materials & Solar Cells, 204*, 110202.

Mack, S., Wolf, A., Brosinsky, C., Schmeisser, S., Kimmerle, A., Saint-cast, P., et al. (2011). Silicon surface passivation by thin thermal oxide/PECVD layer stack systems. *IEEE Journal of Photovoltaics, 1*, 135–145.

Mandal, N. C., Acharya, S., Biswas, S., Panda, T., Sadhukhan, S., Sharma, J. R., et al. (2020). Evolution of PERC from Al-BSF: Optimization based on root cause analysis. *Applied Physics A, 126(7)*, 1–10.

Mandal, N. C., Biswas, S., Acharya, S., Panda, T., Sadhukhan, S., Sharma, J. R., et al. (2020). Study of the properties of SiOx layers prepared by different techniques for rear side passivation in TOPCon solar cells. *Materials Science in Semiconductor Processing, 119*, 105163. https://doi.org/10.1016/j.mssp.2020.105163.

McIntosh, K. R., Provancha, K. M., & Black, L. E. (2011). Surface passivation of crystalline silicon by APCVD aluminium oxide. In *26th European photovoltaic solar energy conference and exhibition* (pp. 1120–1124). Hamburg: WIP.

Mette, A. (2007). *New concepts for front side metallization of industrial silicon solar cells* (Ph.D. dissertation). Freiburg, Universit€at.

Moynihan, M., et al. (2010). In-line and vertical texturing of mono-crystalline solar cells. In *35th IEEE photovoltaic specialists conference*. https://doi.org/10.1109/pvsc.2010.5614629.

Oviroh, P. O., Akbarzadeh, R., Pan, D., Coetzee, R. A. M., & Jen, T.-C. (2019). New development of atomic layer deposition: Processes, methods and applications. *Science and Technology of Advanced Materials, 20(1)*, 465–496. https://doi.org/10.1080/14686996.2019.1599694.

Pasanen, T., Vähänissi, V., Theut, N., & Savin, H. (2017). Surface passivation of black silicon phosphorus emitters with atomic layer deposited SiO2/Al2O3 stacks. *Energy Procedia, 124*, 307–312.

Pawlik, M., Vilcot, J.-P., Halbwax, M., Aureau, D., Etcheberry, A., Slaoui, A., et al. (2014). Electrical and chemical studies on Al2O3 passivation activation process. *Energy Procedia, 60*, 85–89.

Photoemission. http://www.photoemission.com/.

Plummer, J. D., Deal, M. D., & Griffin, P. B. (2009). *Silicon VLSI technology fundamentals, practice and modeling*. Dorling Kindersley India Pvt. Ltd.

Preu, R., Kleiss, G., Reiche, K., & Bucher, K. (1995). PV-module reflection losses. Measurement, simulation and influence on energy yield and performance ratio. In *13th European photovoltaic solar energy conference* (p. 1465).

Preu, R., Lohmüller, E., Lohmüller, S., Saint-Cast, P., & Greulich, J. M. (2020). Passivated emitter and rear cell—Devices, technology, and modeling. *Applied Physics Reviews, 7*(4), 041315.

Preu, R., Rentsch, J., Hofmann, M., Wagenmann, D., Saint-Cast, P., & Kania, D. (2009). Industrial Negatively Charged c-Si Surface Passivation by Inline PECVD AlOx. In *24th European photovoltaic solar energy conference* (pp. 2275–2278). Hamburg: WIP.

Price, J. B. (1973). Semiconductor silicon. *233*, 339.

Puurunen, R. L. (2005). Surface chemistry of atomic layer deposition: A case study for the trimethylaluminum/water process. *Journal of Applied Physics, 97*(12), 121301. https://doi.org/10.1063/1.1940727.

PVEducation. https://www.pveducation.org/.

Rahman, M. Z. (2012). Modeling minority carrier's recombination lifetime of p-Si solar cell. *International Journal of Renewable Energy Research (IJRER), 2*(1), 117–122.

Reed, M. L., & Plummer, J. D. (1988). Chemistry of Si-SiO2 interface trap annealing. *Journal of Applied Physics, 63*, 5776–5793.

Richter, A., Benick, J., Hermle, M., & Glunz, S. W. (2011). Excellent silicon surface passivation with 5 Å thin ALD Al2O3 layers: Influence of different thermal post-deposition treatments. *Physica Status Solidi Rapid Research Letters, 5*, 202–204.

Richter, A., Hermle, M., & Glunz, S. W. (2013). Reassessment of the limiting efficiency for crystalline silicon solar cells. *IEEE Journal of Photovoltaics, 3*, 1184.

Rose, D., Raffaele, K. V. W., Tous, L., Das, J., Dross, F., Claudio, F., et al. (2012). Optimization of rear point contact geometry by means of 3-d numerical simulation. *Energy Procedia, 27*, 197–202.

Rühle, S. (2016). Tabulated values of the Shockley–Queisser limit for single junction solar cells. *Solar Energy, 130*, 139–147.

Rutherford, E. (1914). The structure of the atom. *Philosophical Magazine, 27*, 488–498.

Saraei, A., Eshraghi, M. J., Tajabadi, F., & Massoudi, A. (2018). ECV doping profile measurements in silicon using conventional potentiostat. *Journal of Electronic Materials, 47*(12), 7309–7315.

Schlüter, H., & Shivarova, A. (1998). Advanced technologies on wave and beam generated plasmas. *Nano science series*. Kluwer Academic Publishers.

Schmidt, J., Merkle, A., Bock, R., Altermatt, P. P., Cuevas, A., Harder, N.-P., et al. (2008). Progress in the surface passivation of silicon solar cells. In *23rd European photovoltaic solar energy conference, Valencia* (pp. 974–981).

Schmidt, J., Veith, B., & Brendel, R. (2009). Effective surface passivation of crystalline silicon using ultrathin Al2O3 films and Al2O3/SiNx stacks. *Physica Status Solidi Rapid Research Letters, 289*, 287–289.

Schmidt, J., Werner, F., Veith, B., Dimitri, Z., Steingrube, S., Altermatt, P. P., et al. (2012). Advances in the surface passivation of silicon solar cells. *Energy Procedia, 15*, 30–39.

Schneiderlöchner, E., Preu, R., Lüdemann, R., & Glunz, S. W. (2002). Laser-fired rear contacts for crystalline silicon solar cells. *Progress in Photovoltaics: Research and Applications, 10*(1), 29–34.

Schultz, O., Hofmann, M., Glunz, S. W., & Willeke, G. P. (2005). Silicon oxide/silicon nitride stack system for 20% efficient silicon solar cells. In *31st IEEE photovoltaic specialists conference, Florida* (pp. 872–876).

Sekuler, R., & Blake, R. (1985). *Perception*. New York: Alfred A. Knopf Inc.

Shaffner, T., & Schroder, D. K. (2000). Electrical, physical, and chemical characterization. In *Handbook of semiconductor manufacturing technology* (pp. 889–891).

Sharma, V. (2013). *Study of charges present in silicon nitride thin films and their effect on silicon solar cell efficiencies*. Arizona State University.

128 Sustainable developments by artificial intelligence & machine learning

Shockley, W., & Queisser, H. J. (1961). Detailed balance limit of efficiency of *p-n* junction solar cells. *Journal of Applied Physics, 32*, 510.

Shockley, W., & Read, W. T. (1952). Statistics of the recombinations of holes and electrons. *Physical Review, 87*, 835–842.

Singh, P. K., Kumar, R., Lal, M., Singh, S. N., & Das, B. K. (2001). Effectiveness of anisotropic etching of silicon in aqueous alkaline solutions. *Solar Energy Materials & Solar Cells, 70*, 103–113. https://doi.org/10.1016/s0927-0248(00)00414-1.

Sinton, R. A., & Cuevas, A. A. (2000). Quasi-steady-state open-circuit voltage method for solar cell characterization. In *16th European photovoltaic solar energy conference, Glasgow, Scotland* (pp. 1152–1155).

Sinton, R. A., Cuevas, A., & Stuckings, M. (1996). Quasi-steady-state photoconductance, a new method for solar cell material and device characterization. In *Conference record of the twenty fifth IEEE photovoltaic specialists conference* (pp. 457–460).

Smits, F. M. (1958). Measurement of sheet resistivities with the four-point probe. *Bell System Technical Journal, 34*, 711–718. https://doi.org/10.1002/j.1538-7305.1958.tb03883.x.

Synopsys. (2014). *Sentaurus TCAD user manual*. v. F2014.

Sze, S. M., & Ng, K. K. (2007). *Physics of semiconductor devices* (3rd ed.). Hoboken, NJ: John Wiley & Sons, Inc.

Talló, M. C., & Mcintosh, K. R. (2009). Permeability of TiO2 antireflection coatings to damp heat. In *24th European photovoltaic solar energy conference, Hamburg* (pp. 2037–2040).

Terlinden, N. M., Dingemans, G., Vandalon, V., Bosch, R. H. E. C., & Kessels, W. M. M. (2014). Influence of the SiO2 interlayer thickness on the density and polarity of charges in Si/SiO2/Al2O3 stacks as studied by optical second-harmonic generation. *Journal of Applied Physics, 115*, 033708.

Thorsten, D., Helge, H., Ulrike, B., Tom, F., Karsten, B., Stefan, S., et al. (2012). Fine-line printed 5 busbar PERC solar cells with conversion efficiencies beyond 21%. In *29th European photovoltaic solar energy conference and exhibition, Amsterdam* (pp. 621–626).

Tiedje, T., Yablonovitch, E., Cody, G. D., & Brooks, B. G. (1984). Limiting efficiency of silicon solar cells. *IEEE Transactions on Electron Devices, 31*, 711.

Tyagi, M. S., & Van Overstraeten, R. (1983). Minority carrier recombination in heavily-doped silicon. *Solid-State Electronics, 26*(6), 577–597.

Urrejola, E., Peter, K., Plagwitz, H., & Schubert, G. (2011). Silicon diffusion in aluminum for rear passivated solar cells. *Applied Physics Letters, 98*, 153508. https://doi.org/10.1063/1.3579541.

van de Loo, B. W. H., Ingenito, A., Verheijen, M. A., Isabella, O., Zeman, M., & Kessels, W. M. M. (2017). Surface passivation of n-type doped black silicon by atomic layer-deposited SiO2/Al2O3 stacks. *Applied Physics Letters, 110*, 263106.

van de Loo, B. W. H., Knoops, H. C. M., Dingemans, G., Janssen, G. J. M., Lamers, M. W. P. E., Romijn, I. G., et al. (2015). "Zero-charge" SiO2/Al2O3 stacks for the simultaneous passivation of n+ and p+ doped silicon surfaces by atomic layer deposition. *Solar Energy Materials & Solar Cells, 143*, 450–456.

Veith, B., Dullweber, T., Siebert, M., Kranz, C., Werner, F., Harder, N. P., et al. (2012). Comparison of ICP-AlOx and ALD-Al2O3 layers for the rear surface passivation of c-Si solar cells. *Energy Procedia, 27*, 379–384.

Verhoef, L. A., & Sinke, W. C. (1990). Minority-carrier transport in nonuniformly doped silicon-an analytical approach. *IEEE Transactions on Electron Devices, 37*(1), 210–221.

Vermang, B., Goverde, H., Lorenz, A., Uruena, A., Vereecke, G., Meersschaut, J., et al. (2011). On the blistering of atomic layer deposited Al2O3 as Si surface passivation. In *37th IEEE photovoltaic specialists conference, Seattle* (pp. 3562–3567).

Passivated emitter and rear contact solar cells **Chapter | 4 129**

Wang, A. (1992). *High efficiency PERC and PERL silicon solar cells*. Univ. of New South Wales.

Werner, S., Lohmüller, E., Saint-Cast, P., Greulich, J. M., Weber, J., Maier, S., et al. (2017). Key aspects for fabrication of p-type Cz-Si PERC solar cells exceeding 22% conversion efficiency. In *33rd European photovoltaic solar energy conference.*

Werner, F., Veith, B., Tiba, V., Poodt, P., Roozeboom, F., Brendel, R., et al. (2010). Very low surface recombination velocities on p- and n-type c-Si by ultrafast spatial atomic layer deposition of aluminum oxide. *Applied Physics Letters, 97*, 162103.

Wijekoon, K., Yan, F., Zheng, Y., Wang, D., Mungekar, H., Zhang, L., et al. (2013). Optimization of rear local contacts on high efficiency PERC solar cells structures. *International Journal of Photoenergy, 2013.*

Williams, K. R., & Muller, R. S. (1996). Etch rates for micromachining processing. *Journal of Microelectromechanical Systems, 5*(4), 256–269.

Würfel, P., & Würfel, U. (2016). *Physics of solar cells: From basic principles to advanced concepts* (3rd ed.). Weinheim: Wiley-VCH Verlag GmbH & Co., KGaA.

Xiao, S., & Shuyan, X. (2014). High-efficiency silicon solar cells—Materials and devices physics. *Critical Reviews in Solid State and Materials Sciences, 39*(4), 277–317.

Xu, G., Yang, Y., Zhang, K., Liu, W., Shaoyong, F., Feng, Z., et al (Eds.). (2013). *An improved optical simulation method for crystalline silicon solar cells 2013 IEEE 39th photovoltaic specialists conference (PVSC)* (pp. 2677–2680). IEEE.

Yang, Y., et al. (2016). 22.13% efficient industrial p-type mono PERC solar cell. In F. Ye, W. Deng, W. Guo, R. Liu, D. Chen, & Y. Chen (Eds.), *2016 IEEE 43rd photovoltaic specialists conference (PVSC)* (pp. 3360–3365). IEEE.

Zhao, J., Wang, A., & Green, M. A. (1999). 24·5% Efficiency silicon PERT cells on MCZ substrates and 24·7% efficiency PERL cells on FZ substrates. *Progress in Photovoltaics: Research and Applications, 7*, 471–474.

Zhou, J., Tan, Y., Liu, W., Cai, X., Huang, H., & Cao, Y. (2020). Effect of front surface light trapping structures on the PERC solar cell. *SN Applied Sciences, 2*(5), 1–10.

Chapter 5

Online-based approach for frequency control of microgrid using biologically inspired intelligent controller

Bhola Jha, Manoj Kumar Panda, and Yatindra Kumar
G. B. Pant Institute of Engineering and Technology, Pauri, Uttarakhand, India

5.1 Introduction

The depleting conventional energy resources may create a shortage of power for future generations. In such a situation, renewable resources such as solar, wind, fuel cell, etc., are alternative options that can provide sustainable and pollution-free electricity (Freris & Infield, 2008). Instability and uncertainty are the main problems in renewable power systems because of their dynamic behavior. There is deviation in load frequency because of differences in total generation and demand. In power systems (Ghafouri, Milimonfared, & Gharehpetian, 2015; Golpîra & Bevrani, 2014), load frequency control (LFC) plays a vital role. Practicing engineers decide which type of controllers are necessary for LFC issues (Panda, Panda, & Ardil, 2009; Yousef, Khalfan, Albadi, & Hosseinzadeh, 2014). A wide range of literature regarding LFC issues is reviewed in Pandey, Mohanty, and Kishor (2013), Parmar, Majhi, and Kothari (2012), and Rahmani and Sadati (2013). The research papers (Ansari & Velusami, 2010; Khuntia & Panda, 2012) have used artificial neural network (ANN) and fuzzy logic controller (FLC) in respect of LFC problems. The time taken to train the neurons of an ANN is large, which is one of its drawbacks. Zadeh (1965) introduced the fuzzy set theory in which there are difficulties in determining exact membership functions. Since the penetration of renewables keeps on increasing, so the conventional controller is unable to limit the frequency variations within the required range. Therefore, examples in the literature address that issues by implementing intelligent control or tuning the constants of PI/fuzzy through optimization techniques either in on- or offline mode.

Sustainable Developments by Artificial Intelligence and Machine Learning for Renewable Energies.
https://doi.org/10.1016/B978-0-323-91228-0.00015-X
Copyright © 2022 Elsevier Inc. All rights reserved.

132 Sustainable developments by artificial intelligence & machine learning

There are uncertainties in the membership function of type-1 fuzzy, which can be minimized by interval type-2 fuzzy logic systems (IT2FLS) (Liang & Mendel, 2000). Type-2 fuzzy systems have a three dimensional membership function, which is an advantage over the two dimensional membership function of type-1 fuzzy sets. The extra third dimensional membership function helps to regulate the uncertainty more efficiently. The applications of IT2FLC in different areas of engineering have been presented in Castillo and Melin (2014). LFC of a two-area generation rate constraint (GRC) nonlinearity power system using interval type-2 fuzzy is described (Sudha & Santhi, 2011). The implementation of IT2FLC using feedback error learning (FEL) method is suggested for the LFC (Sabahi, Ghaemi, & Pezeshki, 2014). The studies published are different in terms of their results and claims.

In Saxena (2019), fractional order controller via internal mode control (IMC) for the LFC of a given power system is proposed. The IMC framework is made to acts as a robust controller using the CRONE principle, model-order reduction, and FO filter, which is applied to a single area system and then extended to a two area system. The paper (Xiong, Li, & Wang, 2018) addresses the time delay issues for the LFC of a three area system. The delay causes a potential threat to the stability of power system. To mitigate the threat, robust control with delay margin estimation using a linear matric inequality (LMI) method is introduced. Designing a modified active disturbance rejection control (MADRC) scheme considering extended state observer (ESO) is proposed in Srikanth and Yagaiah (2018) to overcome the difficulties of nonminimum phase dynamics. For this, constraints of multiobjective functions are unified and a teaching-learning-based optimization (TLBO) technique is applied. This MADRC is implemented to a given power system for load frequency.

The paper (Aziz, Than, & Stojcevski, 2018) addresses the frequency sensitivity of wind penetration into the grid and also the grid code compatibility. To support the frequency of a microgrid, the vehicle to grid (V2G) concept is introduced in this paper (Khooban et al., 2017). This concept uses a multiobjective fractional order fuzzy PID (proportional integral derivative) controller (MOFOFPID) whose parameters are tuned using a modified black hole optimization algorithm (MBHOA). In paper (Wang et al., 2016), a sliding mode pitch angle controller is designed for a wind energy generator (WEG) to get proper output power. The sliding mode load frequency controller is designed as a secondary frequency regulation for a diesel power plant to bring back the system frequency to normal. For the improvement of dynamic performance, the sliding mode LFC is redesigned utilizing a disturbance observer (DO). A frequency control approach for the PV generator in a PV-diesel hybrid power system using fuzzy is proposed by Datta et al. (2011). Huge piercing of wind energy causing frequency deviation to the power system is inevitable for which an experimental real time implementation is done by Bervani et al. (2013).

Supplying electricity from a conventional big power plant to remote areas is a challenge due to huge transportation costs, operation, maintenance, transmission line erection, etc. Therefore, this chapter addresses the installation of a hybrid microgrid system, which is feasible, reliable, economically viable, and meets the demand for electricity locally. The durability of this grid system will be enhanced because the life span of individual energy resources is going to be increased. The grid should also ensure the availability of power at a constant frequency, which is usually difficult because of the presence of varying wind speed and solar irradiation supplying microgrid systems. The proposed hybrid test system consists of wind energy, solar PV array, fuel cell, diesel engine generator, battery energy storage system, and flywheel energy storage system. The test system and its parameters are taken to be the same as that of H. Bevrani (Bervani et al., 2012). Since the proposed test system includes WEG and PV, which are the major sources causing the frequency variations, so an intelligent optimized controller, i.e., GWO (gray wolf optimization)-based fuzzy logic controller cum PI (proportional integral) controller, is used for a diesel engine generator, which is a conventional resource in which natural frequency change is not possible. Another intelligent optimized controller for the same engine, i.e., PSO (particle swarm optimization)-based fuzzy logic controller cum PI controller, is employed in which the parameters of PI, i.e., K_p (proportional gain) and K_i (integral gain), are tuned using a PSO technique.

To demonstrate the capability of the proposed methods, the outcomes of results are collated with respect to PI and FLC plus PSO, and PI and FLC plus GWO is investigated.

5.2 Test system description

The dynamic nature of the power system is definitely affected by renewable energy sources. Nowadays, renewables integration into the grid keeps on increasing. Therefore, a standard test system proposed by H. Bevrani (Bervani et al., 2012) in IEEE Transactions on Smart Grid is taken for study, and all parameters are the same. A typical block-diagram of the test system is depicted in Fig. 5.1.

5.2.1 Photovoltaic model

The solar cell is nothing but a simply diode. The simple circuit representation of solar cell is a diode with parallel current source, as shown in Fig. 5.2.

The photoelectric effect is the principle of working of PV. As per this principle, light energy is transformed to electricity. For a photovoltaic module, the equation of ideal solar cell, as described in Onar, Uzunoglu, and Alam (2006) and Qi and Ming (2012), is given by:

$$I = I_L - I_R \left[\exp\left(\frac{V}{aV_t}\right) - 1 \right]$$

(5.1)

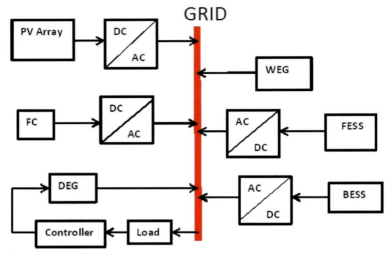

FIG. 5.1 Proposed test system.

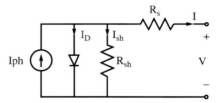

FIG. 5.2 Solar PV circuit.

where I_L = photocurrent (A), I_R = reverse saturation current (A), V = Diode voltage (V), V_t = thermal voltage (V_t = 27.5 mV at 25°C), and a = ideality factor of diode.

The V-I characteristic of a solar cell is rewritten:

$$I = I_L - I_R \left[\exp\left(\frac{V + IR_{Se}}{aV_i}\right) - 1 \right] - \frac{V + IR_{Se}}{R_{Sh}} \tag{5.2}$$

where R_{Se} is series resistance and R_{Sh} is shunt resistance in solar cell.

Since low voltage (approximately 0.5 V) is obtained from a PV cell, large numbers of series and parallel connected PV cells are required to produce the desired output power. Eq. (5.3) gives the V-I characteristic equation of a PV module, where N_s and N_p are series and parallel cells, respectively:

$$I^M = N_p I_L - N_p I_R \left[\exp\left(\frac{V^M/N_s + I^M/N_s}{AV_i}\right) \right] - \frac{(N_p/N_s)V^M + I^M R_{Se}}{R_{Sh}} \tag{5.3}$$

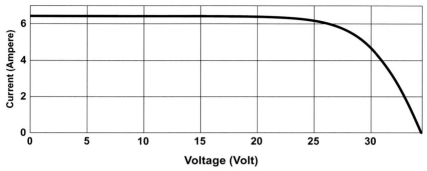

FIG. 5.3 Solar *I-V* characteristics.

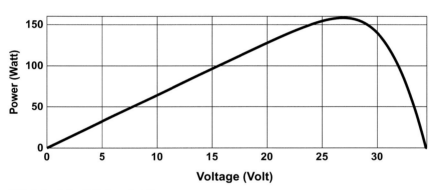

FIG. 5.4 Solar *P-V* characteristics.

Solar *I-V* and *P-V* characteristics are obtained through simulation, which are shown Figs. 5.3 and 5.4, respectively.

5.2.2 Wind energy

The basic equation concerning air or wind velocity as an input and the output power as mechanical force, generated by the rotor blades, is described by Simoes and Ferret (2007), Bevrani and Hiyama (2016), and Onar et al. (2006):

$$P_{mech} = C_P(\omega, \sigma) \frac{1}{2} \tau a v_{wind}^3 \qquad (5.4)$$

where P_{mech} = turbine output power (W), C_P = coefficient of performance of the turbine, ω = tip speed ratio, σ = blade pitch angle (degrees), τ = air density (kg (m^3)$^{-1}$), a = turbine swept area (m^2), and v = wind speed (m s^{-1}).

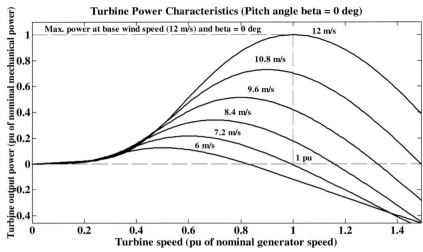

FIG. 5.5 Speed power characteristics of turbine.

Normalization and simplification can be done in Eq. (5.5) for particular values of τ and a. The per unit (p.u.) equation is as follows:

$$P_{mech-pu} = k_p C_{P-pu} v_{wind-pu}^3 \tag{5.5}$$

where $P_{mech\text{-}Pu}$ = per unit power for given values of τ and a (p.u.), K_p = power gain for $C_{P\text{-}pu} = 1$ p.u. and $v_{wind\text{-}pu} = 1$ p.u., $C_{P\text{-}pu}$ = per unit (p.u.) performance coefficient, and $v_{wind\text{-}Pu}$ = per unit of base wind speed. Speed-output power characteristics of a turbine are shown in Fig. 5.5 for various values of air or wind speed. Due to inertia of turbine, output power appears to be zero from 0 to 0.2 p.u. turbine speed.

5.2.3 Diesel engine generator (DEG) model

Transfer function of DEG model is given as follows (Papathanassiou & Papadopoulos, 2001):

$$P_d = -\left(\frac{1}{R} + \frac{K_I}{s}\right)\left(\frac{1}{T_{sg}s+1}\right)\left(\frac{K_d}{T_{\deg}s+1}\right)\Delta f_e \tag{5.6}$$

where P_d = power of diesel engine generator, K_I = constant of integral gain constant, R = speed regulation, T_{sg} = time constant of speed governor, K_d = gain of DEG, and $T_{\deg}$ = DEG time constant; in this, $K_d = 1$.

5.2.4 Fuel cell, BESS, and FESS

The fuel cell (FC) and battery energy storage system (BESS) convert chemical energy to electrical energy, whereas the flywheel energy storage system (FESS) is also known as a mechanical energy storage system. Details and description can be found in the literature (Hawke et al., 2011; Mahmoud et al., 2019; Mousavi et al., 2017; Wang et al., 2019).

The transfer function of FC, BESS, and FESS are given here as:

$$\frac{1}{1+sT_{FC}} \quad \frac{1}{1+sT_{BESS}} \quad \frac{1}{1+sT_{FESS}}$$

where T_{FC}, T_{BESS}, and T_{FESS} are the time constants of FC, BESS, and FESS, respectively; all are considered to be as the first order system.

The detailed block diagram for the frequency response of the proposed microgrid (MG) system is depicted in Fig. 5.6.

Parameters and their values are given in Table 5.1.

5.3 Fuzzy logic controller

Fuzzy logic was developed by Zadeh in 1965 and can be applied to any system where uncertainties and inaccuracies exist. The fuzzy logic controller (FLC) has the following advantages:

(1) A mathematical model of a system is not necessary
(2) It can handle nonlinearity and complexity
(3) It increases the robustness
(4) It does not require the system information
(5) The principle of operation of FLC is linguistic rules, i.e., IF/IF-THEN general instructions or structure.

Apart from these advantages, the FLC has some disadvantages, i.e., high computational burden during hardware and software implementation.

A fuzzy system is made up of three components: fuzzification, fuzzy rule base inference system, and defuzzification.

(A) Fuzzification: Error signals of crisp values are converted into fuzzy variables. The appropriate membership functions (triangular) for the input-output variables are selected. The universe of discourse is established for all inputs and outputs variables. The suitable ranges are selected for bringing all the inputs-outputs to the universe of discourse. Each discourse is split into three overlapping fuzzy sets. Each variable is a member of a subset with a degree of membership ranging from 0 to 1.
(B) Fuzzy rule base inference system: This system represents IF–THEN general instructions governing the input-output variables with membership functions. In this case, the variables are processed through an inference engine executing the rules for GWO and PSO, as shown in Table 5.2.

In an FLC controller, the input and output variables are mapped as below, thus meeting the fitness/objective function dynamically:

$$K = f(\Delta e) \tag{5.7}$$

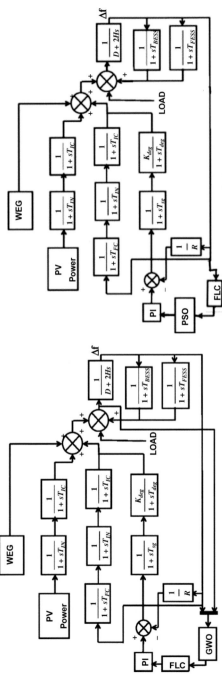

FIG. 5.6 Frequency response schematic diagram of MG using GWO and PSO.

TABLE 5.1 Parameters values.

S.N.	Name/symbol of parameter	Values
1	Damping coefficient, D (p.u./Hz)	0.015
2	Inertia constant, H (p.u. s)	0.835
3	Fuel cell time constant, T_{FC}	0.26
4	Battery energy storage system time constant, T_{BESS}	0.1
5	Flywheel energy storage system time constant, T_{FESS}	0.1
6	Speed governor time constant, T_{sg}	0.08
7	Diesel engine time constant, T_{deg}	0.4
8	Inverter time constant, T_{IN}	0.04
9	Connector time constant, T_{1C}	0.004
10	Speed regulation, R (Hz/p.u.)	3
11	WEG rating (kW)	100
12	PV power rating (kW)	30
13	FC rating (kW)	70
14	DEG rating (kW)	160
15	FESS rating (kW)	45
16	BESS rating (kW)	45
17	Total load (kW)	410

TABLE 5.2 Fuzzy rule base for GWO and PSO.

	$\Delta f \rightarrow$					
$\Delta P_L \downarrow$	*BN*	*MN*	*SN*	*SP*	*MP*	*BP*
S	BN	MN	SN	SP	SP	MP
M	BN	BN	MN	SP	MP	MP
B	BN	BN	BN	MP	MP	MP

	O $\rightarrow$	
$\Delta f \downarrow$	O_1	O_2
S	S	S
M	M	M
B	B	B

Here, BN: Big Negative, MN: Mean Negative, SN: Slight Negative, SP: Slight Positive, MP: Mean Positive, BP: Big Positive, S: Slight, M: Mean and B: Big. O: Output, O_1 = Output 1 and O_2 = Output 2.

140 Sustainable developments by artificial intelligence & machine learning

where K = output of controller, Δe = errors input to the controller, and f = nonlinear function.

Now it is necessary to select the membership function for the input and output variables. In this chapter, triangular membership is selected to reduce the computational burden. For the above mentioned fuzzy rules, Mamdani-type fuzzy inference is used. The membership functions, fuzzy sets, and fuzzy rules are all selected using hit and trial method to obtain optimum performance.

(C) Defuzzification: In this section, a crisp value of output is obtained by using a centroid defuzzification approach in which the center of each membership function for each rule is evaluated first.

5.4 Particle swarm optimization (PSO)

Particle swarm optimization (PSO) (Kennedy & Eberhart, 1995), as introduced by Kennedy and Eberhart in 1995, is a computational intelligence type of algorithm. PSO is simple and easy, making it popular to use, and is efficient to compute. PSO is a flexible and highly balanced mechanism to improve the local and global exploration capabilities. It has been shown that PSO is robust while solving nonlinear, nondifferentiable, multiple optima, and high dimension problems through adaptation, which has been derived from the socio-psychological theory. This method is found from research on swarms of birds and fish. Here, it is assumed that all information is shared inside the swarm.

In a PSO algorithm for an m-variable optimization problem, a swarm of particles are put into the m-dimensional search space with arbitrarily chosen velocities and positions knowing their best values so far (*pbest*). The velocity of each particle is changed as per its self-flying experience and that of others. The properties of the search are summarized below:

- Initial positions of *pbest* and *gbest* are different. But using the different direction of *pbest* and *gbest*, all particles gradually get close to the global optimal value.
- The changed velocity and position of each particle can be found using the current velocity and the distance from the $pbest_{j,g}$ to $gbest_{j,g}$ as shown in the following formulae:

$$v_{j,g}^{(t+1)} = \begin{cases} w * v_{j,g}^{(t)} + c_1 * r_1^0 * \left(pbest_{j,g} - x_{j,g}^{(t)} \right) \\ + c_2 * r_2^0 * \left(gbest_{j,g} - x_{j,g}^{(t)} \right) \end{cases} \tag{5.8}$$

$$x_{j,g}^{(t+1)} = x_{j,g}^{(t)} + v_{j,g}^{(t+1)} \tag{5.9}$$

with $j = 1, 2, \ldots, n$ and $g = 1, 2, \ldots, m$

where, n = number of particles in a group;
m = dimension;
t = number of iterations (generations);
$v_{j,\,g}^{(t)}$ = velocity of particle j at iteration t,
With $v_{j,\,g}^{m-n} \le v_{j,\,g}^{(t)} \le v_{j,\,g}^{max}$;
w = inertia weight factor;
c_1, c_2 = cognitive and social acceleration factors respectively;
r_1, r_2 = random numbers uniformly distributed in the range (0, 1);
$x_{j,\,g}^{(t)}$ = current position of j at iteration t;
$pbest_{j,\,g}$ = $pbest$ (local best) of particle j;
$gbest$ = global best of the group

The j-th particle in the swarm is represented by a g-dimensional vector $x_j = (x_{j,\,1}, x_{j,\,2}, \ldots, x_{j,\,g})$ and its rate of position change (velocity) is denoted by another g-dimensional vector $v_j = (v_{j,\,1}, v_{j,\,2}, \ldots, v_{j,\,g})$. The best previous position of the j-th particle is represented as $pbest_j = (pbest_{j,\,1}, pbest_{j,\,2}, \ldots, pbest_{j,\,g})$. The index of best particle among all of the particles in the group is represented by the $gbest_g$. In the proposed method each particle has three members K_p, K_i, and K_d. The search space has three dimensions, the particles must fly in a three dimensional space. Fig. 5.7 (Panda & Padhy, 2007) shows the velocity and position updates of a particle.

The goal of PSO is to have the optimum values of K_p and K_i for the PI controller. Here, the optimal means to minimize the objective or fitness function. Here, the fitness function is frequency error or frequency deviation. There are various performance indices to select the error, such as integral of absolute error (IAE), integral square error (ISE), integral square of absolute error (ISAE), etc. In this chapter, IAE is selected as a performance index, i.e., $IAE = \int e\,dt$. Here,

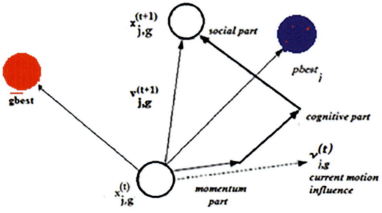

FIG. 5.7 Position and velocity update diagram.

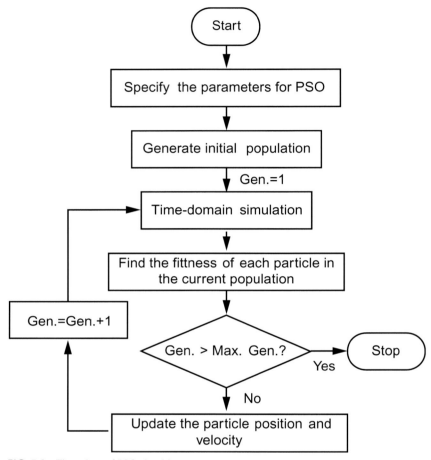

FIG. 5.8 Flow chart of PSO algorithm.

e stands for error. The particle travels with velocity vector and position vector for each iteration and checks for the optimization of fitness function in order to have the best values of PI controller. The computational flow chart of the PSO algorithm is shown in Fig. 5.8. In our PSO-PI controller design, the following parameters are chosen: population size $=5$, number of iterations 5, $c_1 = c_2 = 1$, $W_{max} = 0.8$, $W_{min} = 0.1$, $r_1 = 0.8$, and $r_2 = 0.8$, and the optimum values of the tuning parameters are $K_p = 150$ and $K_I = 120$.

5.5 Gray wolf optimization (GWO)

A GWO algorithm is a metaheuristics search algorithm proposed by Mirjalih. It is based on how wolves hunt prey for their food. The GWO algorithm emulates the chain of command and trapping system of gray wolves in the environment. Four names used to describe the hierarchy of wolf packs, alpha (α), beta (β), delta (δ), and omega (ω), are used to imitate or simulate the chain of command. While designing the mathematical model of GWO, α is considered as the best solution of fittest function, popularly known as the leader of wolves who is guiding others and asking them to participate in hunting. Accordingly, the second and third best results are beta (β) and delta (δ), commonly. The remaining individual outcomes are considered to be omega (ω).

The wolves undertake various activities to get their food, i.e., searching, chasing, tracking, encircling, and then attacking the prey. This behavior can be mathematically modeled as:

$$D = |\vec{C} \cdot \overrightarrow{X_p(t)} - \overrightarrow{X_p(t)}|$$
$$\vec{X}(k+1) = \overrightarrow{X_p(t)} - \vec{A} \cdot \vec{D}$$

where k denotes the number of iterations. D, A, and C indicate the coefficient vectors, and X_p and X indicate the position vector of prey and gray wolf, respectively.

The vectors A and C can be determined as:

$$\vec{A} = 2\vec{a} \cdot \vec{p} - \vec{a}$$
$$\vec{C} = 2 \cdot \vec{q}$$

where $\vec{a}$ vector linearly decreases from 2 to 0 during the iteration process, and $\vec{p}$ and $\vec{q}$ are the random vectors in [0 1].

When the wolves stop moving, it is understood that hunting is over.

Out of many optimization algorithms, GWO is implemented in this chapter because it possesses the following features:

- Since this algorithm uses the concept of probability, it becomes flexible, robust, and has easy convergence.
- It may lead to premature convergence, which increases the search action capability.
- Less time of optimization.
- GWO can be implemented in online/offline mode. In this chapter, the scaling factor of FLC is also tuned in online mode.

The objective of GWO is to obtain the optimum scaling factors of FLC. Here, the optimum means to minimize the objective function or fitness function. The objective/fitness function is frequency error $(\Delta f) = e$. There are various performance indices to select the error, such as integral of absolute error (IAE) $I_{AE} = \int |e(t)| dt$, integral square error (ISE) $I_{SE} = \int e^2(t)dt$, integral square of absolute error (ISAE) $= I_{SAE} = \int |e(t)|^2 dt$, and integral of time multiplied by absolute error (ITAE) $= I_{TAE} = \int t |e(t)| dt$. In this method, integral square error (ISE) $I_{SE} = \int e^2(t)dt$ is selected as a one of the performance indices.

In this way, the search agents (wolves) move and check for the optimization of objective function for each iteration, keeping the constraints of scaling factor lying between 0 and 1. Here, the number of search agents = 10 and number of iterations = 5.

The flow chart for implementing the GWO algorithm is depicted in Fig. 5.9.

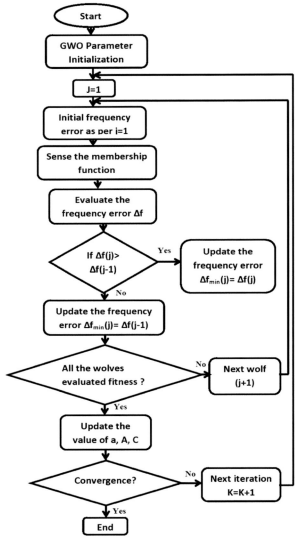

FIG. 5.9 GWO flow chart.

5.6 Results analysis

The load change profile on microgrid is shown in Fig. 5.10. There is increase and decrease in load at 5 s, 10 s, 15 s, etc.

The per unit change in frequencies by the conventional PI controller and two proposed methods, i.e., PI and FLC plus PSO and PI and FLC plus GWO are depicted in Figs. 5.11–5.13, respectively.

5.7 Conclusion

Each component of the microgrid is simulated using MATLAB. The results are illustrated. The comparative evaluation of frequency profile is validated. It is understood from the above figures that the per unit change in frequency decreases/increases in accordance with the per unit change in load by all controllers. The frequency variation is less using the proposed method, i.e., PI, GWO, and FLC.

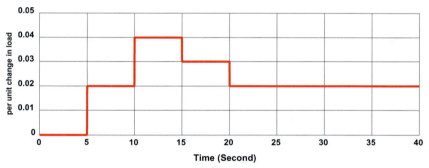

FIG. 5.10 Load profile on microgrid system.

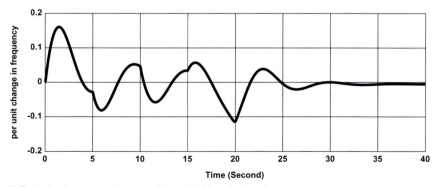

FIG. 5.11 Frequency change profile using PI controller alone.

146 Sustainable developments by artificial intelligence & machine learning

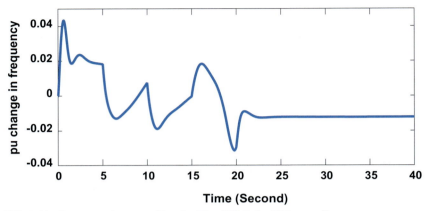

FIG. 5.12 Frequency change profile using PI and FLC plus PSO controller.

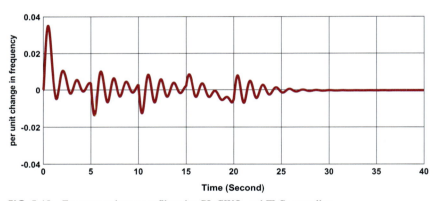

FIG. 5.13 Frequency change profile using PI, GWO, and FLC controller.

References

Ansari, M. M. T., & Velusami, S. (2010). Dual mode linguistic hedge fuzzy logic controller for an isolated wind diesel hybrid power system with superconducting magnetic energy storage unit. *Energy Conversion and Management*, 51(1), 169–181. http://www.sciencedirect.com/science/article/pii/S0196890409003501.

Aziz, A., Than, A., & Stojcevski, A. (2018). Analysis of frequency sensitive wind plant penetration effect on load frequency control of hybrid power system. *International Journal of Electrical Power and Energy Systems*, 99, 603–617. https://doi.org/10.1016/j.ijepes.2018.01.045.

Bervani, H., et al. (2012). Intelligent frequency control in an AC micro grid: Online PSO-based fuzzy tuning approach. *IEEE Transaction on Smart Grid, 3*(4), 1935–1944.

Bervani, H., et al. (2013). Intelligent LFC concerning high penetration of wind power: Synthesis and real time application. *IEEE Transaction on Sustainable Energy, 5*(2), 655–662.

Bevrani, H., & Hiyama, T. (2016). *Intelligent automatic generation control*. CRC Press.

Castillo, O., & Melin, P. (2014). A review on interval type-2 fuzzy logic applications in intelligent control. *Information Sciences, 279*, 615–631. http://www.sciencedirect.com/science/article/pii/S0020025514004629.

Datta, M., et al. (2011). A frequency control approach by photovoltaic generator in a PV-diesel hybrid power system. *IEEE Transaction on Energy Conversion, 26*(2), 559–571.

Freris, L., & Infield, D. (2008). *Renewable energy in power systems* (p. 2008). UK: John Wiley & Sons.

Ghafouri, A., Milimonfared, J., & Gharehpetian, G. B. (2015). Coordinated control of distributed energy resources and conventional power plants for frequency control of power systems. *IEEE Transactions on Smart Grid, 6*(1), 104–114. http://ieeexplore.ieee.org/document/6908022/.

Golpîra, H., & Bevrani, H. (2014). A framework for economic load frequency control design using modified multi-objective genetic algorithm. *Electric Power Components and Systems, 42*(8), 788–797. http://www.tandfonline.com/doi/abs/10.1080/15325008.2014.893545.

Hawke, J., et al. (2011). A modular fuel cell with hybrid energy storage. In *IEEE energy conversion congress & exposition*.

Kennedy, J., & Eberhart, R. (1995). Particle swarm optimization. In *Vol. IV. Proceedings of IEEE International Conference on Neural Networks, Perth, Australia* (pp. 1942–1948).

Khooban, M.-H., et al. (2017). Load frequency control in microgrids based on stochastic non-integral controller. *IEEE Transaction on Sustainable Energy*. https://doi.org/10.1109/TSTE.2017.2763607.

Khuntia, S. R., & Panda, S. (2012). Simulation study for automatic generation control of a multi-area power system by ANFIS approach. *Applied Soft Computing, 12*(1), 333–341. http://www.sciencedirect.com/science/article/pii/S156849461100322X.

Liang, Q., & Mendel, J. M. (2000). Interval type-2 fuzzy logic systems: Theory and design. *IEEE Transactions on Fuzzy Systems, 8*(5), 535–550. http://www.sciencedirect.com/science/article/pii/S0142061515001520.

Mahmoud, T. S., et al. (2019). The role of intelligent generation control algorithms in optimizing battery energy storage system size in microgrids: A case study from Western Australia. *Energy Conversion and Management, 196*, 1335–1352.

Mousavi, S. M., et al. (2017). A comprehensive review of flywheel energy storage system technology. *Renewable and Sustainable energy Reviews, 67*, 477–490.

Onar, O. C., Uzunoglu, M., & Alam, M. S. (2006). Dynamic modeling, design and simulation of a wind/fuel cell/ultra-capacitor-based hybrid power generation system. *Journal of Power Sources, 161*(1), 707–722. https://doi.org/10.1016/j.jpowsour.2006.03.055.

Panda, S., & Padhy, N. P. (2007). Comparison of particle swarm optimization and genetic algorithm of TCSC-based controller design. *International Journal of Electrical and Electronics Engineering, 1*(5), 305–313.

Panda, G., Panda, S., & Ardil, C. (2009). Hybrid neuro fuzzy approach for automatic generation control of two–area interconnected power system. *International Journal of Computational Intelligence, 5*(1), 80–84. http://www.waset.org/publications/6556.

Pandey, S., Mohanty, S. R., & Kishor, N. (2013). A literature survey on load–frequency control for conventional and distribution generation power systems. *Renewable and Sustainable Energy Reviews, 25*, 318–334. http://www.sciencedirect.com/science/article/pii/S1364032113002815.

148 Sustainable developments by artificial intelligence & machine learning

Papathanassiou, S. A., & Papadopoulos, M. P. (2001). Dynamic characteristics of autonomous wind-diesel systems. *Renewable Energy*, *23*(2), 293–311.

Parmar, K. S., Majhi, S., & Kothari, D. P. (2012). Load frequency control of a realistic power system with multi-source power generation. *International Journal of Electrical Power & Energy Systems*, *42*(1), 426–433. http://www.sciencedirect.com/science/article/pii/S0142061512001676.

Qi, C., & Ming, C. Z. (2012). Photovoltaic module simulink model for a stand-alone PV system. Science Direct International Conference on Applied Physics and Industrial Engineering *Physics Procedia*, *24*(A), 94–100.

Rahmani, M., & Sadati, N. (2013). Two-level optimal load–frequency control for multi-area power systems. *International Journal of Electrical Power & Energy Systems*, *53*, 540–547. http://www.sciencedirect.com/science/article/pii/S0142061513002123.

Sabahi, K., Ghaemi, S., & Pezeshki, S. (2014). Application of type-2 fuzzy logic system for load frequency control using feedback error learning approaches. *Applied Soft Computing*, *21*, 1–11. http://www.sciencedirect.com/science/article/pii/S1568494614000921.

Saxena, S. (2019). Load frequency control strategy via fractional-order controller and reduced-order modelling. *International Journal of Electrical Power and Energy Systems*, *104*, 603–614. https://doi.org/10.1016/j.ijepes.2018.07.005.

Simoes, M. G., & Ferret, F. A. (2007). *Alternate energy systems: Design and analysis with induction generator* (2nd ed.). CRC Press-Taylor and Francis Group.

Srikanth, M. V., & Yagaiah, N. (2018). An AHP based optimized tuning of modified active disturbance rejection control: An application to power system load frequency control problems. *ISA Transaction*. https://doi.org/10.1016/j.isatra.2018.07.001.

Sudha, K. R., & Santhi, R. V. (2011). Robust decentralized load frequency control of interconnected power system with generation rate constraint using type-2 fuzzy approach. *International Journal of Electrical Power & Energy Systems*, *33*(3), 699–707. http://www.sciencedirect.com/science/article/pii/S0142061511000135.

Wang, C., et al. (2016). Frequency control of an isolated micro-grid using double sliding mode controllers and disturbance observer. *IEEE Transaction on Smart Grid*. https://doi.org/10.1109/TSG.2016.2571439.

Wang, Y., et al. (2019). Optimization of power plant component size on board a fuel cell/battery hybrid bus for fuel economy and system durability. *International Journal of Hydrogen Energy*, *44*(335), 18283–18292.

Xiong, L., Li, H., & Wang, J. (2018). LMI based robust load frequency control for time delayed power system via delayed margin estimation. *International Journal of Electrical Power and Energy Systems*, *100*, 91–103. https://doi.org/10.1016/j.ijepes.2018.02.027.

Yousef, H. A., Khalfan, A. K., Albadi, M. H., & Hosseinzadeh, N. (2014). Load frequency control of a multi-area power system: An adaptive fuzzy logic approach. *IEEE Transactions on Power Systems*, *29*(4), 1822–1830. http://ieeexplore.ieee.org/document/6717058.

Zadeh, L. A. (1965). Fuzzy sets. *Information and Control*, *8*(3), 338–353.

Chapter 6

Optimal allocation of renewable energy sources in electrical distribution systems based on technical and economic indices

Mohamed Zellagui[a,b], Samir Settoul[c], and Heba Ahmed Hassan[d]

[a]*Department of Electrical Engineering, University of Quebec, Montreal, QC, Canada,* [b]*Department of Electrical Engineering, University of Batna 2, Batna, Algeria,* [c]*Department of Electrotechnic, Mentouri University of Constantine 1, Constantine, Algeria,* [d]*Electrical Power Engineering Department, Cairo University, Giza, Egypt*

6.1 Introduction

6.1.1 Motivation

The rapid development of distributed generators (DGs) in different forms and capacities is transforming the image of classical planning of electrical distribution systems (EDSs). Despite the benefits offered by renewable DG technologies, several economic and technical challenges can result from the inappropriate integration of DG into the existing EDS. As a result, optimizing DG preparation is critical to ensuring that the EDS's output meets the expected power efficiency, voltage stability, power loss reduction, durability, and profitability (Abdmouleh, Gastli, Ben-Brahim, Haouari, & Al-Emadi, 2017; Ehsan & Yang, 2018).

Decentralized power generation from renewable energy sources (RESs) is a long-term solution that addresses the present environmental threats because of its widespread availability, sustainability, nonpolluting generation, and eco-friendliness. The most widely used renewable DG types are the photovoltaic (PV) and wind turbine (WT) systems, where generation is intermittent.

Because of its universal affordability, reliability, nonpolluting generation, and eco-friendliness, decentralized power generation from RESs is a long-term approach that addresses current environmental challenges. PV and WT systems, which generate intermittently, are the most commonly used renewable DG forms. DGs are a secure and cost-effective way to deliver power to consumers,

Sustainable Developments by Artificial Intelligence and Machine Learning for Renewable Energies.
https://doi.org/10.1016/B978-0-323-91228-0.00014-8
Copyright © 2022 Elsevier Inc. All rights reserved.

149

150 Sustainable developments by artificial intelligence & machine learning

and they are usually wired either to the distribution network or on the customer's premises. As a result, it is critical to plan and distribute DGs optimally, in terms of scale, placement, and form, in order to achieve the expected economic, technological, environmental, and regulatory benefits of power systems (Zubo et al., 2017).

6.1.2 Literature review

There are numerous techniques and algorithms to address the optimal DG planning problem. In 2017, determination of optimal location of DG using analytical methods (AMs) was used for the minimization of active power loss (APL), cost of energy losses, and annual investment cost of DG units with a new voltage stability index (VSI) (Kazmi & Shin, 2017). These methods include the particle swarm optimization (PSO) algorithm for various techno-economic and environmental indices (Tanwar & Khatod, 2017), PSO for the maximization of profit and initial investments for distribution companies (Kansal, Tyagi, & Kumar, 2017), symbiotic organism search (SOS) algorithm to minimize APL (Nguyen-Phuoc, Vo-Ngoc, & Tran-The, 2017), flower pollination algorithm (FPA) approach in order to minimize APL and improve the bus voltage (Oda, Abdelsalam, Abdel-Wahab, & El-Saadawi, 2017), krill herd algorithm (KHA) in order to minimize line losses (Chithra Devi, Lakshminarasimman, & Balamurugan, 2017), ant lion optimization algorithm (ALOA) to reduce APL and improve voltage deviation (VD) and VSI (Ali, Elazim, & Abdelaziz, 2017), and the applied gray wolf optimizer (GWO) algorithm for the minimization of active and reactive energy indexes (Sultana, Khairuddin, Mokhtar, Qazi, & Sultana, 2017).

In 2018, a new and effective stochastic fractal search algorithm (SFSA) was applied based on three objectives involving APL, VD, and VSI (Nguyen & Ngoc Vo, 2018). Ant lion optimization algorithm (ALOA) was proposed for reduction of APL and VD (Hadidian-Moghaddam, Arabi-Nowdeh, Bigdeli, & Azizian, 2018), and ALOA for reducing APL and consequently maximizing the net saving (Ali, Elazim, & Abdelazi, 2018). A new AM was applied for the optimization of DG-based wind farms (WFs) considering the voltage stability of the system and minimizing the cost components of WFs (Nikkhah & Rabiee, 2018), grasshopper optimizer algorithm (GOA) to minimize the energy VSI (Sultana et al., 2018), population-based incremental learning (PBIL) algorithm to reduce APL and VD (Grisales-Noreña, Montoya, & Ramos-Paja, 2018), differential learning with biogeography-based optimization (DLBBO) algorithm to minimize the APL (Ravindran & Victoire, 2018), dragonfly algorithm (DA) approach to reduce the APL-based loss sensitivity factors (LSF) (Suresh & Belwin, 2018), a comprehensive teaching learning-based optimization (CTLBO) technique to minimize APL and maximize VSI (Quadri, Bhowmick, & Joshi, 2018), human opinion dynamics (HOD) technique to minimize APL and VD (Mahajan & Vadhera, 2018), and moth-flame optimization (MFO) algorithm to minimize APL (Tolba, Diab, Tulsky, & Abdelaziz, 2018).

In 2019, the chaotic differential evolution (CDE) technique was applied for minimizing various objectives including yearly economic loss, maintenance cost, and APL (Kumar, Mandal, & Chakraborty, 2019a). Opposition-based tuned-chaotic differential evolution (OTCDE) technique was applied considering cost index, voltage deviation index, and line flow capacity index (Kumar, Mandal, & Chakraborty, 2019b), and elephant herding optimization (EHO) algorithm for the reduction of APL and reactive power loss (QPL) for all the buses (Prasad, Subbaramaiah, & Sujatha, 2019). A discrete-continuous hyper-spherical search (DC-HSS) algorithm was applied to minimize APL in EDS with allocated soft open points and DG (Diaaeldin, Abdel Aleem, El-Rafei, Abdelaziz, & Zobaa, 2019), spider monkey optimization (SMO) algorithm to minimize voltage deviation (Deb, Chakraborty, & Deb, 2019), heuristic moment matching (HMM) technique applied for a multiobjective index for minimizing APL and VD (Ehsan, Cheng, & Yang, 2019), moth-flame optimization (MFO) algorithm to reduce the APL and augmented VSI in practical Algerian EDS (Settoul, Zellagui, Abdelaziz, & Chenni, 2019), and MFO algorithm to minimize the ALP (Settoul, Chenni, Hassan, Zellagui, & Kraimia, 2019). Biogeography-based optimization (BBO) algorithm was applied to reduce APL while maintaining voltage harmonic distortion at the limits (Duong, Pham, Nguyen, Doan, & Tran, 2019), unified particle swarm optimization (UPSO) to minimize the APL (Gkaidatzis et al., 2019), and hybrid PSO algorithm considering APL minimization as an objective (Yaprakdal, Baysal, & Anvari-Moghaddam, 2019).

Recently, in 2020, improved Harris Hawks optimization (HHO) algorithm was used with the aim of minimizing the APL, reducing VD, and increasing the VSI (Selim, Kamel, Alghamdi, & Jurado, 2020), improved estimation of distribution algorithm (IEDA) to minimize the APL (Yang, Yang, Wu, & Liu, 2018), modified moth-flame optimization (MMFO) algorithm to minimize the total operating cost, APL, VD, and pollution emission (Elattar & Elsayed, 2020), backtracking search optimization (BTSO) algorithm for minimizing the active and reactive power flows in the lines and maximized microgrid success indicator (Osama, Zobaa, & Abdelaziz, 2020), craziness-based PSO (CRPSO) algorithm based on the game-theoretic strategy to minimize the cost of buying power, and APL (Karimizadeh, Soleymani, & Faghihi, 2020), coyote optimization algorithm (COA) with considerations include the APL and voltage regulator tap changes at different load levels (Chang & Chinh, 2020), and applied spring search algorithm (SSA) to minimize the emissions, VD, APL, and energy cost (Dehghani, Montazeri, & Malik, 2020).

6.1.3 Contribution and chapter organization

This chapter presents a new approach to optimal integration of RES-based DG in electrical power systems using the modified WOA approach called cosine

152 Sustainable developments by artificial intelligence & machine learning

adapted WOA algorithm (CAWOA) for techno-economic profits. The salient contributions and findings of this study can be summarized as follows:

- Optimal integration of various DG-based RES: solar photovoltaic PV-DG and wind WT-DG using CAWOA techniques.
- The CAWOA approach is applied to the standard IEEE 33-, 69-, and 118-bus EDS.
- The optimal integration has been selected to minimize various technical and economic indexes: the total APL, TVV, and TOC for test systems.
- The validity of the proposed algorithm is demonstrated by comparing with existing optimization techniques.
- The chapter studies the impact of different types of DGs on active and reactive branch currents in the distribution lines.
- It studies the impact of voltages profile and power losses under different loadability that varied between 60% and 120% to reflect load uncertainty.

This chapter contains five sections along with references list, and is organized as follows: Section 6.2 presents the mathematical problem formulation; Section 6.3 introduces the CAWOA algorithm applied; Section 6.4 provides the simulation results, discussions, and comparisons of the result; and Section 6.5 presents the conclusions.

6.2 Problem formulation

6.2.1 Multiobjective function

In this chapter, the authors proposed a multiobjective function (MOF) for the EDS and represented by the following equations:

$$MOF = \min \sum_{i=1}^{Nbus} \sum_{j=2}^{Nbus} \alpha_1 \cdot \Delta APL_{ij} + \alpha_2 \cdot \Delta TVV_i + \alpha_3 \cdot \Delta TOC_{ij} \qquad (6.1)$$

$$\sum_{m=1}^{3} \alpha_m = |\alpha_1| + |\alpha_2| + |\alpha_3| = 1 \qquad (6.2)$$

In this chapter, α_1 is taken as 0.5, while α_2 and α_3 are taken as 0.25, based on the practical, technical, and economic indicators. The three indices are presented by:

$$\Delta APL = \frac{P_{Loss}^{After\ DG}}{P_{Loss}^{Before\ DG}} \qquad (6.3)$$

and:

$$P_{Loss} = \sum_{i=1}^{Nbus} \sum_{j=2}^{Nbus} R_{ij} \frac{\left(P_{ij}^2 + Q_{ij}^2\right)}{V_i^2} \qquad (6.4)$$

$$\Delta TVV = \frac{TVV_{After\,DG}}{TVV_{Before\,DG}} \tag{6.5}$$

where:

$$TVV = \sum_{i=2}^{N} |1 - V_i| \tag{6.6}$$

The TOC values for system study with the operation and maintenance costs included the cost of DG integrated into the distribution system (Diaaeldin et al., 2019). The ΔTOC index is calculated by:

$$\Delta TOC = \frac{TOC_{After\,DG}}{TOC_{Before\,DG}} \tag{6.7}$$

The TOC before and after the integration of the DG is calculated by the following expressions (Imran & Kowsalya, 2014):

$$TOC_{Before\,DG} = K_2 \times P_{Loss}^{Before\,DG} \tag{6.8}$$

$$TOC_{After\,DG} = \left(K_1 \times P_{Loss}^{After\,DG} \right) + \left(K_2 \times P_{Loss}^{Before\,DG} \right) \tag{6.9}$$

6.2.2 Equality constraints

The equality constraints represent the load flow balance equations, which can be formulated as (Chang & Chinh, 2020; Dehghani et al., 2020; Elattar & Elsayed, 2020; Hassan & Zellagui, 2019; Imran & Kowsalya, 2014; Karimizadeh et al., 2020; Osama et al., 2020; Yang et al., 2018):

$$P_G + P_{DG} = P_D + P_{Loss} \tag{6.10}$$

$$Q_G + Q_{DG} = Q_D + Q_{Loss} \tag{6.11}$$

6.2.3 Inequality constraints of distribution line

Inequality constraints represent the power system operating limits, which can be given as:

(a) Bus voltage limits:

$$V_{min} \leq |V_i| \leq V_{max} \tag{6.12}$$

(b) Voltage drop limit:

$$|1 - V_i| \leq \Delta V_{max} \tag{6.13}$$

154 Sustainable developments by artificial intelligence & machine learning

(c) Line capacity constraint:

$$|S_{ij}| \leq |S_{max}|$$
(6.14)

6.2.4 Inequality constraints of DG units

In the case of PV, the DG source only delivers active power, while in the case of WT, the DG source delivers active and reactive power (Chang & Chinh, 2020; Dehghani et al., 2020; Hassan & Zellagui, 2019; Imran & Kowsalya, 2014; Karimizadeh et al., 2020; Lasmari et al., 2020).

(a) DG capacity limits:

$$P_{DG}^{min} \leq P_{DG} \leq P_{DG}^{max}$$
(6.15)

$$Q_{DG}^{min} \leq Q_{DG} \leq Q_{DG}^{max}$$
(6.16)

(b) Power Factor of DG:

$$PF_{DG}^{min} \leq PF_{DG} \leq PF_{DG}^{max}$$
(6.17)

$$PF_{DG} = \frac{P_{DG}}{\sqrt{P_{DG}^2 + Q_{DG}^2}}$$
(6.18)

(c) Position of DG:

$$2 \leq DG_{position} \leq N_{bus}$$
(6.19)

(d) Number of DG:

$$N_{DG} \leq N_{DG.}\mathrm{max}$$
(6.20)

(e) Location of DG:

$$n_{DG,i}/Location \leq 1$$
(6.21)

The limits of constraint values are shown in Table 6.1.

6.3 Cosine adapted whale optimization algorithm (CAWOA)

In 2016, Mirjalili developed the basic whale optimization algorithm (WOA) approach (Mirjalili & Andrew, 2016) as an innovative biological-inspired heuristic technique. The natural body motions of humpback whales are the basis of the WOA. Three steps are used to describe the bubble-net hunting process and

TABLE 6.1 Limits values of constraints.

Parameters		Cases studies		
		33-bus	69-bus	118-bus
Bus number		33	69	118
P_{DG} (kW)	min	10	10	10
	max	3000	3000	3000
Q_{DG} (kVar)	min	10	10	10
	max	2000	2000	2000
V_i (p.u.)	min	0.95	0.95	0.90
	max	1.05	1.05	1.10
ΔV_{max} (%)		5	5	10
PF_{DG}	min	0.80	0.80	0.80
	max	1.00	1.00	1.00
Number of DGs		2	3	5

derive the WOA mathematical model. Modeling steps are detailed in the following subsections (Elaziz & Mirjalili, 2019; Mirjalili & Andrew, 2016).

Fig. 6.1 represent the exploration mechanism implemented in basic WOA. This enhancement strategy can modify spiral and circular movements, depending on the measure of p (Elaziz & Mirjalili, 2019).

In the basic WOA approach, the movement between exploration and exploitation phase occurs with a decrease of distance control parameter "a" linearly from 2 to 0. The cosine adapted WOA (CAWOA) approach employs the cosine function to tune the parameter "a" of the WOA for varying exploration and exploitation combinations over the course of iterations (Eid & Abraham, 2018). The CAWOA approach employs a cosine function instead of the linear function for the decay of the parameter "a" over the course of iterations; as given in Eq. (6.22). The use of a cosine feature for decaying "a" result in a differing exploration and exploitation combination (Saha & Panda, 2020). The pseudo code of the CAWOA technique is presented in Algorithm 6.1.

6.4 Results and discussion

6.4.1 Test systems

The proposed WOA algorithm is developed using MATLAB 2017.b and simulations are carried out on a PC that possesses an Intel i5, 2.7 GHz, and 8 GB RAM.

156 Sustainable developments by artificial intelligence & machine learning

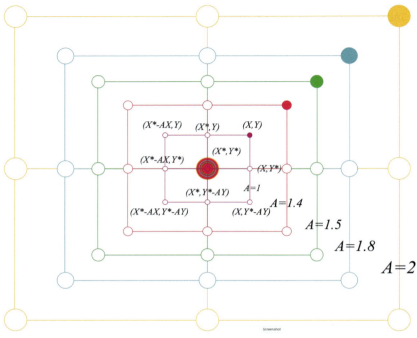

FIG. 6.1 Exploration mechanism implemented in basic WOA (Mirjalili & Andrew, 2016).

The first test system is a standard IEEE 33-bus system and composed of 33 buses, 32 distribution lines with an active power load of 3715 kW, and reactive power load of 2300 kVar with nominal voltage of 12.66 kV, as represented in Fig. 6.2B. The second test system is a standard IEEE 69-bus system, which is composed of 69 buses, 68 distribution lines with active power load of 3791.89 kW, and reactive power load of 2694.10 kVar with nominal voltage of 12.66 kV, as shown Fig. 6.2C. To illustrate the applicability of the proposed algorithm in large-scale distribution systems, it is applied on the standard IEEE 118-bus system, which is composed of 118 buses with active power load of 22.71 MW and reactive power load of 17.04 MVar, as shown Fig. 6.2C. The nominal voltage is 12.66 kV for all test systems.

The convergence optimization characteristics of the proposed CAWOA algorithm for the three considered IEEE distributions systems in the presence of PV-DG and WT-DG are presented in Fig. 6.3.

From the figure, when installing PV-DG, the IEEE 33-bus system takes almost 50 iterations to achieve the optimal solution, whereas the IEEE 69-bus system obtains the optimal solution by iteration 170, and the larger systems IEEE 118-bus, takes a long time to achieve the optimal solution—more than 170 iterations—but for the three systems, a quick convergence in the first iterations can be observed.

Algorithm 6.1 Pseudo-code of CAWOA.

1. Generate the initial population X_i (i = 1, 2, ..., n_p)
2. Evaluate the fitness for each candidate solutions in X_i
3. $X*$ is the best candidate solutions
4. while ($t < t_{max}$)
5.   for i = 1 to n_p (for each search agent)
6. Update the parameter a by equation:

$$a = 1 + 0.5 \cdot \cos\left(\pi \frac{t}{t_{max}}\right) \tag{6.22}$$

   Update A, C, l and p
7. if ($p < 0.5$)
8.   if ($|A| < 1$)
9.   Update the position of the search agent by equations:

$$\vec{D} = \left| \vec{C} \cdot \vec{X}^* (t) - \vec{X}(t) \right| \tag{6.23}$$

$$\vec{X}(t+1) = \vec{X}^*(t) - \vec{A} \cdot \vec{D} \tag{6.24}$$

10.   else if ($|A| \geq 1$)
11.   Select a random search agent (X_{rand})
12.   Update the position of the search agent by equations:

$$\vec{D} = \left| \vec{C} \cdot \vec{X}_{rand} - \vec{X} \right| \tag{6.25}$$

$$\vec{X}(t+1) = \vec{X}_{rand} - \vec{A} \cdot \vec{D} \tag{6.26}$$

13.   end if
14.   else if ($p \geq 0.5$)
15.   Update the position of the current agent by equation:

$$\vec{X}(t+1) = \vec{D'} \cdot e^{bt} \cdot \cos(2\pi l) + \vec{X}^*(t) \tag{6.27}$$

16.   end if
17.   end for
18.   Check if any search agent goes beyond the search space
19.   Calculate the MOF of each search agent
20.   Update $X*$ if there is a better solution
21. $t = t + 1$
22.   end while
23.   Return $X*$

In the case study of installing the WT-DG, IEEE 33-bus and 69-bus provide the optimal solution by 60 iterations and the 118-bus system takes almost 180 iterations to attain the best solution, this is related to the large size of this system.

6.4.2 Analysis of optimal results

The parameters of all distribution systems before DG integration are represented in Table 6.2, the parameters of all DGs installed on EDSs are represented in Table 6.3, and optimization results after integration of PV-DG and WT-DG are represented in Table 6.4.

According to the results illustrated in Table 6.4, for the first test system (IEEE 33-bus), the selected sizes by CAWOA algorithm for PV-DGs are 850.3 and 1158.6 kW, and for WT-DG are 828.0 kW with a power factor of

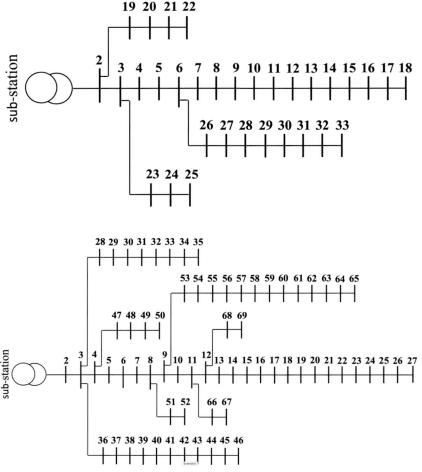

FIG. 6.2 Single line diagram of standard EDSs: (A) IEEE 33-bus, (B) IEEE 69-bus,
(Continued)

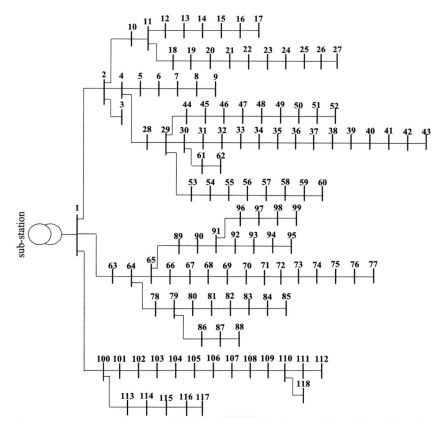

Fig. 6.2—Cont'd (C) IEEE 118-bus.

0.8944 (lagging) and 1243.3 kW with a power factor of 0.8 (lagging). The DGs are connected to their optimal places. They are placed in buses 13 and 30; thus, the ΔAPL is minimized to 41.3132% for PV-DG and 13.9037% for WT-DG.

For the *sec*ond test system, the IEEE 69-bus is considered. After optimizing the sizes of the three DGs, the optimal size is found to be 576.1, 339.3, and 1716.8 kW connected at buses 11, 21, and 61, respectively, which minimizes the ΔAPL to 30.8611%. Correspondingly, the WT-DGs with sizes of 432.7, 315.8, and 1662.2 kW are installed at buses 11, 21, and 61, respectively, and lead to the minimization of AAPL to 2.0168%.

For the third test system, the IEEE 118-bus, this larger system needed five PV-DGs to minimize the ΔAPL to 44.3794% placed in buses 39, 74, 80, 96, and 110 with a defined size for each PV-DG equal to 2971.4, 2257.4, 2382.4, 1782.6, and 2911.1 kW. On the other hand, five WT-DGs units are suggested

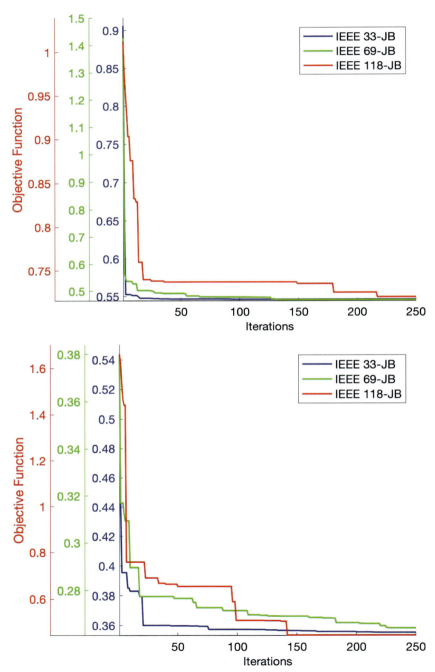

FIG. 6.3 Convergence curve of CAWOA for EDSs studies: (A) PV-DG, (B) WT-DG.

Optimal allocation of renewable energy sources **Chapter | 6** **161**

TABLE 6.2 Parameters of EDSs before DG.

Parameters	IEEE 33-bus	IEEE 69-bus	IEEE 118-bus
$\Sigma\,P_{Loss}$ (kW)	210.9875	224.9480	1297.5000
$\Sigma\,Q_{Loss}$ (kVar)	143.1284	102.1406	978.4956
V_{min} (p.u.)	0.9038	0.9092	0.8688
$\Sigma\,TVV$ (p.u)	1.8047	1.8704	5.2877
TOC (k$)	1054.9375	1124.740	6487.500

TABLE 6.3 Parameters of DGs installed.

DGs type	DGs Parameters	IEEE 33-bus	IEEE 69-bus	IEEE 118-bus
PV-DG				2971.4 (39)
			576.1 (11)	2257.4 (74)
	P_{DG}, kW (Bus)	850.3 (13)	339.3 (21)	2382.4 (80)
		1158.6 (30)	1716.8 (61)	1782.6 (96)
				2911.1 (110)
	$\Sigma\,P_{DG}$ (MW)	2.0089	2.6322	12.3049
WT-DG				2.9038 (39)
			432.7 (12)	2.1597 (47)
	P_{DG}, kW (Bus)	828.0 (13)	315.8 (21)	2.7667 (72)
		1243.3 (30)	1662.2 (61)	2.3067 (80)
				2.9171 (118)
	$\Sigma\,P_{DG}$ (MW)	2.0713	2.4107	13.0540
				0.8321 (39)
			0.8000 (12)	0.8869 (47)
		0.8944 (13)	0.8000 (21)	0.8321 (72)
	PF_{DG}	0.8000 (30)	0.8000 (61)	0.8028 (80)
				0.8321 (118)

TABLE 6.4 Optimization results after integration of DG.

Parameters	PV-DG	WT-DG	PV-DG	WT-DG	PV-DG	WT-DG
	IEEE 33-bus		*IEEE 69-bus*		*IEEE 118-bus*	
ΣP_{Loss} (kW)	87.1657	29.3351	69.4215	4.5367	575.8223	233.3590
ΔAPL (%)	41.3132	13.9037	30.8611	2.0168	44.3794	17.9853
V_{min} (p.u.)	0.9685	0.9804	0.9790	0.9943	0.9554	0.9662
ΣTVV (p.u.)	0.6775	0.1862	1.0277	0.1087	3.3633	1.4590
ΔTVV (%)	37.541	10.318	54.946	5.812	63.6061	27.5923
TOC (k$)	1403.60	1172.28	1402.43	1142.89	8790.79	7420.94
ΔTOC (%)	133.051	111.123	124.689	101.613	135.503	114.388
$CPU\ TIME$ (s)	6.0750	7.7052	11.6577	13.2048	21.6061	23.3794

to be placed in buses 39, 47, 72, 80, 118, the appropriate output of active power injected from WT-DGs is 2.9038, 2.1597, 2.7667, 2.3067, and 2.9171 kW, respectively, with the location mentioned before, which allow the minimization of the ΔAPL to 17.9853%.

From this analysis, it is interesting to note that there is a significant improvement after the installation of DGs in terms of the minimization of the power losses, the enhancement of voltage profile, and also the minimization of the operating cost of the tested systems, but the main observation is that the contribution of WT-DG is higher than PV-DGs. This is related not just to the optimal reactive power defined by the proposed algorithm, but also to the optimal lagging power factors attain by the algorithm, which leads to the optimal choice of the active and reactive powers being injected by the WT-DGs installed in their optimal places.

Fig. 6.4 shows the bus voltage profiles for different case studies performed for all distribution systems. The voltage profiles have improved across the tested power systems after the installation of DGs units. It can be seen that for different voltage profiles for each type of DG unit, the effect of the optimal

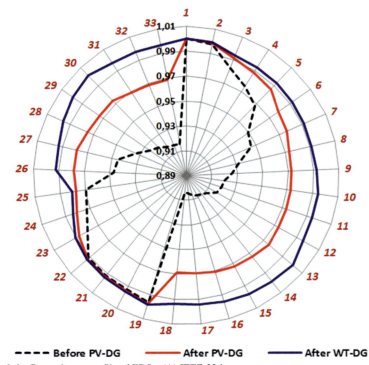

FIG. 6.4 Bus voltages profile of EDSs: (A) IEEE 33-bus,

(Continued)

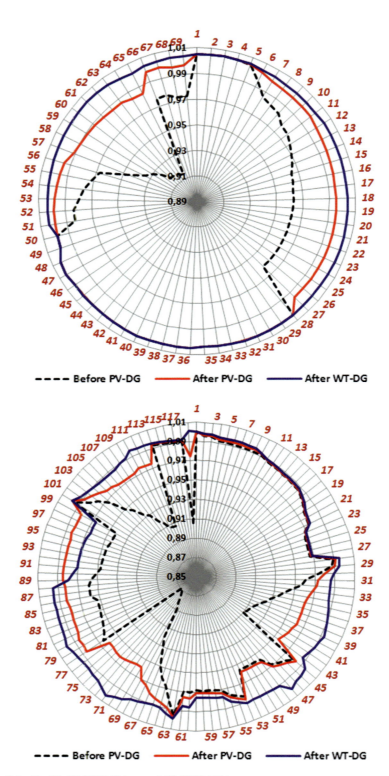

Fig. 6.4—Cont'd (B) IEEE 69-bus, and (C) IEEE 118-bus.

Optimal allocation of renewable energy sources **Chapter | 6** 165

factor obtained by the proposed algorithm allows the WT-DGs to maintain a better improvement to the voltage profiles and make them almost flat.

Fig. 6.5 illustrates the impact of DG installation on active power losses for the three test systems. Analysis of Fig. 6.5 reveals that after installation of the DG, there is significant minimization on the amount of the active power losses.

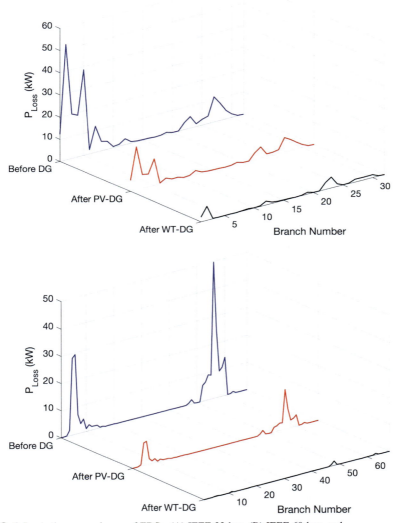

FIG. 6.5 Active power losses of EDSs: (A) IEEE 33-bus, (B) IEEE 69-bus, and

(Continued)

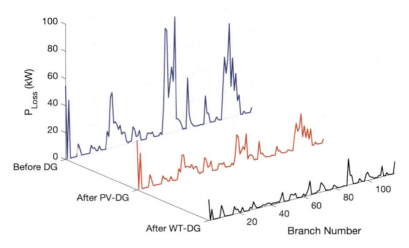

Fig. 6.5—Cont'd (C) IEEE 118-bus.

It is observed that the WT-DG installation contributes to more minimization of the power losses than the installation of PV-DG. These results are due to the contribution of the reactive power injected from WT-DG, and also the optimal power factors defined by the proposed algorithm. As mentioned above in Table 6.3, the power factors are set between 0.8 and 0.9 (lagging).

Fig. 6.6 demonstrates the influence of DG installation on the bus voltage deviations of the three test systems. In general, the installation of the two types of DGs, i.e., PV-DG and WT-DG, will reduce the voltage deviation in all the buses. However, from the voltage variation curve, it can be recognized that the active and reactive powers injected from the integration of WT-DGs into the tested systems are more effective in minimizing the voltage deviations. Moreover, it can be noted that WT-DGs provide more improvement than PV-DGs in terms of system voltage level.

6.4.3 Comparison results

Table 6.5 contains the comparison results of various optimization algorithms against the proposed CAWOA algorithm for all test EDS systems.

The comparison shown in Table 6.5 is carried out to demonstrate the performance of the proposed CAWOA algorithm. The table contains the details of comparison of the proposed algorithms with other benchmark algorithms from the literature in terms of active power losses, ΔAPL, ΔTVV, and ΔTOC for the three test systems with PV-DG and WT-DG integration.

Optimal allocation of renewable energy sources **Chapter | 6 167**

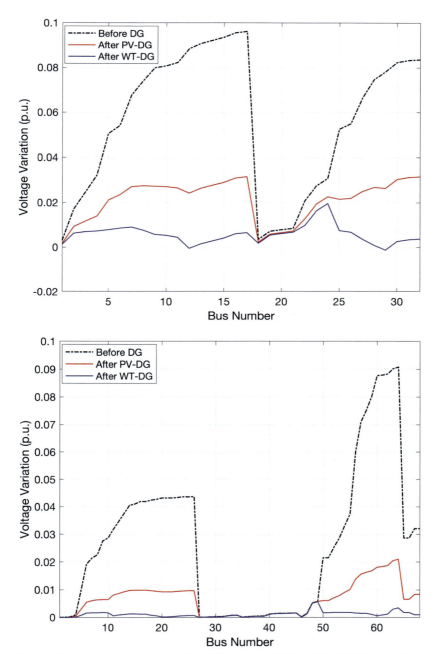

FIG. 6.6 Bus voltage variation of standard EDSs: (A) IEEE 33-bus, (B) IEEE 69-bus, and
(Continued)

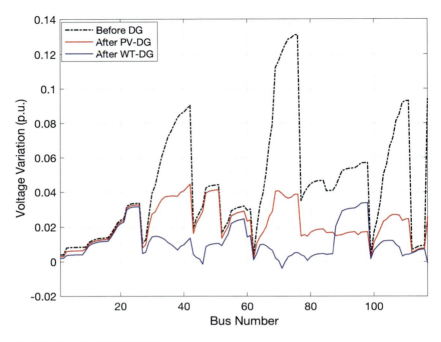

Fig. 6.6—Cont'd (C) IEEE 118-bus.

For the first test system, IEEE 33-bus, the CAWOA was compared with three algorithms namely BSOA, FPA, and KHA for PV-DGs and PSO, BSOA, and ALOA for WT-DGs, in the presence of PV-DGs. The compared algorithms reducing the amount of active power losses to 89.340 kW, 89.2000 kW, and 87.4260 kW, which in turn reduced the ΔAPL to 42.3437%, 42.2774%, and 41.4366% and the ΔTOC to 133.875%, 133.822%, and 133.149%, respectively. On the other hand, the CAWOA reduced the active power losses to 87.1657 kW. This leads to the minimization of the ΔAPL and ΔTOC to 41.3132% and 133.051%, respectively. The compared algorithms based on the integration of WT-DGs are also able to determine the optimal placement and sizing needed to reduce the power losses to 39.1000 kW, 31.9800 kW, and 30.9251 kW, respectively, as mentioned previously. In this instance, the ΔAPL has been minimized to 18.5319%, 15.1573%, and 14.6573%. The ATOC has been reduced to 114.826%, 112.126%, 111.726%, and 111.123%.

The second system, IEEE 69-bus, is compared with three existing algorithms called CDE, CSFS, and SFSA. In the case study of PV-DG installation, it can be observed that after installation of three PV-DG in their best locations obtained by these algorithms, the power loss has been reduced to 69.4360 kW, 69.4300 kW, and 69.4280 kW, and ΔTOC to 124.694%, 124.692%, and 124.691%, respectively; at the same time, the minimum power loss and ΔTOC

TABLE 6.5 System results given by various optimization algorithms.

Test system		Algorithms applied	ΣP_{Loss} (kW)	ΔAPL (%)	ΔTVV (%)	ΔTOC (%)
IEEE 33-bus	PV-DG	BSOA (El-Fergany, 2015)	89.3400	42.3437	41.0040	133.875
		FPA (Oda et al., 2017)	89.2000	42.2774	35.5738	133.822
		KHA (Chithra Devi et al., 2017)	87.4260	41.4366	37.4577	133.149
		CAWOA	**87.1657**	**41.3132**	**37.5409**	**133.051**
	WT-DG	PSO (Kaur, Kumbhar, & Sharma, 2014)	39.1000	18.5319	10.6721	114.826
		BSOA (El-Fergany, 2015)	31.9800	15.1573	32.6148	112.126
		ALOA (Ali et al., 2017)	30.9251	14.6573	9.2037	111.726
		CAWOA	**29.3351**	**13.9037**	**10.3175**	**111.123**
IEEE 69-bus	PV-DG	CDE (Kumar et al., 2019a)	69.4360	30.8676	54.9401	124.694
		CSFS (Nguyen & Vo, 2019)	69.4300	30.8649	54.9401	124.692
		SFSA (Nguyen & Ngoc Vo, 2018)	69.4280	30.8640	54.9348	124.691
		CAWOA	**69.4215**	**30.8611**	**54.9455**	**124.689**
	WT-DG	LSFSA (Injeti & Kumar, 2013)	16.2600	7.2283	15.0075	105.783
		BFOA (Imran & Kowsalya, 2014)	12.9000	5.7347	25.9356	104.588
		GWO (Nowdeh et al., 2019)	7.3100	3.2496	7.4476	102.600
		CAWOA	**4.5367**	**2.0168**	**5.8116**	**101.613**

Continued

TABLE 6.5 System results given by various optimization algorithms—cont'd

Test system		Algorithms applied	ΣP_{Loss} (kW)	ΔAPL (%)	ΔTVV (%)	ΔTOC (%)
IEEE 118-bus	PV-DG	CSFS (Nguyen & Vo, 2019)	581.580	44.8231	69.0849	135.858
		SFSA (Nguyen & Ngoc Vo, 2018)	578.741	44.6043	70.1780	135.683
		KHA (Nowdeh et al., 2019)	576.490	44.4308	69.6220	135.545
		CAWOA	**575.822**	**44.3794**	**63.6061**	**135.503**
	WT-DG	LSFSA (Injeti & Kumar, 2013)	684.028	52.7189	49.2880	142.175
		SFSA (Nguyen & Ngoc Vo, 2018)	236.528	18.2295	41.4244	114.584
		KHA (Sultana & Roy, 2016)	233.383	17.9871	40.9649	114.390
		CAWOA	**233.359**	**17.9853**	**27.5923**	**114.388**

Optimal allocation of renewable energy sources **Chapter | 6** **171**

obtained by the proposed algorithm are 69.4215 kW and 124.689%, respectively. On the other side, the results of the WT-DGs case study has been compared with LSFSA, BFOA, and GWO; as reported, the power losses it provides are 16.2600 kW, 12.9000 kW, and 7.3100 kW, and the ΔAPL and ΔTOC are 7.2283%, 5.7347%, and 3.2496%, and 105.783%, 104.588%, and 102.600%, respectively. The power loss amount attained by CAWOA is 4.5367, so ΔAPL and ΔTOC are 2.0168% and 101.613%, respectively.

The third test system (IEEE 118-bus) is equipped with PV-DGs. The obtained results have been compared with CSFS, SFSA, and KHA, which made power losses of 581.580, 578.741, and 576.490 kW, and ΔTOC of 142.175%, 114.584%, and 114.390%, respectively. In the same manner as PV-DGs, the WT-DG have been compared with LSFSA, SFSA, and KHA; based on these algorithms, the power losses are reduced to 684.028 kW, 236.528 kW, and 233.383 kW, and the ΔTOC to 142.175%, 114.584%, and 114.390%, respectively, whereas the power losses and ATOC attained by CAWOA are 233.359 kW, and 114.388%.

Considering the holistic view of the above results and the comparison of the literature benchmark algorithms against the proposed CAWOA algorithm, a great power loss reduction has been achieved by the literature algorithms. However, the CAWOA algorithm provides better results in terms of power losses minimization and the total operating cost for the three tested systems with PV-DGs and WT-DGs; this is an indication for the superior performance of the CAWOA algorithm in determination of the best location and sizing of PV-DGs and WT-DGs.

Fig. 6.7 shows the bar chart for ΔTOC and ΔAPL of the three standard IEEE systems in the presence of PV-DG and WT-DG. This is to show the effectiveness of the proposed algorithm versus the standard literature algorithms.

The results shown in Fig. 6.7 suggest that in all three IEEE test scenarios, while applying CAWOA to optimize the siting and sizing of DGs for different systems with the aim to reduce the active power losses index and the total operating cost, the proposed algorithm proves its ability to handle this kind of the proposed objective function. Furthermore, the comparison of CAWOA with the other algorithms shows that the CAWOA algorithm performs well and is able to obtain the best and optimum results in terms of either the minimization of power loss or operating cost enhancement based on different types of DG units.

6.4.4 Impact of DG on branch currents

Fig. 6.8 shows the influence of DG installation on active and reactive branch currents on all three test systems. As shown in the figure, the active current is plotted with dashed lines and the reactive current is plotted with continuous lines; the negative sign of reactive currents indicates that the nature of loads is inductive. One of the objectives of installing DGs is reducing the branch

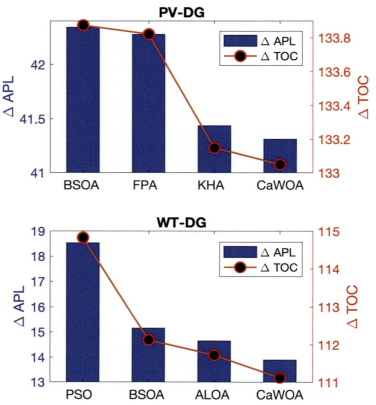

FIG. 6.7 Comparison of CAWOA with other algorithms in the presence different type of DG: (A) IEEE 33-bus,

(Continued)

current, as knowing the losses of the branches in the system is dependent in the first place on the current flowing through these branches. As can be noted, after the installation of DGs, the branch current has clearly reduced for the three tested systems; thus, the power losses are consequently reduced.

Because of the reactive power injected from the WT-DGS, the branch current is affected by WT-DGs more than the PV-DGs in the three tested systems.

6.4.5 Impact of loadability variation on EDS

To validate the effectiveness and performance of the proposed algorithm, it has been implemented on standard IEEE distribution systems. For all the cases of test systems, the power system feeder loads are linearly varied from 60% (light load) to 120% (peak load) in steps of 10% load; at the same time, the optimal size of PV-DG and WT-DG are calculated by CAWOA algorithm for each load

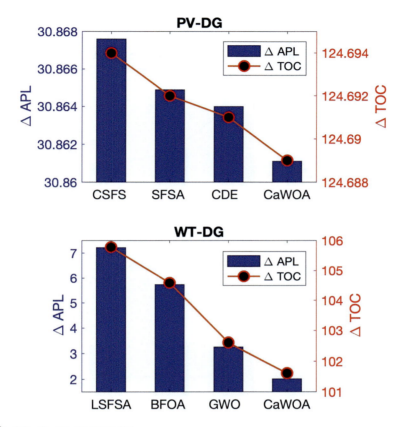

Fig. 6.7—Cont'd (B) IEEE 69-bus,

(Continued)

step variation. The voltage profile under different load variations of IEEE 33-, 69-, and 118-bus are represented in Fig. 6.9.

The accepted upper and lower voltage limits for each test system are plotted with dashed red lines in Fig. 6.9. Before DG-installation, some of the bus voltages broke the permissible limits with specific loads. For example, in the IEEE 33-bus system, the buses from bus number 6 to bus 18 and from bus 26 to bus 33 had the minimum values of the voltages—these values were less than 0.95 p.u.—but after installing the PV-DGs and taking into account the changes in consumption, it is observed that the voltage profiles have significantly improved, and the minimum value has become 0.95 p.u. in bus number 18. Similarly, in WT-DGs, the voltage profiles have improved but better than PV-DGs. The same observation can be made for IEEE 69-bus and IEEE 118-bus systems—the best voltage profiles for different consumption loads are obtained based on PV-DGs and WT-DGs. Also, it can be noted that the voltages profiles for the system in light loads consumption is always better than medium and peak loads.

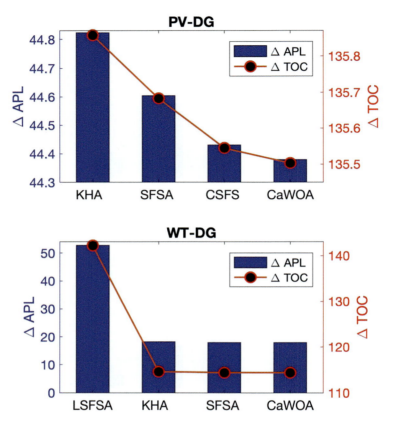

FIG. 6.7—Cont'd (C) IEEE 118-bus.

Fig. 6.10 is presented to investigate the impact of the loadability on the total active and reactive powers before and after the integration of DG units for the three test systems, in the same way as the variation of loads done previously. The loads vary from 60% to 120%, with steps of 10%, and the active and reactive power losses are calculated simultaneously.

As shown in the figure, the minimum of active and reactive distribution line losses are achieved in the case of light loads consumption (60% of loads). On the other side, the highest power losses are obtained in the case of high load consumption. Whether before or after DG unit installation, it is seen that the power losses are proportional with the increase in loads.

Another observation is that the best profile is obtained after the integration of DG units—the injected power from the DGs units always made the losses up

Optimal allocation of renewable energy sources **Chapter | 6** 175

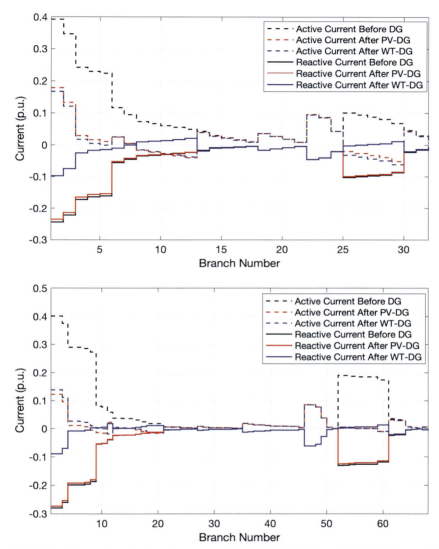

FIG. 6.8 Active and reactive branch currents of EDSs: (A) IEEE 33-bus, (B) IEEE 69-bus, and
(Continued)

to their optimal minimum value in the different loadabilities. Moreover, the active and reactive power losses measured in the presence of WT-DGs and by the linear variation in loads are better than those obtained in the presence of PV-DGs; this observation is valid for the three tested systems.

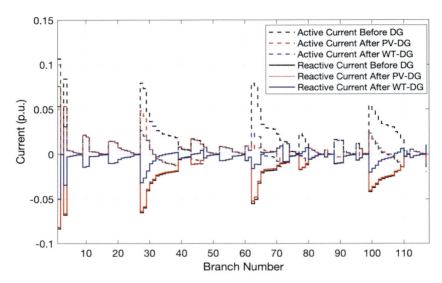

Fig. 6.8—Cont'd (C) IEEE 118-bus.

6.5 Conclusions

In this chapter, the CAWOA algorithm has been proposed to solve the techno-economic analysis problem of optimal location and sizing of renewable DG in the EDSs. Thus, a multiobjectives function has been used for the optimal integration of PV-DG and WT-DG sources with the optimum power factors. The optimal power factors obtained by the proposed algorithms show a positive and clear effect on power losses and bus voltages, unlike many techniques proposed in the literature, which suggest a predefined power factor. In general, the best results are obtained by installing the WT-DGs.

In this study, the proposed algorithm is applied to minimize the active power losses index, voltage variation index, and the total operation cost index subject to the quality and inequality constraints within a predefined tolerance to avoid violation of these constraints. The active and reactive branch currents are calculated, and the results show that the currents are minimized after the installation of DGs and the profile of the current has greater reduction thru the presence of WT-DGs. Careful and realistic results have been attained by a linear change in loads; the loadability results show that the total power losses and the voltage profiles are proportional with the increasing or decreasing of loads.

To show the performance of the proposed algorithms, the obtained results have been compared with other excising results of algorithms from the literature. The comparative study shows that CAWOA is better than the compared algorithms in terms of minimizing the power losses of the studied systems;

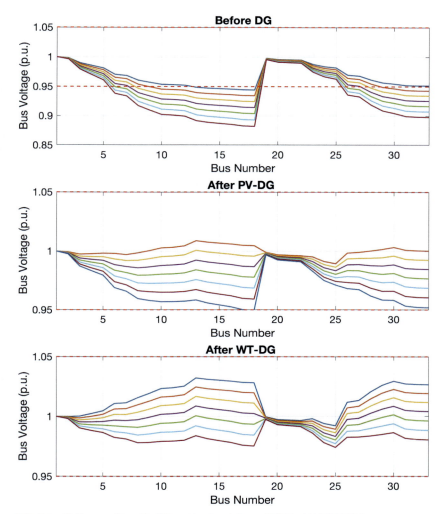

FIG. 6.9 Voltage profile with different load variations of EDSs: (A) IEEE 33-bus,

(Continued)

furthermore, it performs well and obtains better results considering the same number of DGs installed in the systems. In general, the results show that WAO can handle this type of proposed objective function.

This research work gives guidelines and hints for power distribution companies to reducing power losses. This work will also help the power distribution companies in incorporating small-sized renewable energy sources to distribution networks easily and more reliably.

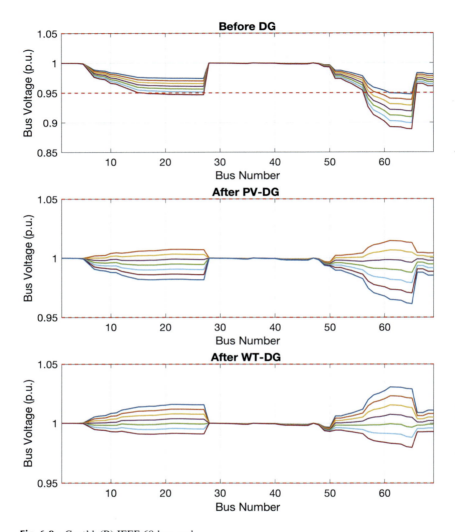

Fig. 6.9—Cont'd (B) IEEE 69-bus, and

(Continued)

Abbreviations and symbols

Abbreviations

APL	active power loss
EDS	electrical distribution system
OPF	optimal power flow
PV-DG	PV source-based DG
VD	voltage deviation
VSI	voltage stability index
WT-DG	WT source-based DG

Optimal allocation of renewable energy sources **Chapter | 6** 179

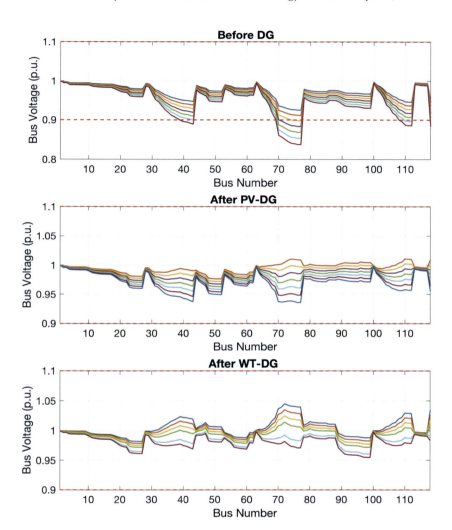

Fig. 6.9—Cont'd (C) IEEE 118-bus.

Symbol objective function

K_1	cost coefficient of real power supplied by the substation and DG (4 $/kW)
K_2	the maintenance and installation costs coefficient of DGs (5 $/kW)
N_{bus}	number of buses
OF	objective function
P_{ij}, Q_{ij}	active and reactive powers of the distribution line

180 Sustainable developments by artificial intelligence & machine learning

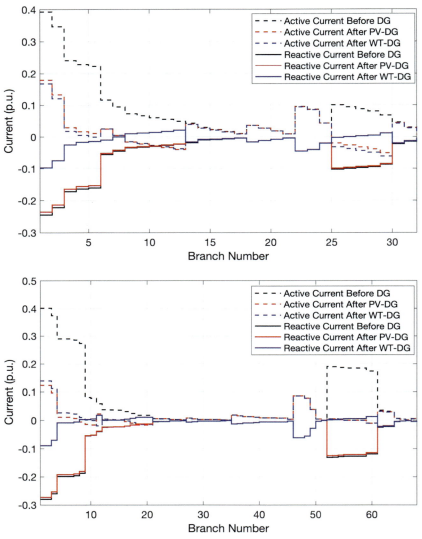

FIG. 6.10 Total active and reactive power losses under different loadability value of EDSs: (A) IEEE 33-bus, (B) IEEE 69-bus, and

(Continued)

P_{Loss}	total active loss on the distribution
R_{ij}	resistance of the distribution line
TOC	total operating cost
TVV	total voltage variation
V_i	the voltage at bus i

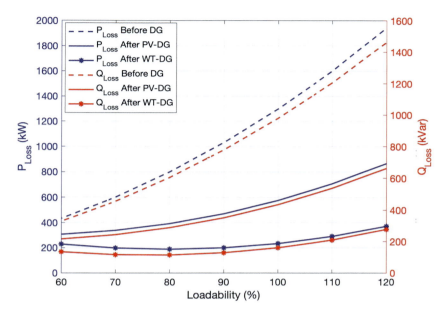

Fig. 6.10—Cont'd (C) IEEE 118-bus.

α_1, α_2, and α_3	weighting factors
ΔAPL	active power loss index
ΔTOC	TOC index
ΔTVV	TVV index

Symbols of constraints

$DG_{Position}$	position of DG units
N_{DG}	number of DG units
$n_{DG,i}$	location of PV-DG units at bus i
P_D, Q_D	total active and reactive of demand load
P_{DG}, Q_{DG}	total active and reactive injection from DG
$P_{DG}^{min}, P_{DG}^{max}$	minimum and maximum active power output limits of the DG
PF_{DG}	power factor of DG
$PF_{DG}^{min}, PF_{DG}^{max}$	minimum and maximum value PF_{DG}
P_G, Q_G	total active and reactive power of generator
P_{Loss}, Q_{Loss}	total active and reactive powers losses
$Q_{DG}^{min}, Q_{DG}^{max}$	minimum and maximum reactive power output limits of the DG
S_{ij}	apparent power in bus i to bus j branch
S_{max}	maximum apparent power

182 Sustainable developments by artificial intelligence & machine learning

V_1 voltage at the generating station is equal to 1.0 p.u.

V_{min}, V_{max} limited specified voltages

ΔV_{max} maximum voltage drops

Parameters of CA WOA algorithm

A and C	the coefficient vectors
a	control parameter
b	constant to represent the state of the logarithmic helix
D	distance between ith whale and the prey
D'	determines the location or distance of whales i to the prey
l	arbitrary number in the range -1, and 1
p	arbitrary number having a range between 0 and 1.
r	random vector in the interval of 0 and 1
t	indicates the current iteration
t_{max}	maximum number of iterations
X	position vector
$X*$	position vector of the best solution obtained so far
X_{rand}	arbitrary position vector (a random whale)

References

Abdmouleh, Z., Gastli, A., Ben-Brahim, L., Haouari, M., & Al-Emadi, N. A. (2017). Review of optimization techniques applied for the integration of distributed generation from renewable energy sources. *Renewable Energy, 113*, 266–280. https://doi.org/10.1016/j.renene.2017. 05.087.

Ali, E. S., Elazim, S. A., & Abdelazi, A. Y. (2018). Optimal allocation and sizing of renewable distributed generation using ant lion optimization algorithm. *Electrical Engineering, 100*(1), 99–109. https://doi.org/10.1007/s00202-016-0477-z.

Ali, E. S., Elazim, S. A., & Abdelaziz, A. Y. (2017). Ant lion optimization algorithm for optimal location and sizing of renewable distributed generations. *Renewable Energy, 101*, 1311–1324. https://doi.org/10.1016/j.renene.2016.09.023.

Chang, G. W., & Chinh, N. C. (2020). Coyote optimization algorithm-based approach for strategic planning of photovoltaic distributed generation. *IEEE Access, 8*, 36180–36190. https://doi.org/ 10.1109/ACCESS.2020.2975107.

Chithra Devi, S. A., Lakshminarasimman, L., & Balamurugan, R. (2017). Study krill herd algorithm for multiple DG placement and sizing in a radial distribution system. *Engineering Science and Technology, an International Journal, 20*(2), 748–759. https://doi.org/10.1016/j.jestch. 2016.11.009.

Deb, G., Chakraborty, K., & Deb, S. (2019). Spider monkey optimization technique-based allocation of distributed generation for demand side management. *International Transactions on Electrical Energy Systems, 29*(5), e12009. https://doi.org/10.1002/2050-7038.12009.

Dehghani, M., Montazeri, Z., & Malik, O. P. (2020). Optimal sizing and placement of capacitor banks and distributed generation in distribution systems using spring search algorithm. *International Journal of Emerging Electric Power Systems, 21*(1), 1–9. https://doi.org/10.1515/ ijeeps-2019-0217.

Diaaeldin, I., Abdel Aleem, S., El-Rafei, A., Abdelaziz, A. Y., & Zobaa, A. F. (2019). Optimal network reconfiguration in active distribution networks with soft open points and distributed generation. *Energies, 12*(21), 4172. https://doi.org/10.3390/en12214172.

Duong, M. Q., Pham, T. D., Nguyen, T. T., Doan, A. T., & Tran, H. V. (2019). Determination of optimal location and sizing of solar photovoltaic distribution generation units in radial distribution systems. *Energies, 12*(1), 174. https://doi.org/10.3390/en12010174.

Ehsan, A., Cheng, M., & Yang, Q. (2019). Scenario-based planning of active distribution systems under uncertainties of renewable generation and electricity demand. *CSEE Journal of Power and Energy Systems, 5*(1), 56–62. https://doi.org/10.17775/CSEEJPES.2018.00460.

Ehsan, A., & Yang, Q. (2018). Optimal integration and planning of renewable distributed generation in the power distribution networks: A review of analytical techniques. *Applied Energy, 210*, 44–59. https://doi.org/10.1016/j.apenergy.2017.10.106.

Eid, H. F., & Abraham, A. (2018). Adaptive feature selection and classification using modified whale optimization algorithm. *International Journal of Computer Information Systems and Industrial Management Applications, 10*, 174–182.

Elattar, E. E., & Elsayed, S. K. (2020). Optimal location and sizing of distributed generators based on renewable energy sources using modified moth flame optimization technique. *IEEE Access, 8*, 109625–109638. https://doi.org/10.1109/ACCESS.2020.3001758.

Elaziz, M. A., & Mirjalili, S. (2019). A hyper-heuristic for improving the initial population of whale optimization algorithm. *Knowledge-Based Systems, 172*, 42–63. https://doi.org/10.1016/j.knosys.2019.02.010.

El-Fergany, A. (2015). Optimal allocation of multi-type distributed generators using backtracking search optimization algorithm. *International Journal of Electrical Power and Energy Systems, 64*, 1197–1205. https://doi.org/10.1016/j.ijepes.2014.09.020.

Gkaidatzis, P. A., Bouhouras, A. S., Sgouras, K. I., Doukas, D. I., Christoforidis, G. C., & Labridis, D. P. (2019). Efficient RES penetration under optimal distributed generation placement approach. *Energies, 12*(7), 1250. https://doi.org/10.3390/en12071250.

Grisales-Noreña, L. F., Montoya, D. G., & Ramos-Paja, C. A. (2018). Optimal sizing and location of-distributed generators based on PBIL and PSO techniques. *Energies, 11*(4), 1018. https://doi.org/10.3390/en11041018.

Hadidian-Moghaddam, M. J., Arabi-Nowdeh, S., Bigdeli, M., & Azizian, D. (2018). A multi-objective optimal sizing and siting of distributed generation using ant lion optimization technique. *Ain Shams Engineering Journal, 9*(4), 2101–2109. https://doi.org/10.1016/j.asej.2017.03.001.

Hassan, H. A., & Zellagui, M. (2019). MVO algorithm for optimal simultaneous integration of DG and DSTATCOM in standard radial distribution systems based on technical-economic indices. In *21st international Middle East power systems conference (MEPCON), 17–19 December*. Egypt: Tanta University. https://doi.org/10.1109/MEPCON47431.2019.9007995.

Imran, A. M., & Kowsalya, M. (2014). Optimal size and siting of multiple distribution generators in distribution system using bacterial foraging optimization. *Swarm and Evolutionary Computation, 15*, 58–65. https://doi.org/10.1016/j.swevo.2013.12.001.

Injeti, S. K., & Kumar, N. P. (2013). A novel approach to identify optimal access point and capacity of multiple DGs in a small, medium and large-scale radial distribution systems. *International Journal of Electrical Power & Energy Systems, 45*(1), 142–151. https://doi.org/10.1016/j.ijepes.2012.08.043.

Kansal, S., Tyagi, B., & Kumar, V. (2017). Cost-benefit analysis for optimal distributed generation placement in distribution systems. *International Journal of Ambient Energy, 38*(1), 45–54. https://doi.org/10.1080/01430750.2015.1031407.

Karimizadeh, K., Soleymani, S., & Faghihi, F. (2020). Optimal placement of DG units for the enhancement of MG networks performance using coalition game theory. *IET Generation, Transmission & Distribution*, *14*(5), 853–862. https://doi.org/10.1049/iet-gtd.2019.0070.

Kaur, S., Kumbhar, G., & Sharma, J. (2014). A MINLP technique for optimal placement of multiple DG units in distribution systems. *International Journal of Electrical Power & Energy Systems*, *63*, 609–617. https://doi.org/10.1016/j.ijepes.2014.06.023.

Kazmi, S. A. A., & Shin, D. R. (2017). DG placement in loop distribution network with new voltage stability index and loss minimization condition-based planning approach under load growth. *Energies*, *10*(8), 1203. https://doi.org/10.3390/en10081203.

Kumar, S., Mandal, K. K., & Chakraborty, N. (2019a). Optimal DG placement by multi-objective opposition based chaotic differential evolution for techno-economic analysis. *Applied Soft Computing*, *78*, 70–83. https://doi.org/10.1016/j.asoc.2019.02.013.

Kumar, S., Mandal, K. K., & Chakraborty, N. (2019b). A novel opposition-based tuned-chaotic differential evolution technique for techno-economic analysis by optimal placement of distributed generation. *Engineering Optimization*, *51*, 1–20. https://doi.org/10.1080/0305215X.2019. 1585832.

Lasmari, A., Zellagui, M., Chenni, R., Semaoui, S., El-Bayeh, C. Z., & Hassan, H. A. (2020). Optimal energy management system for distribution systems using simultaneous integration of PV-based DG and DSTATCOM units. *Energetika*, *66*(1), 1–14. https://doi.org/10.6001/energetika. v66i1.4294.

Mahajan, S., & Vadhera, S. (2018). Optimal location and sizing of distributed generation unit using human opinion dynamics optimization technique. *Distributed Generation and Alternative Energy Journal*, *33*(2), 38–57. https://doi.org/10.1080/21563306.2018.12002410.

Mirjalili, S., & Andrew, L. (2016). The whale optimization algorithm. *Advances in Engineering Software*, *95*, 51–67. https://doi.org/10.1016/j.advengsoft.2016.01.008.

Nguyen, T. P., & Ngoc Vo, D. (2018). A novel stochastic fractal search algorithm for optimal allocation of distributed generators in radial distribution systems. *Applied Soft Computing Journal*, *70*, 773–796. https://doi.org/10.1016/j.asoc.2018.06.020.

Nguyen, T. P., & Vo, D. N. (2019). Improved stochastic fractal search algorithm with chaos for optimal determination of location, size, and quantity of distributed generators in distribution systems. *Neural Computing and Applications*, *31*(11), 7707–7732. https://doi.org/10.1007/ s00521-018-3603-1.

Nguyen-Phuoc, T., Vo-Ngoc, D., & Tran-The, T. (2017). Optimal number, location, and size of distributed generators in distribution systems by symbiotic organism search-based method. *Advances in Electrical and Electronic Engineering*, *15*(5), 724–735. https://doi.org/10. 15598/aeee.v15i5.2355.

Nikkhah, S., & Rabiee, A. (2018). Optimal wind power generation investment, considering voltage stability of power systems. *Renewable Energy*, *115*, 308–325. https://doi.org/10.1016/j. renene.2017.08.056.

Nowdeh, S. A., Davoudkhani, I. F., Moghaddam, M. H., Najmi, E. S., Abdelaziz, A. Y., Ahmadi, A., et al. (2019). Fuzzy multi-objective placement of renewable energy sources in distribution system with objective of loss reduction and reliability improvement using a novel hybrid method. *Applied Soft Computing*, *77*, 761–779. https://doi.org/10.1016/j.asoc.2019.02.003.

Oda, E. S., Abdelsalam, A. A., Abdel-Wahab, M. N., & El-Saadawi, M. M. (2017). Distributed generations planning using flower pollination algorithm for enhancing distribution system voltage stability. *Ain Shams Engineering Journal*, *8*(4), 593–603. https://doi.org/10.1016/j.asej.2015. 12.001.

Osama, R. A., Zobaa, A. F., & Abdelaziz, A. Y. (2020). A planning framework for optimal partitioning of distribution networks into microgrids. *IEEE Systems Journal, 14*(1), 916–926. https://doi.org/10.1109/JSYST.2019.2904319.

Prasad, C. H., Subbaramaiah, K., & Sujatha, P. (2019). Cost–benefit analysis for optimal DG placement in distribution systems by using elephant herding optimization algorithm. *Renewables, 6* (2), 1–12. https://doi.org/10.1186/s40807-019-0056-9.

Quadri, I. A., Bhowmick, S., & Joshi, D. A. (2018). Comprehensive technique for optimal allocation of distributed energy resources in radial distribution systems. *Applied Energy, 211*, 1245–1260. https://doi.org/10.1016/j.apenergy.2017.11.108.

Ravindran, S., & Victoire, T. A. A. (2018). A bio-geography-based algorithm for optimal siting and sizing of distributed generators with an effective power factor model. *Computers & Electrical Engineering, 72*, 482–501. https://doi.org/10.1016/j.compeleceng.2018.10.010.

Saha, N., & Panda, S. (2020). Cosine adapted modified whale optimization algorithm for control of switched reluctance motor. *Computational Intelligence*, 1–41. https://doi.org/10.1111/coin.12310.

Selim, A., Kamel, S., Alghamdi, A. S., & Jurado, F. (2020). Optimal placement of DGs in distribution system using an improved Harris Hawks optimizer based on single- and multi-objective approaches. *IEEE Access, 8*, 52815–52829. https://doi.org/10.1109/ACCESS.2020.2980245.

Settoul, S., Chenni, R., Hassan, H. A., Zellagui, M., & Kraimia, M. N. (2019). MFO algorithm for optimal location and sizing of multiple photovoltaic distributed generations units for loss reduction in distribution systems. In *7th international renewable and sustainable energy conference (IRSEC), Agadir, Morocco, 27–30 November*. https://doi.org/10.1109/IRSEC48032.2019.9078241.

Settoul, S., Zellagui, M., Abdelaziz, A. Y., & Chenni, R. (2019). Optimal integration of renewable distributed generation in practical distribution grids based on moth-flame optimization algorithm. In *International conference on advanced electrical engineering (ICAEE), Algiers, Algeria, 19–21 November*. https://doi.org/10.1109/ICAEE47123.2019.9014662.

Sultana, U., Khairuddin, A., Mokhtar, A. S., Qazi, S. H., & Sultana, B. (2017). An optimization approach for minimizing energy losses of distribution systems based on distributed generation placement. *Jurnal Teknologi, 79*(4), 87–96. https://doi.org/10.11113/jt.v79.5574.

Sultana, U., Khairuddin, A. B., Sultana, B., Rasheed, N., Qazi, S. H., & Malik, N. R. (2018). Placement and sizing of multiple distributed generation and battery swapping stations using grasshopper optimizer algorithm. *Energy, 165*, 408–421. https://doi.org/10.1016/j.energy.2018.09.083.

Sultana, S., & Roy, P. K. (2016). Krill herd algorithm for optimal location of distributed generator in radial distribution system. *Applied Soft Computing, 40*, 391–404. https://doi.org/10.1016/j.asoc.2015.11.036.

Suresh, M. C. V., & Belwin, E. J. (2018). Optimal DG placement for benefit maximization in distribution networks by using Dragonfly algorithm. *Renewables, 5*(4), 1–8. https://doi.org/10.1186/s40807-018-0050-7.

Tanwar, S. S., & Khatod, D. K. (2017). Techno-economic and environmental approach for optimal placement and sizing of renewable DGs in distribution system. *Energy, 127*, 52–67. https://doi.org/10.1016/j.energy.2017.02.172.

Tolba, M. A., Diab, A. A. Z., Tulsky, V. N., & Abdelaziz, A. Y. (2018). LVCI approach for optimal allocation of distributed generations and capacitor banks in distribution grids based on moth–flame optimization algorithm. *Electrical Engineering, 100*(3), 2059–2084. https://doi.org/10.1007/s00202-018-0684-x.

Yang, L., Yang, X., Wu, Y., & Liu, X. (2018). Applied research on distributed generation optimal allocation based on improved estimation of distribution algorithm. *Energies*, *11*(9), 2363. https://doi.org/10.3390/en11092363.

Yaprakdal, F., Baysal, M., & Anvari-Moghaddam, A. (2019). Optimal operational scheduling of reconfigurable microgrids in presence of renewable energy sources. *Energies*, *12*(10), 1858. https://doi.org/10.3390/en12101858.

Zubo, R. H. A., Mokryani, G., Rajamani, H. H., Aghaei, J., Niknam, T., & Pillai, P. (2017). Operation and planning of distribution networks with integration of renewable distributed generators considering uncertainties: A review. *Renewable and Sustainable Energy Reviews*, *72*, 1177–1198. https://doi.org/10.1016/j.rser.2016.10.036.

Chapter 7

Optimization of renewable energy sources using emerging computational techniques

Aman Kumar[a], Krishna Kumar[b], and Nishant Raj Kapoor[a]

[a]*CSIR-CBRI, AcSIR—Academy of Scientific and Innovative Research, Roorkee, India,* [b]*Research & Development Unit, Uttarakhand Jal Vidyut Nigam (UJVN) Ltd., Dehradun, Uttarakhand, India*

Abbreviations

AMPE	absolute mean percentage error
ANFIS	adaptive neuro-fuzzy inference system
ANN	artificial neural network
ARIMA	autoregressive integrated moving average
ARMA	autoregressive and moving average
BPNN	feed-forward back propagation neural network
BPS	back propagation algorithm
CGP	Pola-Ribiere conjugate gradient
DNN	dynamic neural network
EMVS	emission virtual sensors
ENVS	engine virtual sensors
FAWNN	fuzzy arithmetic wavelet neural networks
FC	fuel cell
FFNN	feed forward neural network
GA	genetic algorithm
GMDH	group method of data handling
GP	genetic programming
GRNN	generalized regression neural network
GSR	global solar radiations
IDBN	improved deep belief network
IEA	International Energy Agency
IVMD	improved variational mode decomposition
KNN	k-nearest neighbor

Sustainable Developments by Artificial Intelligence and Machine Learning for Renewable Energies.
https://doi.org/10.1016/B978-0-323-91228-0.00012-4
Copyright © 2022 Elsevier Inc. All rights reserved.

LM	Levenberg-Marguardt
LR	linear regression
LSM	least-squares method
MFC	microbial fuel cell
ML	machine learning
MLR	multiple linear regression
MSE	mean square error
NFIS	neuro-fuzzy interface system
NN	neural network
NWP	numerical weather predictions
OLF	on-line fuzzy
PCR	principal component regression
PLS	partial least squares regression
RANN	recurrent artificial neural network
RBFNN	radial basis function neural networks
RDA	regularized discriminant analysis
RF	random forest
RNN	recurrent neural network
RRMSE	relative root mean squared error
SCG	scaled conjugate gradient
SVD	singular value decomposition
SVM	support vector machines
VGCHP	vertical ground coupled heat pump

7.1 Introduction

Energy generation and consumption can precisely define the economic development of any country. The share of greenhouse gases generated by human activities related to energy generation by nonrenewable sources is 73%, and significantly responsible for climate change (UNDP, 2021a). Out of 17 sustainability goals laid out by the United Nations (UN-SDG), the seventh goal is "affordable and clean energy" (UNDP, 2021b). To fulfill energy demand together with achieving the sustainability goal in the energy sector, it is necessary to increase the percentage of renewable energy (RE) generation; as reported by UNDP, only 17.5% of power was generated using renewable resources in the year 2017 (UNDP, 2021a). Conventional sources of energy are responsible for climate change and environmental damage. According to the report of US IEA, "Total energy-related greenhouse gas (GHG) emission would lead to considerable climate degradation with an average 6°C global warming" (International Energy Agency, 2020). Clean energy (hydroelectricity, geothermal, tidal energy, solar, and wind energy) is the only practicable and feasible alternative to these conventional sources of energy. In 2003, the US government started an initiative developing hydrogen fuel to protect the environment from harmful gases and ensure the future requirements of clean energy (Chalk & Miller, 2006) with an expected outcome by 2030.

In electricity production, the share of wind and solar energy reached only 8.5% in the year 2019 (Global Energy Statistical Yearbook, 2015). The total energy consumption by the world's top 10 leading countries is shown in Fig. 7.1. China accounts for a large proportion of consumption in the total energy among all the countries, and produces only 27% of its electricity through RE sources. The United States is the second-largest energy consumption country in the world and produces only 17.9% from solar and wind sources. India holds the third position and has a 20.7% share in total electricity generation by RE (IRENA, 2021b).

In the Asian continent, China is the leading country in electricity generation from renewable sources (biofuels, geothermal, hydropower, ocean, solar, and wind energy). The second place in RE electricity generation is held by India. The largest proportion of installed capacity is also held by these two countries. In the African continent, Egypt has held first place from the year 2016 to now in RE electricity generation and Mozambique stands in second place. South Africa is in first place in installed capacity and Egypt in second place. Germany, Norway, and Italy are the countries which hold the first, second, and third

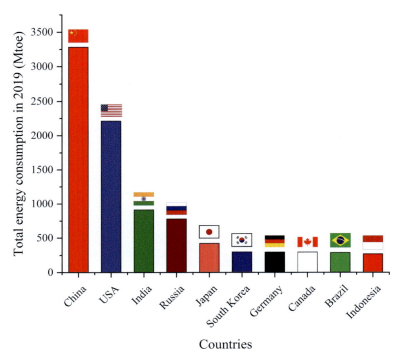

FIG. 7.1 Energy consumption data of world-leading countries.

positions in RE electricity generation in Europe, respectively. But there is a small change in the installed capacity: the first rank is held by Germany, second place is taken by Italy, and France holds third place. In Oceania, Australia is in first place and New Zealand in second place in RE electricity generation. In installed capacity, the same places are held by these two countries. In the North America continent, the United States and Canada hold the top places in RE electricity generation and in installed capacity. In the South American continent, Brazil is in first place and Paraguay holds second place in RE electricity generation. The installed capacity varies, in that first place is held by Brazil but Venezuela is in second position. In the Middle East (Asia), Iran has first place and second place is held by Jorden in energy generation and the installed capacity values of Iran are higher than Iraq. In Eurasia, the first place is held by Russia and Turkey is in second place in RE electricity generation; they also hold the same places in installed capacity. In Central America and the Caribbean, Costa Rica and Guatemala are in the first and second place in RE electricity generation and also hold the same places in installed capacity, respectively, as shown in Fig. 7.2 (IRENA, 2021b).

Within RE sources, hydropower energy plays a vital role and contributes 63% in energy generation; however, its installed capacity is only 49.8%. The second most used source of energy is wind energy and its contribution to energy generation is 19.1%. The installed wind energy capacity is 23.9%. The third most commonly used source of energy is solar energy and provides an 8.5% role in energy generation. Solar energy has an installed capacity of near about 20.7%, which means only 41.06% of the solar systems are in working condition. The fourth place is occupied by bioenergy and its contribution to energy generation is 7.8% with only 5% of installed capacity. The contribution of geothermal energy in energy generation is only 1.3% and its installed capacity is 0.6% as shown in Fig. 7.3 (Tableau Public, 2021).

The equation of clean sustainable energy is given as:

$$ACE + CAP = RE + AI \qquad (7.1)$$

where ACE = affordable clean energy, CAP = control accuracy and precision, RE = renewable energy, and AI = artificial intelligence.

7.2 Sources of renewable energy

The main sources of renewable energies are bioenergy, geothermal, hydropower energy, hydrogen energy, solar energy, wind energy, and ocean/marine energy (Chalk & Miller, 2006; Johansson, Kelly, Reddy, & Williams, 1992). A schematic chart of the different sources of energies is shown in Fig. 7.4. A brief explanation of renewable energies follows.

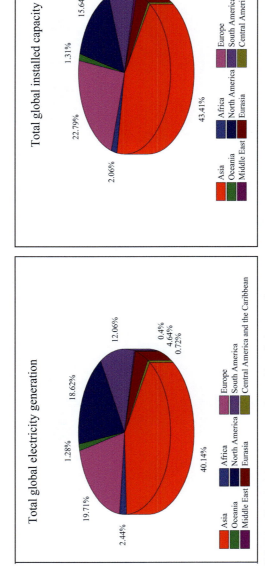

FIG. 7.2 Global total energy generation and installation capacity.

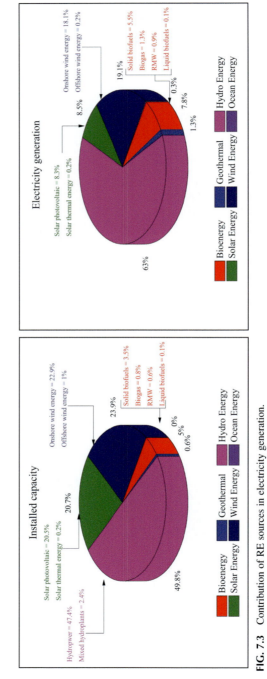

FIG. 7.3 Contribution of RE sources in electricity generation.

7.2.1 Bioenergy (BE)

Bioenergy is energy in which electric power is generated using biologically decomposable and degradable materials such as agricultural wastes, wood, vegetables, organic wastes, and microorganisms (Dahiya, 2015; Khanal et al., 2010). Broadly, bioenergy is categorized as modern and traditional technologies. Modern bioenergy technology includes liquid biofuels, which are generated from the bagasse plants (bio-refiners, biogas) produced through anaerobic digestion of residues. The traditional techniques are the combustion of biomass such as animal waste, wood, and charcoal. The categorization of bioenergy is shown in Fig. 7.4.

For developing a balanced sustainable society, biofuels are the cardinal form among the various RE sources. The twofold benefits of this energy combine the degradation of biological materials and simultaneously the generation of energy during the degradation process. Extensive research is ongoing on energy extracted from degradable biomass, especially focused on developing liquid mobility fuels. Lignocellulose, heterotrophic, and autotrophic algal biofuel production are gaining attention to generate energy. Furthermore, oilseed crops like Jatropha that are of the nonfood category are used for the production of biodiesel. Biofuels reduce the dependency on petroleum fuels and strengthen the local, state, and national economy by supporting the agriculture sector. The bioenergy production chain is shown in Fig. 7.5. The status of electricity generation from bioenergy and installed bioenergy capacity is shown in Fig. 7.6, and the world's top countries' contribution in electricity generation from bioenergy and installed bioenergy capacity is presented in Fig. 7.7 (IRENA, 2021a).

7.2.2 Geothermal energy (GE)

Geothermal energy is generated and stored beneath the surface of the Earth (Archer, 2020). The geothermal gradient is created between the Earth's core

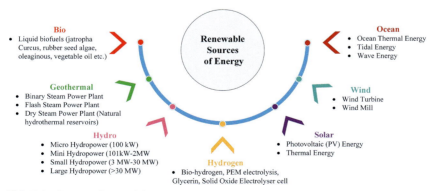

FIG. 7.4 Sources of renewable energy.

194 Sustainable developments by artificial intelligence & machine learning

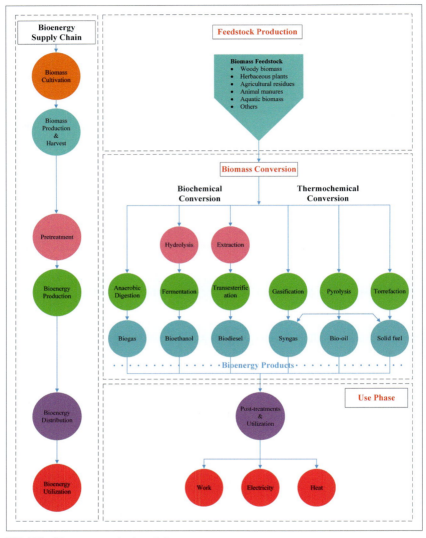

FIG. 7.5 Bioenergy production chain.

and the Earth's surface due to differences in temperatures. Lava is formed due to the steady decay of radioactive elements within the Earth's surface. This lava is disintegrated by the movement of tectonic plates and produces large geothermal reservoirs, which is the source of geothermal energy. Geothermal reservoirs mean fumaroles, geysers, steam vents and hot springs, altered ground, etc. The reservoirs are defined as "the hot part of the geothermal system that can be exploited either by extracting the fluid contained (water, steam, or gases)"

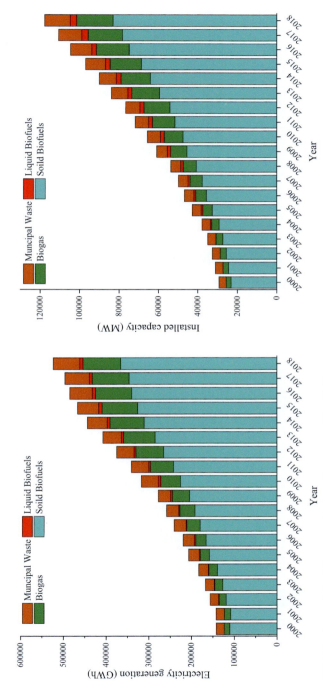

FIG. 7.6 Status of electricity generation from bioenergy and installed bioenergy capacity.

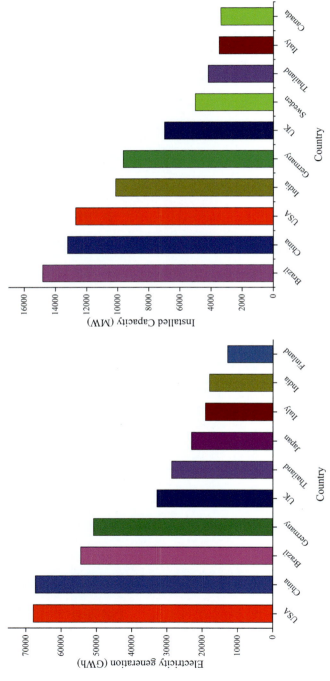

FIG. 7.7 The world's top countries' contribution in electricity generation from bioenergy and installed bioenergy capacity.

(Grant, Donaldson, & Bixley, 1982). These geothermal reservoirs occur in places where the flow of heat to the surface is high. This energy can be extracted economically and used in heating, agricultural, and electricity generation applications. Geothermal energy is divided into three parts, as per the electricity generation chart shown in Fig. 7.4.

Geothermal technologies are directly used in geothermal heat pumps, greenhouses, and district heating (heat networks or teleheating). These geothermal heating technologies first came into use in 1913. Medium-temperature fields are preferred to generate electricity through binary cycle technology. In the future, enhanced geothermal systems (EGS) can be introduced to increase the efficiency and production of clean geothermal energy (Geothermal Energy, 2021). The status of electricity generation from geothermal and installed geothermal capacity is shown in Fig. 7.8, and the world's top countries' contribution in electricity generation from geothermal and installed geothermal capacity in Fig. 7.9 (Geothermal Energy, 2021).

The various locations of geothermal energy in India are shown in Fig. 7.10.

7.2.3 Hydropower energy (HPE)

For thousands of years, water has been used for mechanical energy to rotate sawmills, flour mills, watermills, etc. (Hamududu & Killingtveit, 2012). In the late 17th century and early 18th century, hydraulic turbines were developed and used for many purposes. In England in 1870, a technique to generate electricity from water was first discovered. Due to this invention, the first hydropower plant was installed in Appleton, Wisconsin in 1882, which could produce 12.5 kW of electricity—enough to power 250 lamps (Milly, Dunne, & Vecchia, 2005). Hydroelectric power, latterly called hydropower, is a source of RE, producing electrical energy by utilizing natural sources of water like waterfalls, etc., or the controlled fall of water on turbines, as in dams. Hydropower has a noble role in socioeconomic development and its promise of sustainability for the future. In 1897, India started its first hydropower station in Darjeeling with a capacity of 130 kW (Hydropower Potential in India, 2021). The main limitation of hydropower plants is silt erosion. There are various technologies that are available for the optimization of hydropower energy generation. Artificial intelligence and machine learning are being utilized for operation and maintenance optimization (Kumar & Saini, n.d.; Kumar & Saini, 2021; Kumar, Singh, Ranjan, & Kumar, 2021).

The energy extracted from flowing water is in accordance with following equation (Roy & Majumder, 2016):

$$P = \eta \times \rho \times g \times Q \times H \qquad (7.2)$$

where P = power potential, η = efficiency of the conversion process, Q = discharge of water, H = head, g = gravitational force, and ρ = density of water.

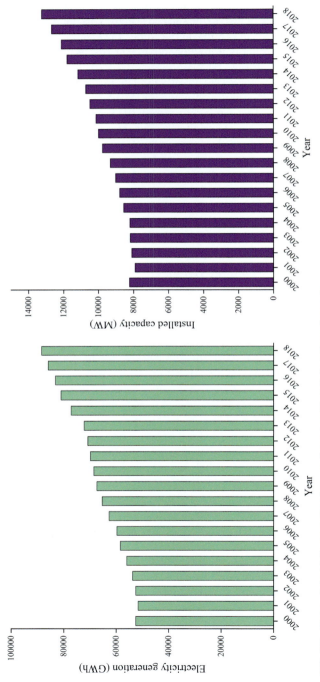

FIG. 7.8 Status of electricity generation from geothermal and installed geothermal capacity.

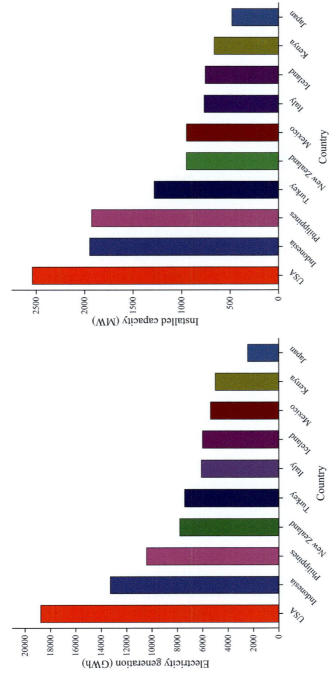

FIG. 7.9 The world's top countries' contribution in electricity generation from geothermal and installed geothermal capacity.

200 Sustainable developments by artificial intelligence & machine learning

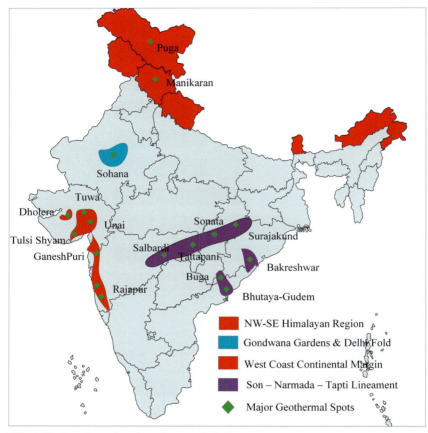

FIG. 7.10 Locations of geothermal energy in India.

The status of electricity generation from hydropower energy and installed hydropower capacity is shown in Fig. 7.11, and the world's top countries' contribution in electricity generation from hydropower and installed capacity in Fig. 7.12 (Hydropower Energy, 2021).

7.2.4 Hydrogen energy (HE)

Hydrogen is the fundamental constituent of all fuel products. The electrolysis process of water and other biomass-derived composites such as organic wastes or the biological process of bacteria produces hydrogen, further higher energy is produced from burning and can be utilized as an RE source for electricity production (Kothari, Buddhi, & Sawhney, 2008; McDowall, 2012; Stern, 2018). Innovations and adopting the latest technologies help to improve the overall process efficiency.

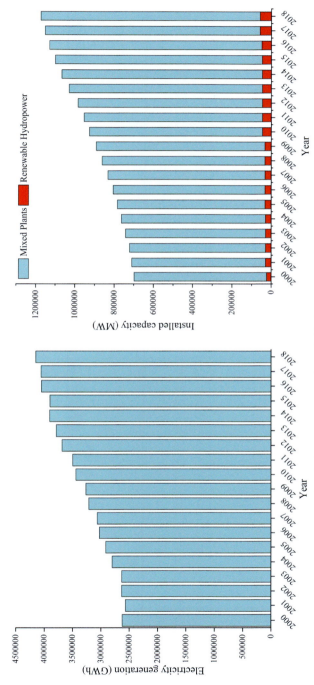

FIG. 7.11 Status of electricity generation from hydropower energy and installed hydropower capacity.

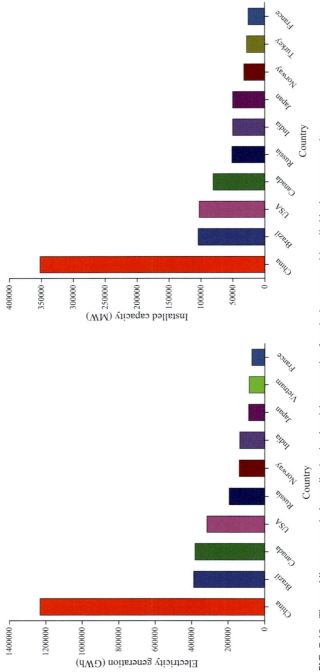

FIG. 7.12 The world's top countries' contribution in electricity generation from hydropower and installed hydropower capacity.

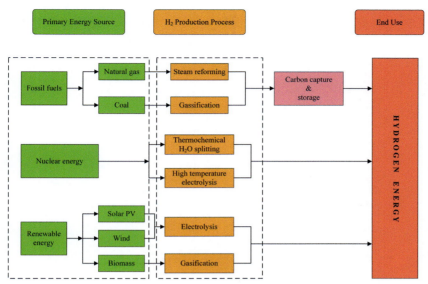

FIG. 7.13 Various methods of hydrogen production with their production process.

Hydrogen exists in many forms on the surface of the Earth but it is not freely available. Among all the energy fuels, H_2 has high energy per unit mass. In the present scenario, hydrogen is produced from various sources such as fossil-based natural coal and gas, coal gasification, and steam reforming, etc. Energy produced by H_2 is mainly divided into three categories: (i) thermal (conversion of biomass and fossil to H_2), (ii) photonic (split light to H_2, e.g., H_2O or organic molecules), and (iii) electrolytic (split electricity to H_2O) (Dincer & Acar, 2017; Kothari et al., 2008). The various methods of hydrogen production with their production process are shown in Fig. 7.13.

> *Hydrogen is today enjoying unprecedented momentum. The world should not miss this unique chance to make hydrogen an important part of our clean and secure energy future.*
>
> IEA (2021)

Hydrogen has a higher energy density per mass than any other fuel. The key drivers of the hydrogen economy are shown in Fig. 7.14. The sustainable development scenario (SDS) of hydrogen generation is presented in Fig. 7.15.

7.2.5 Solar energy (SE)

The sun is the major source of energy on the Earth's surface for living creatures. Solar radiation has been used by human beings for different purposes for many centuries (N'Tsoukpoe, Liu, Le Pierrès, & Luo, 2009; Yadav & Chandel, 2013;

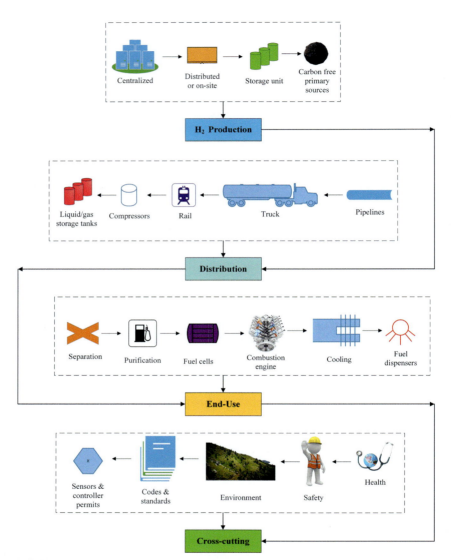

FIG. 7.14 Key driver and challenges for the hydrogen economy.

Zhang, Zhao, Deng, Xu, & Zhang, 2017). In the 7th century BCE, the first evidence was recorded of the use of the sun rays to create fire. The elaborated history of solar energy is mentioned in Kreider and Kreith (1981). The categorization of solar energy is shown in Fig. 7.4.

The research into solar elements helps to improve the efficiency and quality in the areas of global policies, application, design, novel and efficient materials, storage, and low energy buildings. Flow of solar energy to the Earth's surface is shown in Fig. 7.16.

Optimization of renewable energy sources **Chapter | 7** 205

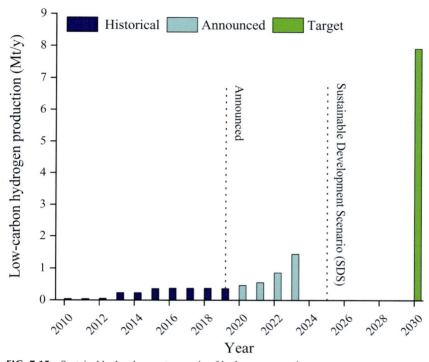

FIG. 7.15 Sustainable development scenario of hydrogen generation.

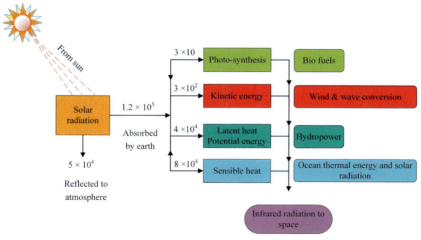

FIG. 7.16 Flow of solar energy to the Earth's surface (units: terawatts).

206 Sustainable developments by artificial intelligence & machine learning

The status of electricity generation from solar energy and installed solar capacity is shown in Fig. 7.17, and the world's top countries' contribution in electricity generation from solar energy and installed solar capacity in Fig. 7.18 (Solar Energy, 2021).

7.2.6 Wind energy (WE)

Wind is generated due to the Earth's rotation and the sun's rays on the Earth's surface. Wind energy is generated using windmills and wind turbines by converting kinetic energy into mechanical energy. In the year 1887, the first electricity-producing windmill was installed in Ohio, United States, with a capacity of 12 kW (World Wind Energy Reports 2006–2015 (2021)). Wind energy is the second largest source of RE after hydropower energy. Electricity production through wind energy increased by 11.41% in 2018, as compared to 2017 statistical data.

The foremost advantage of a wind power station is that it is easy to install, it is scalable, and it has low carbon emission throughout the life of the project. The status of electricity generation from wind energy and installed wind energy capacity is shown in Fig. 7.19, and the world's top countries' contribution in electricity generation from wind energy and installed wind capacity shown in Fig. 7.20 (Wind Energy, 2021).

7.2.7 Ocean energy (OE)

The energy generated from the ocean is also considered as part of hydropower energy and is sometimes called marine energy. Ocean energy is categorized into three parts, as shown in Fig. 7.4. The working mechanisms of water turbines and wind turbines are quite similar, but the relative density and viscosity of the air and water make for differences—the viscosity and density of water are 1000 and 100 times greater than air, respectively. At deep ocean levels, the temperature differences have enough energy to produce thermal energy, and this is called ocean thermal energy (OTE). Wave energy is accumulated through floating bodies that perform elliptic movement under the action of waves. The status of electricity generation from ocean energy and installed ocean capacity is shown in Fig. 7.21, and the world's top countries' contribution in electricity generation from ocean energy and installed ocean capacity in Fig. 7.22 (Ocean Energy, 2021).

7.3 Artificial intelligence (AI)

Nowadays, artificial intelligence is used in every sector from industries to agriculture. It is a technique to solve complex problems by training a computer system. The development of AI reduces the burden of manual calculations. AI is based on numerous learning techniques, among which some are: neural

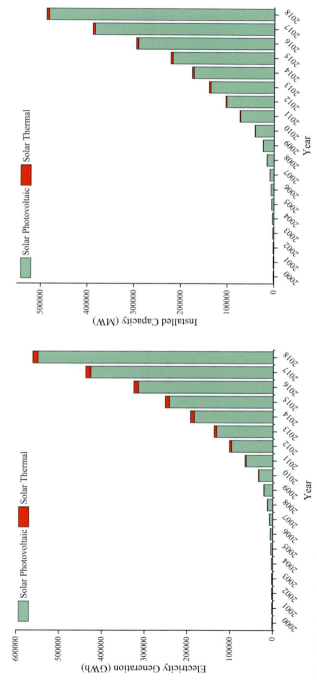

FIG. 7.17 Status of electricity generation from solar energy and installed solar capacity.

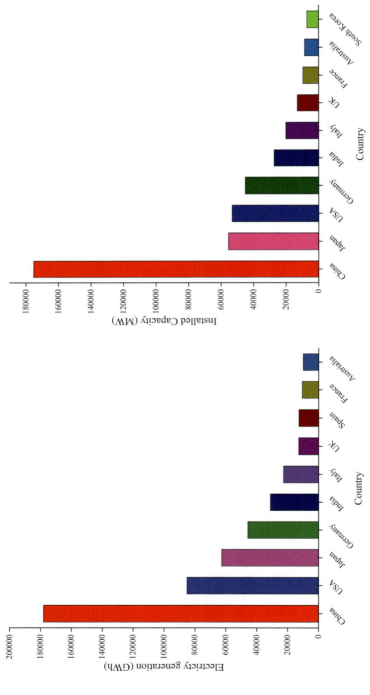

FIG. 7.18 The world's top countries' contribution in electricity generation from solar and installed solar capacity.

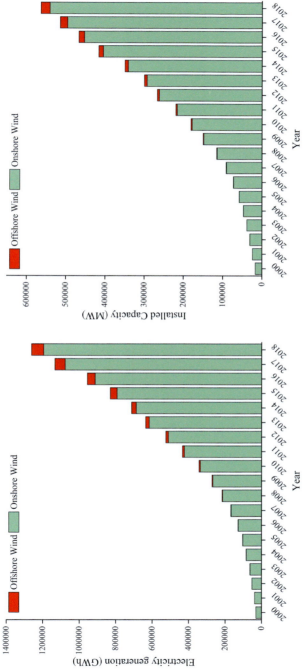

FIG. 7.19 Status of electricity generation from wind energy and installed wind capacity.

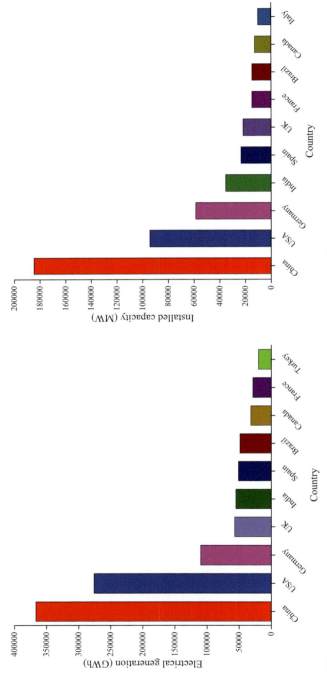

FIG. 7.20 The world's top countries' contribution in electricity generation from wind energy and installed wind capacity.

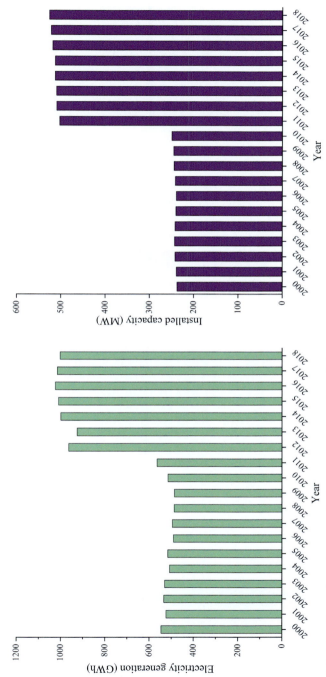

FIG. 7.21 Status of electricity generation from ocean energy and installed ocean capacity.

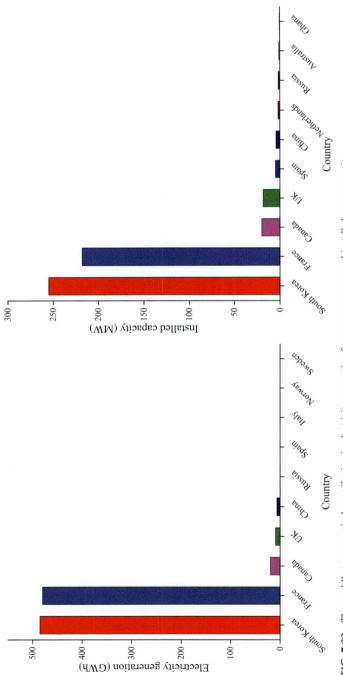

FIG. 7.22 The world's top countries' contribution in electricity generation from ocean energy and installed ocean capacity.

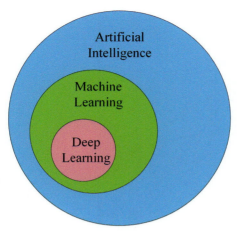

FIG. 7.23 Foundation of artificial intelligence.

learning, statistical learning, evolutionary learning, etc. AI contributes in the engineering and medical fields by simulating complex problems, and optimizing and predicting the best possible economic solution to the problem. AI will work as a catalyst for achieving the sustainability goals in the energy sector. Machine learning (ML) is a subfield of AI, as shown in Fig. 7.23 (Kumar & Mor, 2021). ML contains algorithms that are capable of learning a complicated task and generating predictive models from the sample data. The positive response of ML to solving complex problems leads to innovation in deep learning.

Deep learning is a subfield of machine learning and, working with technologies like neural networks, etc., DL eliminates the learning portion of the set of data in the engineering features. The architectural model of an artificial neural network (ANN) (Kumar & Mor, 2021) is presented in Fig. 7.24. This ANN model contains three layers. The first layer is the input parameters and depends on the type of data to be optimized and predict; the second layer is the hidden layer, in which the behind-the-scenes processing work is done (it may also have further subcategories, depending on the amount of data); and the third layer is the output layer, providing the final outcome after prediction and optimization.

Use of artificial intelligence for different sources of RE is depicted in Fig. 7.25. AI is used in the design, prediction, optimization, control, supply, administration, and policymaking of RE sources.

7.3.1 Artificial intelligence in bioenergy

AI is the predominant computational technique used in bioenergy to predict biomass feedstock properties, the biomass conversion pathway (thermochemical conversion and biochemical conversion technologies), and is also applied

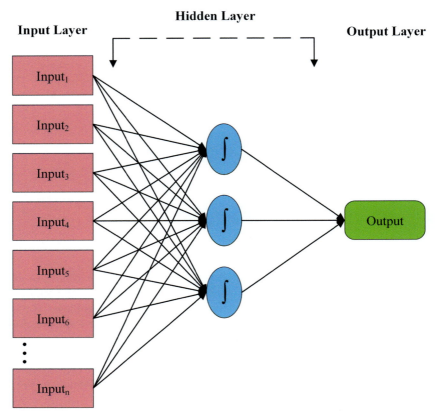

FIG. 7.24   Sample model of ANN.

for the end-users. In predicting biomass properties, AI affects the quality of the end product and the conversion operability of biomass. AI also helps in the prediction of higher heating values (HHV) using data from ultimate analysis. In the common context, proximate analysis data are used to predicting the HHV because it is time-saving and economic as compared to the ultimate analysis.

Researchers have reviewed stochastic and deterministic mathematical modeling for optimizing forest biomass to generate RE. Various single and some hybrid approaches have been described in a number of reports and studies, as listed in Table 7.1. The hybrid methods used in controlling, prediction, and optimization are also shown in the table.

7.3.2   Artificial intelligence in geothermal energy

AI in geothermal energy was started during the last decade. The AI techniques with smart sensors are making it possible to predict and optimize the design

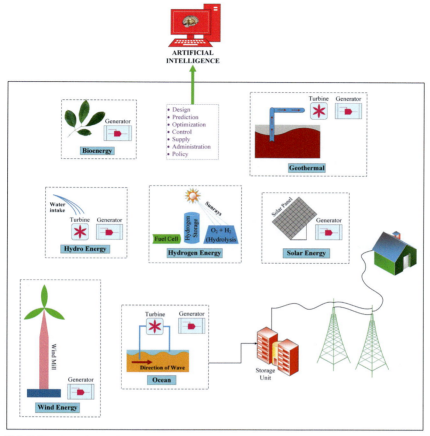

FIG. 7.25 Use of artificial intelligence in RE.

control, geothermal reservoir and drilling process, etc. Researchers from the literature used various AI methodologies, as shown in Table 7.2. But ANN (BPNN) is used in almost every geothermal sector and provides good accuracy in prediction and enhances optimization. The input parameters used by the various researchers are: change in temperature, time variation, ammonia fraction, pressure, density and viscosity of water, and surface area of the geothermal source, as mentioned in Table 7.2.

7.3.3 Artificial intelligence in hydro energy

The design and control of hydropower plants using modern techniques like those shown, along with their input, in Table 7.3. The hybrid AI methods are also used in the prediction of hydropower energy.

TABLE 7.1 AI approach to bioenergy.

Authors	AI methodology	Application	Input parameters	Results
Yang, Liu, and Li (2011)	GRNN	Cetane number and density prediction	Cetane number, paraffins, density, dicycloparaffins, monocycloparaffins, tricycloparaffins, benzocycloalkanes, alkylbenzenes, fluorenes, benzodicycloalkanes, aromatic sulfur, and triaromatics	$R^2 = 0.93$
Strik, Domnanovich, Zani, Braun, and Holubar (2005)	BPNN	H_2S and NH_3 detection in biogas	Sulfate and nitrogen loading rate, H_2S and NH_3 in biogas, sulfides, inorganic nitrogen, ammonia, and ammonium in reactor, biogas productivity, and organic loading rate	$R^2 = 0.91$ for H_2S and 0.83 for NH_3
Ramadhas, Jayaraj, Muraleedharan, and Padmakumari (2006)	BPNN, RNN, GRNN, RBFNN	Cetane number prediction	—	3.4%, 3.6%, 3.8%, and 5.0% accuracy, respectively
Ozkaya, Demir, and Bilgili (2007)	BPNN	Prediction of CH_4 percentage in landfill biogas	pH, alkalinity, COD, sulfate, conductivity, chloride, waste temperature, and refuse age	$R^2 = 0.951-0.957$
(Balabin & Safieva, 2011)	BPNN, MLR, PLS, PCR	Prediction of biodiesel specifications	Kinematic viscosity, density, methanol percentage, and water content	RMSE = 0.42–51
Kumar, Srinivasa Pai, and Shrinivasa Rao (2012)	RBFNN	Biodiesel engine performance prediction	Compression ratio, load percentage, injection timing, injection pressure, and blend percentage	MSE = 0.001985–0.0011
Balabin and Safieva (2011)	SVM, KNN, RDA, PLS	Biodiesel classification	Coconut, sunflower, palm, soy, castor, rapeseed, Jatropha, linseed, cottonseed, and used frying oil	Support vector machine accuracy 95%

Ghugare, Tiwary, Elangovan, and Tambe (2014)	GP, BPNN	Higher heating value prediction	—	Correlation 0.95
Koutroumanidis, Ioannou, and Arabatzis (2009)	ANN, ARIMA, ARIMA-ANN	Prediction of price of fuelwood	—	RMSE $=0.05$
Romeo and Gareta (2009)	ANN-Fuzzy logic (Hybrid method)	Controlling biomass boiler	Steam mass flow, fuel flow, drum pressure, final steam temperature, grate load, grate surface, total airflow, and furnace load	Output of turbine increases 3.5%
Abu Qdais, Bani Hani, and Shatnawi (2010)	BPNN-GA (Hybrid method)	Prediction of CH_4 from waste	% Total solids, % volatile solids, pH value, temperature	$R^2 = 0.8703$
Gueguim, Oloke, Lateef, and Adesiyan (2012)	BPNN-GA (Hybrid method)	Optimizing the biogas generation	Rice bran, stem of banana, waste paper, sawdust, cow dung	8.64% increased production

TABLE 7.2 AI approach to geothermal energy.

Authors	AI methodology	Application	Input parameters	Results
Esen and Inalli (2009)	VGCHP	Prediction of performance	—	$R^2 = 0.9998$
	BPNN (SCG, CGP, LM)			
Bassam, Santoyo, Andaverde, Hernández, and Espinoza-Ojeda (2010)	BPNN (LM)	Geothermal well SFT prediction	Temperature, time, dT/dt	$R^2 > 0.95$
Arslan (2011)	BPNN (SCG, CGP, LM)	Prediction of geothermal power	Outlet temperature, ammonia fraction, and the vapor fraction of geothermal water	$R^2 = 0.9987$
Arslan and Yetik (2011)	BPNN (SCG, CGP, LM)	Prediction of geothermal power	Vapor fraction of geothermal water, the outlet temperature of working fluid from the condenser, working fluids, outlet temperature, outlet pressure, circulation pump power, and generated power	$R^2 = 0.9999$
Kalogirou et al. (2012)	BPNN	Generating geothermal map	—	Correlation 0.9253
Keçebaş and Yabanova (2012)	BPNN (LM)	Performance prediction of AGDHS	Ambient temperature, temperatures, pressures, and the volume flow rates	$R^2 = 0.9999$
Álvarez del Castillo, Santoyo, and García-Valladares (2012)	BPNN (LM)	VF prediction	Wellbore diameter, steam quality, wellhead pressure, viscosity, fluid density, and the dimensionless numbers Froude, Weber, and Reynolds	MPE 0.17

Reference	Method	Application	Inputs	Results
Yabanova and Keçebaş (2013)	BPNN	PID controller efficiency prediction	Network inputs, the mass flow rate of the fluid, enthalpy, rate of energy destruction, net energy transfer by heat at temperature T, heat transfer rate, rate of entropy generation	Correlation 0.9986
Arslan and Yetik (2014)	BPNN (LM)	Modeling of the geothermal plant	Outlet temperature, working fluids, vapor fraction of geothermal water, outlet pressure, condenser temperature	$R^2 = 0.99$
Yeo and Yee (2014)	BPNN (LM, SCG)	Site location modeling	Altitude, inclination, direction degree, distance to the road (the shortest distance), ground coverage	$R^2 = 0.85$
Kalogirou, Florides, Pouloupatis, Christodoulides, and Joseph-Stylianou (2015)	BPNN	Conductivity map generation	Lithology class, minimum annual temperature, elevation, rainfall, east coordinate, north coordinate, mean annual temperature, and maximum annual temperature	Correlation 0.9553
Farghally, Atia, El-madany, and Fahmy (2014b)	Fuzzy logic	Design of RAS system	Surface area, resource temperature, relative humidity, air temperature, water temperature	Error zero error
Farghally, Atia, El-madany, and Fahmy (2014a)	Fuzzy logic	Design of RAS system	Surface area, resource temperature, relative humidity, air temperature, water temperature	Max. RAS production at 2°C
Porkhial, Salehpour, Ashraf, and Jamali (2015)	GMDH-GA-SVD (Hybrid method)	Prediction of temperature	Angle, northing, and easting (top point of wells), azimuth, and major depth	$R^2 = 0.9899$

TABLE 7.3 AI approach in hydropower energy.

Authors	AI methodology	Application	Input parameters	Results
Dolling and Varas (2002)	BPNN (LM)	Prediction of streamflow	Mean temperature, month number, number of cloudy days, sunshine hours, relative humidity, snow depth, wind velocity, and precipitation	MAPE < 5%
Firat and Güngör (2007)	ANN, ANFIS, MR	Prediction of river flow	—	RMSE for ANFIS = 7.1
Molina, Isasi, Berlanga, and Sanchis (2000)	LVQ-ART-MAP (Hybrid method)	Maintenance and acoustic prediction	—	False alarm rate < 10%
Toro, Gómez Meire, Gálvez, and Fdez-Riverola (2013)	HC-FFT-ANN (Hybrid method)	Prediction of river flow	—	Std. 26. 48
Uzlu, Akpinar, Özturk, Nacar, and Kankal (2014)	ANN-ABC (Hybrid method)	Prediction of hydraulic energy	Average yearly temperature, energy consumption, population, and gross electricity energy demand	Mean absolute percentage error = 4.6%
Kumar and Saini (n.d.)	SOM	Prediction of daily silt pattern	Daily silt data	Detected the outliers for accurate forecasting of silt pattern
Uzlu, Akpınar, Özturk, Nacar, and Kankal (2014)	ANN	Prediction of daily plant load	Daily plant load data of various power plants	Daily load forecasting

7.3.4 Artificial intelligence in hydrogen energy

The area of research to generate electricity from hydrogen energy is not the newest, but in this area, the percentage of energy generation is lowest among all the energy sources. The area of AI is fruitful in the risk management system of the hydrogen power plant. The AI methodologies which are able to predict the performance and control the Power Entry Module and Filter (PEMF) are shown in Table 7.4.

7.3.5 Artificial intelligence in solar energy

In solar energy, the AI technique ANN is used to optimize the heating load and prediction in the design process. ANN techniques are also helpful in the prediction and optimization of the weather forecast, temperature, solar radiation, humidity, UV, pressure, wind speed, precipitation, temperature range, and sunshine hours. The other various methodologies for the prediction and optimization of solar energy are tabulated in Table 7.5 along with input parameters.

7.3.6 Artificial intelligence in wind energy

In wind energy, AI is used to predict the power of the wind, the life span of the wind turbine, and to optimize the loss of energy using the methodologies listed in Table 7.6. Hybrid technologies that boost the prediction and optimization with greater accuracy are also shown in the table. The input parameters consist of relative humidity, wind speed, generation hours, site weather, operation aspects, and geography of the wind site, etc. Complete information on wind energy and associated AI methods is given in Table 7.6.

7.3.7 Artificial intelligence in ocean energy

In ocean energy, AI plays a unique role in the optimization of the energy, wind velocity, and temperature conditions. The various AI technologies used in the prediction of ocean waves and sea-level, etc., along with methodologies and input parameters are presented in Table 7.7. The input parameters used by various researchers are wave period, sea level, depth of wave flow, and speed of the wave, etc.

7.4 Conclusion

The chapter has concisely presented the state of the art methods of artificial intelligence used in the RE sector with detailed energy installation, generation, and consumption data of the top 10 countries and different continents. Recent developments aligned with AI in SE, WE, HPE, HE, OE, GE, and BE are presented effectively. Additionally, the presented worldwide data of installed

TABLE 7.4 AI approach in hydrogen energy.

Authors	AI methodology	Application	Input parameters	Results
Garg, Vijayaraghavan, Mahapatra, Tai, and Wong (2014)	MGGP (GP), SVR, ANN	Predict the performance of microbial fuel cell	Ferrous sulfate and temperature	$R^2 = 0.9872, 0.9568, 0.9588$
Prakasham, Sathish, and Brahmaiah (2011)	ANN, GA	Optimization of bio-hydrogen	pH, glucose, xylose, inoculum size, age of inoculum	$R^2 = 0.9999$
Amirinejad et al. (2013)	RBFNN, MLPNN	Prediction of the voltage of proton exchange membrane fuel cell	Cell temperature, current density, inlet gas temperature, and inorganic additives	$R^2 = 0.9955, 0.9961$
Entchev and Yang (2007)	ANN, ANFIS	Prediction of solid oxide fuel cell performance	Stack fuel flow, stack temperature, stack airflow, burner fuel flow, stack air inlet temperature, burner combustion temperature, burner airflow	Both are good, but FC voltage ANN performs better
Yap, Ho, and Karri (2012)	ANN (EMVS and ENVS)	Optimization of exhaust emission and engine parameters	Throttle position, lambda, injection angle, and ignition advanced	RMSE = 4.1% and 6.5%
Tardast et al. (2014)	ANN	Prediction of bioelectricity generation in MFC	pH, temperature and electron acceptor	$R^2 = 0.98868$ and MSE = 0.0024

Caux, Hankache, Fadel, and Hissel (2010)	OLF	Management of hybrid fuel cell	Storage component and propulsion power	Provide good results
Hatti and Tioursi (2009)	DNN	Controlling the Proton-exchange membrane fuel cell	Current drawn, H_2 pressure, O_2, and cell temperature	Accurate and robust
Vichard, Harel, Ravey, Venet, and Hissel (2020)	RNN	Degradation prediction of PEM cell	Ambient temperature and operating time	Performance losses 9% with respect to original life after 5000 h
Kenanoğlu, Baltacıoğlu, Demir, and Erkınay Özdemir (2020)	ANN	Performance and emission of diesel engine	Motor speed, fuel type, and fuel consumption	Accuracy $=96.07\%$, 95.82%, and 92.35% for motor power, motor torque, and NO_x emission, respectively

TABLE 7.5 AI approach in solar energy.

Authors	AI methodology	Application	Input parameters	Results
Rehman and Mohandes (2008)	ANN	Estimation of GSR	Relative humidity and air temperature	For mean temperature, AMPE is 11.8% and for maximum temperature its value is 10.3%
Tasadduq, Rehman, and Bubshait (2002)	BPNN	Prediction of temperature	Temperature value	Temperature error between predicted and measured values is within 5%
Ben Ammar, Ben Ammar, and Oualha (2020)	FFNN, Hybrid model: NFIS, ANFIS (BPS, LSM)	Photovoltaic power prediction	Power load, photovoltaic power, pump load	NMBE and NRMSE=0.0023% and 0.08% on cloudy day
Zhang, Tan, and Wei (2020)	Hybrid model: IVMD, ARIMA, IDBN	Output power prediction of photovoltaic	Solar power	Accuracy improved from 9% to 96%
Shaker, Manfre, and Zareipour (2020)	NWP, FAWNN	Engine prediction	Solar power	RMSE=3%
Kosovic, Mastelic, and Ivankovic (2020)	ML (LR, RF, NN)	Estimation of solar radiation	Solar radiation, humidity, UV, wind speed, temperature	RRMSE=0.0393
Mellit, Benghanem, Arab, and Guessoum (2005)	MLP, MLP-IIR, RBF-IIR	Optimization of the photovoltaic system	Solar radiations	$R^2=0.98$

TABLE 7.6 AI approach in wind energy.

Authors	AI methodology	Application	Input parameters	Results
Carolin Mabel and Fernandez (2008)	BPNN	Prediction of wind power	Relative humidity, wind speed, and generation hours	RMSE = 0.0065
Li and Shi (2010)	ADALINE, RBFNN, and BPNN	Prediction of wind speed	Wind speed	RMSE for BPNN = 1.254
Mabel and Fernandez (2009)	BPNN	Prediction of wind power	Site weather, operation aspects, and geography	MSE = 7.6×10^{-3}
Kariniotakis, Stavrakakis, and Nogaret (1996)	NB and RANN	Prediction of wind power	Wind speed	RMSE = 4.2
Öztopal (2006)	TPCSV and BPNN	Prediction of wind speed	Wind speed and temperature	Correlation for BPNN = 0.95
Alexiadis, Dokopoulos, and Sahsamanoglou (1999)	BPNN	Prediction of wind power and speed	Wind speed and wind generation capacity	20%–40% improved accuracy
Li, Shi, and Zhou (2011)	RBFNN, BPNN, ADALINE, and BC	Wind speed prediction	Wind speed and temperature	RSME for BC = 1.5
Sfetsos (2000)	NLN, ARMA, ANN	Prediction of wind speed	Last hour wind speed data	RMSE for NLN = 4.9%
Cadenas, Jaramillo, and Rivera (2010)	BPNN, and SES	Prediction of wind speed	Wind speed, generation hours, and relative humidity	MAE for BPNN = 0.5251
Simoes, Bose, and Spiegel (1997)	Fuzzy method	Wind generation system design	Wind generation and wind speed	3500 W

Continued

TABLE 7.6 AI approach in wind energy—cont'd

Authors	AI methodology	Application	Input parameters	Results
Sideratos and Hatziargyriou (2007)	Fuzzy methods, RBFNN, and ANN	Prediction of wind power	Wind speed and wind direction	Planning 1–48 h ahead
Monfared, Rastegar, and Kojabadi (2009)	Fuzzy methods and BPNN	Prediction of wind speed	Wind speed	RMSE for BPNN = 3.30
Juban, Siebert, and Kariniotakis (2007)	Probabilistic method	Prediction of wind power	—	Reliability (2%–4%)
Mohandes, Halawani, Rehman, and Hussain (2004)	BPNN, and SVM	Prediction of wind speed	Wind speed	MSE for BPNN = 0.0078
Potter and Negnevitsky (2006)	ANFIS	Prediction of wind power	Wind speed	MAE < 8
Mohandes et al. (2004)	ANFIS	Prediction of wind speed	Wind speed and air density	MAPE 3% at 40 m
Yang et al. (2011)	ANFIS	Interpolation of missing wind data	Wind speed previous data	RMSE = 0.230
Meharrar, Tioursi, Hatti, and Boudghène Stambouli (2011)	ANFIS	Wind production system designing	Wind speed	MSE = 0.05
Yang, Li, and Wang (2008)	WT + BPNN	Fault diagnosis of wind turbine	—	Detection of eight conditions
Jursa and Rohrig (2008)	BPNN + PSO	Prediction of wind power	Power data of wind farms and weather	Accuracy improved by 2.8%

Guo, Zhao, Lu, & Wang, 2012	FNN+EMD	Prediction of wind speed	Wind speed	MSE=0.1296
Pourmousavi Kani and Ardehali (2011)	MC+ANN	Prediction of wind speed	Wind speed transition probability values and wind speed	94.84 Error
Damousis and Dokopoulos (2001) and Hu, Wang, and Zeng (2013)	Fuzzy-GA (Hybrid method)	Prediction of wind power and speed	Hourly wind speed data	Accuracy improved by 29.7%
Damousis and Dokopoulos (2001) and Hu et al. (2013)	EEMD-SVM (Hybrid method)	Prediction of wind speed	Wind speed and air density	MAE=0.12
Cadenas and Rivera (2010)	ARIMA-BPNN (Hybrid method)	Prediction of wind speed	Hourly wind speed	MSE=0.49
Salcedo-Sanz et al. (2009)	MM5-ANN (Hybrid method)	Prediction of wind speed	Wind speed, wind temperature, and direction of the wind	MAE=1.45–2.2 m/s
Liu, Niu, Wang, and Fan (2014)	WT-SVM-GA (Hybrid method)	Prediction of wind speed	Wind temperature and wind speed	MAE=0.6169
Kong, Liu, Shi, and Lee (2015)	SVR-PSO (Hybrid method)	Prediction of wind speed	Wind speed and direction of the wind	Effective accuracy
Rahmani, Yusof, Seyedmahmoudian, and Mekhilef (2013)	ACO-PSO (Hybrid method)	Prediction of wind power	Ambient temperature and wind speed	MAPE=3.5%
Pousinho, Mendes, and Catalão (2011)	ANFIS-WT-PSO (Hybrid method)	Optimization of risk in trading energy	Wind power and energy market prices	Estimation of profit for risk level (0.0–0.1)

228 Sustainable developments by artificial intelligence & machine learning

TABLE 7.7 AI approach in ocean energy.

Authors	AI methodology	Application	Input parameters	Results
Makarynskyy and Makarynska (2007)	BPNN	Wave parameters prediction	Geographic coordinates, training, gaps, validation, gaps	RMSE 0.53
Toprak and Cigizoglu (2008)	BPNN, RBFNN, GRNN	Prediction of dispersion coefficient	Bed shear-stress velocity, depth of the flow, mean cross-sectional velocity of the channel, and channel width	RMSE 27.9 BPNN
Chen, Lin, Lee, and Chen (2010)	Fuzzy logic	Reducing the effect of ocean wave	Wave period, wave amplitude, water depth, gravitational force, mass, draft, width, young's modulus, mass moment of inertia, time delay	Small amplitudes stability is 0.0
Ghorbani, Makarynskyy, Shiri, and Makarynska (2010)	BPNN (LM), GP	Prediction of sea level	Sea level at various times	$R^2 = 0.972-0.973$
Karimi, Kisi, Shiri, and Makarynskyy (2013)	ARMA, BPNN, ANFIS	Prediction of sea level	Time series of hourly sea level	RMSE 0.055 for ANFIS

capacity, energy generation, and consumption enriches the knowledge available in this literature.

The bioenergy production chain, geothermal energy locations in India, methods of hydrogen production with their production process, key drivers and challenges for the hydrogen economy, flow of solar energy to the Earth's surface, and foundation of artificial intelligence with a sample model of an ANN are summarized effectively.

Furthermore, an interesting fact explored during the preparation of this chapter is that ANN (BPNN, GRNN, RNN, DNN, RBFNN, SCG, CGP, LM, and MLPNN) is used frequently by researchers to design, predict, optimize, control, supply, administrate, and make policy in all type of RE sources. The current chapter explains the obviousness of hybrid AI methods along with single AI methods in different activities. AI methods retain infinite potential and the hour demands that hybrid and novel methods should be properly utilized in the future for fruitful outcomes with increased sustainability and prosperity.

To sum up, the junction of technological developments in RE with artificial intelligence will increase the overall efficiency of the energy sector by accurate forecasting, precise optimization, and hassle-free accessibility and sustainably, resulting in low climatic impact by the energy industry.

References

Abu Qdais, H., Bani Hani, K., & Shatnawia, N. (2010). Modeling and optimization of biogas production from a waste digester using artificial neural network and genetic algorithm. *Resources, Conservation and Recycling*, *54*(6), 359–363. https://doi.org/10.1016/j.resconrec.2009.08.012.

Alexiadis, M. C., Dokopoulos, P. S., & Sahsamanoglou, H. S. (1999). Wind speed and power forecasting based on spatial correlation models. *IEEE Transactions on Energy Conversion*, *14*(3), 836–842. https://doi.org/10.1109/60.790962.

Álvarez del Castillo, A., Santoyo, E., & García-Valladares, O. (2012). A new void fraction correlation inferred from artificial neural networks for modeling two-phase flows in geothermal wells. *Computers & Geosciences*, *41*, 25–39. https://doi.org/10.1016/j.cageo.2011.08.001.

Amirinejad, M., Tavajohi-Hasankiadeh, N., Madaeni, S. S., Navarra, M. A., Rafiee, E., & Scrosati, B. (2013). Adaptive neuro-fuzzy inference system and artificial neural network modeling of proton exchange membrane fuel cells based on nanocomposite and recast Nafion membranes. *International Journal of Energy Research*, *37*(4), 347–357. https://doi.org/10.1002/er.1929.

Archer, R. (2020). Geothermal energy. In *Future energy: Improved, sustainable and clean options for our planet* (pp. 431–445). Elsevier. https://doi.org/10.1016/B978-0-08-102886-5.00020-7.

Arslan, O. (2011). Power generation from medium temperature geothermal resources: ANN-based optimization of Kalina cycle system-34. *Energy*, *36*(5), 2528–2534. https://doi.org/10.1016/j.energy.2011.01.045.

Arslan, O., & Yetik, O. (2011). ANN based optimization of supercritical ORC-binary geothermal power plant: Simav case study. *Applied Thermal Engineering*, *31*(17–18), 3922–3928. https://doi.org/10.1016/j.applthermaleng.2011.07.041.

Arslan, O., & Yetik, O. (2014). ANN modeling of an orc-binary geothermal power plant: Simav case study. *Energy Sources, Part A: Recovery, Utilization, and Environmental Effects*, *36*(4), 418–428. https://doi.org/10.1080/15567036.2010.542437.

Bassam, A., Santoyo, E., Andaverde, J., Hernández, J. A., & Espinoza-Ojeda, O. M. (2010). Estimation of static formation temperatures in geothermal wells by using an artificial neural network approach. *Computers & Geosciences*, *36*(9), 1191–1199. https://doi.org/10.1016/j.cageo.2010.01.006.

Ben Ammar, R., Ben Ammar, M., & Oualha, A. (2020). Photovoltaic power forecast using empirical models and artificial intelligence approaches for water pumping systems. *Renewable Energy*, *153*, 1016–1028. https://doi.org/10.1016/j.renene.2020.02.065.

230 Sustainable developments by artificial intelligence & machine learning

Cadenas, E., Jaramillo, O. A., & Rivera, W. (2010). Analysis and forecasting of wind velocity in chetumal, quintana roo, using the single exponential smoothing method. *Renewable Energy*, *35*(5), 925–930. https://doi.org/10.1016/j.renene.2009.10.037.

Cadenas, E., & Rivera, W. (2010). Wind speed forecasting in three different regions of Mexico, using a hybrid ARIMA-ANN model. *Renewable Energy*, *35*(12), 2732–2738. https://doi.org/10.1016/j.renene.2010.04.022.

Carolin Mabel, M., & Fernandez, E. (2008). Analysis of wind power generation and prediction using ANN: A case study. *Renewable Energy*, *33*(5), 986–992. https://doi.org/10.1016/j.renene.2007.06.013.

Caux, S., Hankache, W., Fadel, M., & Hissel, D. (2010). On-line fuzzy energy management for hybrid fuel cell systems. *International Journal of Hydrogen Energy*, *35*(5), 2134–2143. https://doi.org/10.1016/j.ijhydene.2009.11.108.

Chalk, S. G., & Miller, J. F. (2006). Key challenges and recent progress in batteries, fuel cells, and hydrogen storage for clean energy systems. *Journal of Power Sources*, *159*(1), 73–80. https://doi.org/10.1016/j.jpowsour.2006.04.058.

Chen, C.-Y., Lin, J.-W., Lee, W.-I., & Chen, C.-W. (2010). Fuzzy control for an oceanic structure: A case study in time-delay TLP system. *Journal of Vibration and Control*, *16*(1), 147–160. https://doi.org/10.1177/1077546309339424.

Dahiya, A. (2015). *Bioenergy biomass to biofuels*.

Damousis, I. G., & Dokopoulos, P. (2001). *A fuzzy expert system for the forecasting of wind speed and power generation in wind farms* (pp. 63–69). IEEE. https://doi.org/10.1109/PICA.2001.932320.

Dincer, I., & Acar, C. (2017). Innovation in hydrogen production. *International Journal of Hydrogen Energy*, *42*(22), 14843–14864. https://doi.org/10.1016/j.ijhydene.2017.04.107.

Dolling, O. R., & Varas, E. A. (2002). Utilisation des réseaux des neurones artificielles pour la prédiction des écoulements. *Journal of Hydraulic Research*, *40*(5), 547–554. https://doi.org/10.1080/00221680209499899.

Entchev, E., & Yang, L. (2007). Application of adaptive neuro-fuzzy inference system techniques and artificial neural networks to predict solid oxide fuel cell performance in residential micro-generation installation. *Journal of Power Sources*, *170*(1), 122–129. https://doi.org/10.1016/j.jpowsour.2007.04.015.

Esen, H., & Inalli, M. (2009). Modelling of a vertical ground coupled heat pump system by using artificial neural networks. *Expert Systems with Applications*, *36*(7), 10229–10238. https://doi.org/10.1016/j.eswa.2009.01.055.

Farghally, H. M., Atia, D. M., El-madany, H. T., & Fahmy, F. H. (2014a). Control methodologies based on geothermal recirculating aquaculture system. *Energy*, *78*, 826–833. https://doi.org/10.1016/j.energy.2014.10.077.

Farghally, H. M., Atia, D. M., El-madany, H. T., & Fahmy, F. H. (2014b). Fuzzy logic controller based on geothermal recirculating aquaculture system. *Egyptian Journal of Aquatic Research*, *40*(2), 103–109. https://doi.org/10.1016/j.ejar.2014.07.004.

Firat, M., & Güngör, M. (2007). River flow estimation using adaptive neuro fuzzy inference system. *Mathematics and Computers in Simulation*, *75*(3–4), 87–96. https://doi.org/10.1016/j.matcom.2006.09.003.

Garg, A., Vijayaraghavan, V., Mahapatra, S. S., Tai, K., & Wong, C. H. (2014). Performance evaluation of microbial fuel cell by artificial intelligence methods. *Expert Systems with Applications*, *41*(4), 1389–1399. https://doi.org/10.1016/j.eswa.2013.08.038.

IRENA. (2021b). *Trends in Renewable Energy by Region*. Retrieved from: https://www.irena.org/Statistics/View-Data-by-Topic/Capacity-and-Generation/Regional-Trends.

Ghorbani, M. A., Makarynskyy, O., Shiri, J., & Makarynska, D. (2010). Genetic programming for sea level predictions in an island environment. *The International Journal of Ocean and Climate Systems*, 27–35. https://doi.org/10.1260/1759-3131.1.1.27.

Global Energy Statistical Yearbook. (2015). https://yearbook.enerdata.net/.

IRENA. (2021a). *Bioenergy*. Retrieved from: https://www.irena.org/bioenergy (Accessed 25 March 2021).

IEA. (2021). *The Future of Hydrogen*. Retrieved from: https://www.iea.org/reports/the-future-of-hydrogen (Accessed 22 April 2021).

UNDP. (2021b). *Sustainable Development Goals*. Retrieved from: https://www.undp.org/content/undp/en/home/sustainable-development-goals.html (Accessed 20 March 2021).

Geothermal Energy. (2021). Retrieved April 10, 2021, from https://www.irena.org/geothermal (Accessed 20 March 2021).

Grant, M. A., Donaldson, I., & Bixley, P. F. (1982). *Geothermal reservoir engineering. Vol. 370*.

Gueguim Kana, E. B., Oloke, J. K., Lateef, A., & Adesiyan, M. O. (2012). Modeling and optimization of biogas production on saw dust and other co-substrates using Artificial Neural network and Genetic Algorithm. *Renewable Energy*, *46*, 276–281. https://doi.org/10.1016/j.renene.2012.03.027.

Guo, Z., Zhao, W., Lu, H., & Wang, J. (2012). Multi-step forecasting for wind speed using a modified EMD-based artificial neural network model. *Renewable Energy*, *37*(1), 241–249. https://doi.org/10.1016/j.renene.2011.06.023.

Hamududu, B., & Killingtveit, A. (2012). Assessing climate change impacts on global hydropower. *Energies*, *5*(2), 305–322. https://doi.org/10.3390/en5020305.

Hatti, M., & Tioursi, M. (2009). Dynamic neural network controller model of PEM fuel cell system. *International Journal of Hydrogen Energy*, *34*(11), 5015–5021. https://doi.org/10.1016/j.ijhydene.2008.12.094.

Hu, J., Wang, J., & Zeng, G. (2013). A hybrid forecasting approach applied to wind speed time series. *Renewable Energy*, *60*, 185–194. https://doi.org/10.1016/j.renene.2013.05.012.

Tableau Public. (2021). *Trends in Renewable Energy. All Renewable Sources of Energy*. Retrieved from: https://public.tableau.com/views/IRENARETimeSeries/Charts?:embed=y&:showVizHome=no&publish=yes&:toolbar=no (Accessed 24 March 2021).

Hydropower Energy. (2021). Retrieved from https://www.irena.org/hydropower (Accessed 14 April 2021).

Johansson, T. B., Kelly, H., Reddy, A. K. N., & Williams, R. H. (1992). Renewable fuels and electricity for a growing world economy: Defining and achieving the potential. *Energy Studies Review*, *4*(3), 201–212. https://doi.org/10.15173/esr.v4i3.284.

Juban, J., Siebert, N., & Kariniotakis, G. N. (2007). Probabilistic short-term wind power forecasting for the optimal management of wind generation. In *2007 IEEE Lausanne POWERTECH, proceedings* (pp. 683–688). https://doi.org/10.1109/PCT.2007.4538398.

Jursa, R., & Rohrig, K. (2008). Short-term wind power forecasting using evolutionary algorithms for the automated specification of artificial intelligence models. *International Journal of Forecasting*, *24*(4), 694–709. https://doi.org/10.1016/j.ijforecast.2008.08.007.

Kalogirou, S. A., Florides, G. A., Pouloupatis, P. D., Christodoulides, P., & Joseph-Stylianou, J. (2015). Artificial neural networks for the generation of a conductivity map of the ground. *Renewable Energy*, *77*, 400–407. https://doi.org/10.1016/j.renene.2014.12.033.

Kalogirou, S. A., Florides, G. A., Pouloupatis, P. D., Panayides, I., Joseph-Stylianou, J., & Zomeni, Z. (2012). Artificial neural networks for the generation of geothermal maps of ground temperature at various depths by considering land configuration. *Energy*, *48*(1), 233–240. https://doi.org/10.1016/j.energy.2012.06.045.

232 Sustainable developments by artificial intelligence & machine learning

Karimi, S., Kisi, O., Shiri, J., & Makarynskyy, O. (2013). Neuro-fuzzy and neural network techniques for forecasting sea level in Darwin Harbor, Australia. *Computers & Geosciences, 52*, 50–59. https://doi.org/10.1016/j.cageo.2012.09.015.

Kariniotakis, G. N., Stavrakakis, G. S., & Nogaret, E. F. (1996). Wind power forecasting using advanced neural networks models. *IEEE Transactions on Energy Conversion, 11*(4), 762–767. https://doi.org/10.1109/60.556376.

Keçebaş, A., & Yabanova, I. (2012). Thermal monitoring and optimization of geothermal district heating systems using artificial neural network: A case study. *Energy and Buildings, 50*, 339–346. https://doi.org/10.1016/j.enbuild.2012.04.002.

Kenanoğlu, R., Baltacıoğlu, M. K., Demir, M. H., & Erkınay Özdemir, M. (2020). Performance & emission analysis of HHO enriched dual-fuelled diesel engine with artificial neural network prediction approaches. *International Journal of Hydrogen Energy, 45*(49), 26357–26369. https://doi.org/10.1016/j.ijhydene.2020.02.108.

Khanal, S. K., Surampalli, R. Y., Zhang, T. C., Lamsal, B. P., Tyagi, R. D., & Kao, C. M. (2010). Bioenergy and biofuel from biowastes and biomass. In *Bioenergy and biofuel from biowastes and biomass* (pp. 1–505). American Society of Civil Engineers (ASCE). https://doi.org/10.1061/9780784410899.

Kong, X., Liu, X., Shi, R., & Lee, K. Y. (2015). Wind speed prediction using reduced support vector machines with feature selection. *Neurocomputing, 169*, 449–456. https://doi.org/10.1016/j.neucom.2014.09.090.

Kosovic, I. N., Mastelic, T., & Ivankovic, D. (2020). Using artificial intelligence on environmental data from internet of things for estimating solar radiation: Comprehensive analysis. *Journal of Cleaner Production, 266*. https://doi.org/10.1016/j.jclepro.2020.121489.

Kothari, R., Buddhi, D., & Sawhney, R. L. (2008). Comparison of environmental and economic aspects of various hydrogen production methods. *Renewable and Sustainable Energy Reviews, 12*(2), 553–563. https://doi.org/10.1016/j.rser.2006.07.012.

Koutroumanidis, T., Ioannou, K., & Arabatzis, G. (2009/09/01/2009). Predicting fuelwood prices in Greece with the use of ARIMA models, artificial neural networks and a hybrid ARIMA–ANN model. *Energy Policy, 37*(9), 3627–3634. https://doi.org/10.1016/j.enpol.2009.04.024.

Kreider, J. F., & Kreith, F. (1981). Solar energy handbook. *Journal of Solar Energy Engineering*, 362–363. https://doi.org/10.1115/1.3266267.

Kumar, A., & Mor, N. (2021). *An approach-driven: Use of artificial intelligence and its applications in civil engineering* (pp. 201–221). Springer Science and Business Media LLC. https://doi.org/10.1007/978-981-33-6400-4_10.

Kumar, K., & Saini, R. (2021). Application of artificial intelligence for the optimization of hydropower energy generation. *EAI Endorsed Transactions on Industrial Networks and Intelligent Systems*, 170560. https://doi.org/10.4108/eai.6-8-2021.170560.

Hydropower Potential in India. (2021). Retrieved from: http://117.252.14.242/rbis/india_ information/hydropower.htm#:~:text=A%20project%20with%20capacity%20of,first% 20hydropower%20installation%20in%20India%20 (Accessed 12 April 2021).

Kumar, K., & Saini, R. P. (n.d.). Application of machine learning for hydropower plant silt data analysis. Materials Today: Proceedings. doi: https://doi.org/10.1016/j.matpr.2020.09.375.

Kumar, K., Singh, R. P., Ranjan, P., & Kumar, N. (2021). *Daily plant load analysis of a hydropower plant using machine learning* (pp. 819–826). Springer Science and Business Media LLC. https://doi.org/10.1007/978-981-33-4604-8_65.

Li, G., & Shi, J. (2010). On comparing three artificial neural networks for wind speed forecasting. *Applied Energy, 87*(7), 2313–2320. https://doi.org/10.1016/j.apenergy.2009.12.013.

Li, G., Shi, J., & Zhou, J. (2011). Bayesian adaptive combination of short-term wind speed forecasts from neural network models. *Renewable Energy, 36*(1), 352–359. https://doi.org/10.1016/j.renene.2010.06.049.

Liu, D., Niu, D., Wang, H., & Fan, L. (2014). Short-term wind speed forecasting using wavelet transform and support vector machines optimized by genetic algorithm. *Renewable Energy, 62*, 592–597. https://doi.org/10.1016/j.renene.2013.08.011.

Mabel, M. C., & Fernandez, E. (2009). Estimation of energy yield from wind farms using artificial neural networks. *IEEE Transactions on Energy Conversion, 24*(2), 459–464. https://doi.org/10.1109/TEC.2008.2001458.

Makarynskyy, O., & Makarynska, D. (2007). Wave prediction and data supplementation with artificial neural networks. *Journal of Coastal Research, 23*(4), 951–960. https://doi.org/10.2112/04-0407.1.

McDowall, W. (2012). Technology roadmaps for transition management: The case of hydrogen energy. *Technological Forecasting and Social Change, 79*(3), 530–542. https://doi.org/10.1016/j.techfore.2011.10.002.

Meharrar, A., Tioursi, M., Hatti, M., & Boudghène Stambouli, A. (2011). A variable speed wind generator maximum power tracking based on adaptative neuro-fuzzy inference system. *Expert Systems with Applications, 38*(6), 7659–7664. https://doi.org/10.1016/j.eswa.2010.12.163.

Mellit, A., Benghanem, M., Arab, A. H., & Guessoum, A. (2005). An adaptive artificial neural network model for sizing stand-alone photovoltaic systems: Application for isolated sites in Algeria. *Renewable Energy, 30*(10), 1501–1524. https://doi.org/10.1016/j.renene.2004.11.012.

Milly, P. C. D., Dunne, K. A., & Vecchia, A. V. (2005). Global pattern of trends in streamflow and water availability in a changing climate. *Nature, 438*(7066), 347–350. https://doi.org/10.1038/nature04312.

Mohandes, M. A., Halawani, T. O., Rehman, S., & Hussain, A. A. (2004). Support vector machines for wind speed prediction. *Renewable Energy, 29*(6), 939–947. https://doi.org/10.1016/j.renene.2003.11.009.

Molina, J. M., Isasi, P., Berlanga, A., & Sanchis, A. (2000). Hydroelectric power plant management relying on neural networks and expert system integration. *Engineering Applications of Artificial Intelligence, 13*(3), 357–369. https://doi.org/10.1016/S0952-1976(00)00009-9.

Monfared, M., Rastegar, H., & Kojabadi, H. M. (2009). A new strategy for wind speed forecasting using artificial intelligent methods. *Renewable Energy, 34*(3), 845–848. https://doi.org/10.1016/j.renene.2008.04.017.

N'Tsoukpoe, K. E., Liu, H., Le Pierrès, N., & Luo, L. (2009). A review on long-term sorption solar energy storage. *Renewable and Sustainable Energy Reviews, 13*(9), 2385–2396. https://doi.org/10.1016/j.rser.2009.05.008.

Öztopal, A. (2006). Artificial neural network approach to spatial estimation of wind velocity data. *Energy Conversion and Management, 47*(4), 395–406. https://doi.org/10.1016/j.enconman.2005.05.009.

Porkhial, S., Salehpour, M., Ashraf, H., & Jamali, A. (2015). Modeling and prediction of geothermal reservoir temperature behavior using evolutionary design of neural networks. *Geothermics, 53*, 320–327. https://doi.org/10.1016/j.geothermics.2014.07.003.

Potter, C. W., & Negnevitsky, M. (2006). Very short-term wind forecasting for Tasmanian power generation. *IEEE Transactions on Power Systems, 21*(2), 965–972. https://doi.org/10.1109/TPWRS.2006.873421.

Pourmousavi Kani, S. A., & Ardehali, M. M. (2011). Very short-term wind speed prediction: A new artificial neural network-Markov chain model. In *Vol. 52. Energy conversion and management* (pp. 738–745). Elsevier Ltd. https://doi.org/10.1016/j.enconman.2010.07.053. Issue 1.

234 Sustainable developments by artificial intelligence & machine learning

Pousinho, H. M. I., Mendes, V. M. F., & Catalão, J. P. S. (2011). A risk-averse optimization model for trading wind energy in a market environment under uncertainty. *Energy, 36*(8), 4935–4942. https://doi.org/10.1016/j.energy.2011.05.037.

Prakasham, R. S., Sathish, T., & Brahmaiah, P. (2011). Imperative role of neural networks coupled genetic algorithm on optimization of biohydrogen yield. *International Journal of Hydrogen Energy, 36*(7), 4332–4339. https://doi.org/10.1016/j.ijhydene.2011.01.031.

Rahmani, R., Yusof, R., Seyedmahmoudian, M., & Mekhilef, S. (2013). Hybrid technique of ant colony and particle swarm optimization for short term wind energy forecasting. *Journal of Wind Engineering and Industrial Aerodynamics, 123*, 163–170. https://doi.org/10.1016/j.jweia.2013.10.004.

Ramadhas, A. S., Jayaraj, S., Muraleedharan, C., & Padmakumari, K. (2006/12/01/2006). Artificial neural networks used for the prediction of the cetane number of biodiesel. *Renewable Energy, 31*(15), 2524–2533. https://doi.org/10.1016/j.renene.2006.01.009.

Rehman, S., & Mohandes, M. (2008). Artificial neural network estimation of global solar radiation using air temperature and relative humidity. *Energy Policy, 36*(2), 571–576. https://doi.org/10.1016/j.enpol.2007.09.033.

Roy, U., & Majumder, M. (2016). In U. Roy, & M. Majumder (Eds.), *Impact of climate change on small scale hydro-turbine selections* (1st ed.). https://doi.org/10.1007/978-981-287-239-5.

Salcedo-Sanz, S., Ángel, M. P. B., Ortiz-García, E. G., Portilla-Figueras, A., Prieto, L., & Paredes, D. (2009). Hybridizing the fifth generation mesoscale model with artificial neural networks for short-term wind speed prediction. *Renewable Energy, 34*(6), 1451–1457. https://doi.org/10.1016/j.renene.2008.10.017.

Sfetsos, A. (2000). A comparison of various forecasting techniques applied to mean hourly wind speed time series. *Renewable Energy, 21*, 125–131. https://doi.org/10.1016/S0960-1481(99.

Shaker, H., Manfre, D., & Zareipour, H. (2020). Forecasting the aggregated output of a large fleet of small behind-the-meter solar photovoltaic sites. *Renewable Energy, 147*, 1861–1869. https://doi.org/10.1016/j.renene.2019.09.102.

Sideratos, G., & Hatziargyriou, N. D. (2007). An advanced statistical method for wind power forecasting. *IEEE Transactions on Power Systems, 22*(1), 258–265. https://doi.org/10.1109/TPWRS.2006.889078.

Simoes, M. G., Bose, B. K., & Spiegel, R. J. (1997). Design and performance evaluation of a fuzzy-logic-based variable-speed wind generation system. *IEEE Transactions on Industry Applications, 33*(4), 956–965. https://doi.org/10.1109/28.605737.

Stern, A. G. (2018). A new sustainable hydrogen clean energy paradigm. *International Journal of Hydrogen Energy, 43*(9), 4244–4255. https://doi.org/10.1016/j.ijhydene.2017.12.180.

Solar Energy. (2021). Retrieved from: https://www.irena.org/solar (Accessed 25 April 2021).

UNDP. (2021a). *Sustainable Development Goals. Goal 7: Affordable and Clean Energy.* Retrieved from: https://www.undp.org/content/undp/en/home/sustainable-development-goals/goal-7-affordable-and-clean-energy.html (Accessed 20 March 2021).

Tardast, A., Rahimnejad, M., Najafpour, G., Ghoreyshi, A., Premier, G. C., Bakeri, G., et al. (2014). Use of artificial neural network for the prediction of bioelectricity production in a membrane less microbial fuel cell. *Fuel, 117*, 697–703. https://doi.org/10.1016/j.fuel.2013.09.047.

Tasadduq, I., Rehman, S., & Bubshait, K. (2002). Application of neural networks for the prediction of hourly mean surface temperatures in Saudi Arabia. *Renewable Energy, 25*(4), 545–554. https://doi.org/10.1016/S0960-1481(01)00082-9.

Toprak, Z. F., & Cigizoglu, H. K. (2008). Predicting longitudinal dispersion coefficient in natural streams by artificial intelligence methods. *Hydrological Processes, 22*(20), 4106–4129. https://doi.org/10.1002/hyp.7012.

Toro, C. H. F., Gómez Meire, S., Gálvez, J. F., & Fdez-Riverola, F. (2013). A hybrid artificial intelligence model for river flow forecasting. *Applied Soft Computing*, *13*(8), 3449–3458. https://doi.org/10.1016/j.asoc.2013.04.014.

Uzlu, E., Akpinar, A., Özturk, H. T., Nacar, S., & Kankal, M. (2014). Estimates of hydroelectric generation using neural networks with the artificial bee colony algorithm for Turkey. *Energy*, *69*, 638–647. https://doi.org/10.1016/j.energy.2014.03.059.

Vichard, L., Harel, F., Ravey, A., Venet, P., & Hissel, D. (2020). Degradation prediction of PEM fuel cell based on artificial intelligence. *International Journal of Hydrogen Energy*, *45*(29), 14953–14963. https://doi.org/10.1016/j.ijhydene.2020.03.209.

World Wind Energy Reports 2006–2015. (2021). Retrieved from: https://library.wwindea.org/ (Accessed 12 April 2021).

Wind Energy. (2021). Retrieved from https://www.irena.org/wind (Accessed 17 April 2021).

International Energy Agency. (2020). *Global Energy Review 2020*. Retrieved from IEA https://www.iea.org/reports/global-energy-review-2020 (Accessed 21 March 2021).

Yabanova, I., & Keçebaş, A. (2013). Development of ANN model for geothermal district heating system and a novel PID-based control strategy. *Applied Thermal Engineering*, *51*(1–2), 908–916. https://doi.org/10.1016/j.applthermaleng.2012.10.044.

Yadav, A. K., & Chandel, S. S. (2013). Tilt angle optimization to maximize incident solar radiation: A review. *Renewable and Sustainable Energy Reviews*, *23*, 503–513. https://doi.org/10.1016/j.rser.2013.02.027.

Yang, S., Li, W., & Wang, C. (2008). The intelligent fault diagnosis of wind turbine gearbox based on artificial neural network. In *Proceedings of 2008 international conference on condition monitoring and diagnosis, CMD 2008* (pp. 1327–1330). IEEE Computer Society. https://doi.org/10.1109/CMD.2008.4580221.

Yang, Z., Liu, Y., & Li, C. (2011). Interpolation of missing wind data based on ANFIS. *Renewable Energy*, *36*(3), 993–998. https://doi.org/10.1016/j.renene.2010.08.033.

Yap, W. K., Ho, T., & Karri, V. (2012). Exhaust emissions control and engine parameters optimization using artificial neural network virtual sensors for a hydrogen-powered vehicle. *International Journal of Hydrogen Energy*, *37*(10), 8704–8715. https://doi.org/10.1016/j.ijhydene.2012.02.153.

Yeo, I. A., & Yee, J. J. (2014). A proposal for a site location planning model of environmentally friendly urban energy supply plants using an environment and energy geographical information system (E-GIS) database (DB) and an artificial neural network (ANN). *Applied Energy*, *119*, 99–117. https://doi.org/10.1016/j.apenergy.2013.12.060.

Zhang, J., Tan, Z., & Wei, Y. (2020). An adaptive hybrid model for day-ahead photovoltaic output power prediction. *Journal of Cleaner Production*, *244*. https://doi.org/10.1016/j.jclepro.2019.118858.

Zhang, J., Zhao, L., Deng, S., Xu, W., & Zhang, Y. (2017). A critical review of the models used to estimate solar radiation. *Renewable and Sustainable Energy Reviews*, *70*, 314–329. https://doi.org/10.1016/j.rser.2016.11.124.

Strik, D. P. B. T. B., Domnanovich, A. M., Zani, L., Braun, R., & Holubar, P. (2005). Prediction of trace compounds in biogas from anaerobic digestion using the MATLAB Neural Network Toolbox. *Environmental Modelling & Software*, *20*(6), 803–810. https://doi.org/10.1016/j.envsoft.2004.09.006.

Ozkaya, B., Demir, A., & Bilgili, M. S. (2007). Neural network prediction model for the methane fraction in biogas from field-scale landfill bioreactors. *Environmental Modelling & Software*, *22*(6), 815–822. https://doi.org/10.1016/j.envsoft.2006.03.004.

236 Sustainable developments by artificial intelligence & machine learning

Kumar, S., Srinivasa Pai, P., & Shrinivasa Rao, B. R. (2012). Radial-basis-function-network-based prediction of performance and emission characteristics in a bio diesel engine run on WCO ester. *Advances in Artificial Intelligence*, 610487. https://doi.org/10.1016/j.envsoft.2006.03.004.

Balabin, R. M., & Safieva, R. Z. (2011). Biodiesel classification by base stock type (vegetable oil) using near infrared spectroscopy data. *Analytica Chimica Acta*, *689*(2), 190–197. https://doi.org/10.1016/j.aca.2011.01.041.

Ghugare, S. B., Tiwary, S., Elangovan, V., & Tambe, S. S. (2014). Prediction of higher heating value of solid biomass fuels using artificial intelligence formalisms. *BioEnergy Research*, *7*(2), 681–692. https://doi.org/10.1007/s12155-013-9393-5.

Romeo, L. M., & Gareta, R. (2009). Fouling control in biomass boilers. *Biomass and Bioenergy*, *33*(5), 854–861. https://doi.org/10.1016/j.biombioe.2009.01.008.

Ocean Energy. (2021). Retrieved from https://www.irena.org/ocean (Accessed 18 April 2021).

Chapter 8

Advanced renewable dispatch with machine learning-based hybrid demand-side controller: The state of the art and a novel approach

Yuekuan Zhou[a,b]

[a]*Sustainable Energy and Environment Thrust, Function Hub, The Hong Kong University of Science and Technology, Guangzhou, China,* [b]*Department of Mechanical and Aerospace Engineering, The Hong Kong University of Science and Technology, Clear Water Bay, Hong Kong SAR, China*

8.1 Introduction

The high dependence on traditional fossil fuels accelerates the energy shortage crisis, carbon emissions, and global warming. Deployment of renewable systems is one of the effective strategies to satisfy ever-increased energy demand. However, the intermittence and vulnerability of renewable power systems will lead to power instability, power damage, and power failure, in respect of extreme weather conditions. For example, in early 2021, affected by the extreme climate, the extreme cold current (temperature: $-17.78°C$) in the south of the United States led to the paralysis of the thermal power system in Texas, and around 4 million people in the South suffered from the forced blackouts under the severe cold weather conditions. Development of advanced energy dispatch strategies for renewable energy management is of great significance to improve the reliability, resilience, and flexibility of multienergy systems, in respect of fluctuations of power supply and stochastic demand profiles.

Building energy demand forecasting based on advanced mathematical tools with high accuracy and efficiency has attracted worldwide interest. Models for dynamic performance prediction can be classified into physics-based and machine learning (ML) models. Compared to physics-based models, the machine learning models are much simpler in modeling process, more

Sustainable Developments by Artificial Intelligence and Machine Learning for Renewable Energies.
https://doi.org/10.1016/B978-0-323-91228-0.00006-9
Copyright © 2022 Elsevier Inc. All rights reserved.

238 Sustainable developments by artificial intelligence & machine learning

computational efficient, and more adaptive to capture dynamics of nonlinear systems. Researchers focus on machine learning models for predictions of cooling/heating/electrical loads, with respect to supervised, unsupervised, and reinforcement learning. The forecast model development for energy consumption prediction includes original data source and preparation (data redundancy and data cleaning, missing data adjunction, and data normalization), feature extraction and classification (dimensionless distance metrics, cluster generation, energy consumption pattern, and schedule-based feature), trained model development, and model accuracy assessment.

Demand-side response and management can improve self-consumption of renewable energy and load coverage ratio, increase the reliability of power supply in response to intermittent renewable generation, and decrease import/export pressure on the power grid, without sacrificing indoor thermal comfort of building owners. Energy flexibility provided by demand-side response can meet the grid requirement in response to climate conditions, stochastic demand, and intermittent renewable integration. Flexible demand-side management strategies mainly include smart appliances (Hulst et al., 2015), heating, ventilation, and air conditioning (HVAC) systems (Stavrakas & Flamos, 2020), and plug-in loads and storages (Frendo, Graf, Gaertner, & Stuckenschmidt, 2020).

In this chapter, a holistic overview on ML applications in building energy demand prediction is conducted, for cooling, heating, and electrical energy systems. The underlying mechanism for ML to characterize dynamic behaviors of nonlinear systems is studied. Based on the predicted energy demands, flexible demand-side management strategies have been reviewed, including smart appliances, HVAC systems, plug-in loads, and storages. A generic approach for the development of advanced controllers is proposed, including dynamic performance prediction, machine learning-based surrogate model, controller strategies, and control objectives. This chapter can promote the application of ML for the development of hybrid controllers for energy flexible buildings.

8.2 Building energy demand forecasting with machine learning

Machine learning techniques have been applied in building energy demand forecasting. Compared to the traditional physical model, the data-driven model is more computationally efficient and effective, and can be applied for performance predictions on thousands of scenarios with uncertainty. In this section, predictions on cooling/heating/electrical loads have been conducted with ML techniques, and ML-based approaches are comprehensively reviewed, including supervised, unsupervised, and reinforcement learning. Aslam et al. (2021) comprehensively reviewed deep learning approaches for predictions of electric load and renewable generation. The review can provide comparative analysis on each method, in terms of the statistical indicators.

8.2.1 Predictions on cooling/heating/electrical loads

Table 8.1 summarizes the ML applications in predictions of cooling/heating/electrical loads. Researchers are mainly focused on advanced learning algorithms for nonlinear dynamic characterization (Li & Yao, 2021) and hybrid model with composite learning algorithms (Geysen, De Somer, Johansson, Brage, & Vanhoudt, 2018; Tan et al., 2020; Zhou, Zi, et al., 2019). In terms of the cooling load prediction, Ngo (2019) applied a cross-validation algorithm to predict the thermodynamics of an office building. Zhou, Zi, et al. (2019) developed a hybrid model for the cooling load forecasting of a commercial building. They concluded that the hybrid model is more efficient and accurate. To predict the heating load, Luo et al. (2019) adopted k-means clustering and artificial neural network to predict the heating load of an office. The mean absolute error is 3% and 8% for training and testing phases, respectively. Geysen et al. (2018) developed a combined data-driven method, to improve the forecasting accuracy on thermal load. The electrical load prediction has also been studied. Wang and Ahn (2020) developed a hybrid one-step-ahead load predictor for electrical load prediction of a residential community. Compared to previous methods, the proposed approach can generate better predictive and detective accuracy.

8.2.2 Machine learning modeling techniques

Fig. 8.1 demonstrates the development of machine learning models for demand prediction, including original data source and preparation, feature extraction and classification, ML model development, and model accuracy assessment. The stage of original data source and preparation includes data redundancy and data cleaning, missing data adjunction, and data normalization. Feature extraction and classification include dimensionless distance metrics, cluster generation, energy consumption pattern, and schedule-based feature. The original dataset is classified into training, validation, and testing datasets and, for the trained model, hyperparameter tuning and model accuracy assessment, respectively. Statistical indicators are shown in Fig. 8.1 for the accuracy assessment.

Fig. 8.2 shows the learning principles of supervised and unsupervised learning. In respect to supervised learning, as shown in Fig. 8.2A, the original dataset is classified into training, validation, and testing datasets. In the training dataset, researchers develop training algorithms and parameters, for model training and optimization. Based on the dynamic iteration on feature extraction, training algorithm, and parameter setting, well-trained models can be developed with high prediction accuracy in the testing dataset. The well-trained model is thereafter applied in the new building data for energy performance prediction. In respect to unsupervised learning, as shown in Fig. 8.2B, the model is trained from unlabeled data for performance prediction with insufficient information,

TABLE 8.1 A summary of ML applications in cooling/heating/electrical loads.

Types of energy demands	Studies	Building types	Algorithms	Results
Cooling	Ngo (2019)	Office building	Stratified 10-fold cross-validation algorithm	ML model agrees well with the physics-based model.
	Li and Yao (2021)	Residential and nonresidential buildings	Ten different machine learning models	Different algorithms are suitable for buildings in different scales.
	Zhou, Zi, et al. (2019)	Commercial building	A hybrid chaos-support vector regression and wavelet decomposition	The hybrid model is more efficient and accurate.
	Luo (2020)	Office building	Adaptive artificial neural network	The mean absolute errors are 3.59% and 4.71% for training and testing phases.
	Feng, Duan, Chen, Yakkali, and Wang (2021)	Residential building	eXtreme gradient boosting (XGBoost) model	The root-mean-square error and relative absolute error for this model were 0.294 and 0.153, respectively.
Heating	Luo et al. (2019)	Office building	k-means clustering and artificial neural network	Mean absolute errors are 3% and 8% for training and testing phases.
	Bünning, Heer, Smith, and Lygeros (2020)	A complex building with district heating	Autocorrelation forecast correction	Prediction accuracy can be ensured through the approach.
	Idowu, Saguna, Åhlund, and Schelén (2016)	Residential and commercial buildings	Different training algorithms	Support vector machine is the most accurate.

	Geysen et al. (2018)	Residential buildings	Linear regression, extremely randomized trees regression, feed-forward neural network, and support vector machine	The combination of data-driven methods performs better than individual method in thermal load prediction.
	Tan et al. (2020)	Industrial park	An integrated forecasting model	The hybrid model is more accurate and efficient in load prediction.
Electrical demand	Wang and Ahn (2020)	Residential community	Support vector machine (SVM), the k-nearest neighbors (k-NN) method, and the cross-entropy loss function	Better predictive and detective accuracy can be achieved by the proposed approach than previous methods.
	Chitalia, Pipattanasomporn, Garg, and Rahman (2020)	Commercial buildings	Deep recurrent neural networks	The deep learning can provide accurate predictions on electrical load.
	Skomski et al. (2020)	Commercial office buildings	Sequence-to-sequence recurrent neural network	Models perform best for three to 12 h of prior data.
	Rafati, Joorabian, and Mashhour, (2020)	–	Multilayer perceptron neural network with RReliefF for feature selection	The model is accurate with respect to different evaluation criteria.
	Alipour et al. (2020)	National electrical load	Hybrid deep neural networks and wavelet transform integration	The proposed forecasting model is accurate for electrical net-load prediction.

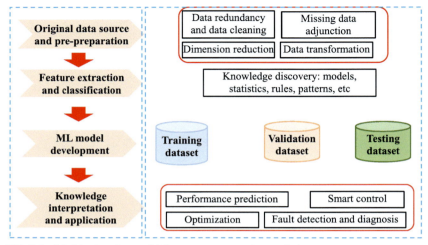

FIG. 8.1 Machine learning model for energy consumption prediction.

but with similar features. Unsupervised learning techniques mainly include k-means and hierarchical clustering methods.

Machine learning modeling techniques have been widely applied in building load predictions, in respect to three machine learning types, as listed in Table 8.2. In terms of supervised learning, Ahmad and Zhang (2020) developed a Gaussian Kernel regression model and nonparametric-based k-NN models to predict the demand of utility companies and an office building. The model is superior in prediction accuracy and performance stability. Zhou and Zheng (2020a) developed a surrogate model with supervised learning to forecast the electric demand of a high-rise office building. The developed data-driven model is accurate, with a normalized mean bias error (NMBE) less than 10%. The unsupervised learning is mainly applied for identification of energy patterns (Bourdeau et al., 2021) and spatiotemporal characteristics (Somu, Gauthama Raman, & Ramamritham, 2021). Furthermore, reinforcement learning is applied for HVAC system control (Azuatalam, Lee, de Nijs, & Liebman, 2020) and prediction accuracy improvement (Liu, Tan, Xu, Chen, & Li, 2020).

8.3 Flexible demand-side management strategies

Demand-side response and management are effective strategies to improve self-consumption of renewable energy and load coverage ratio, increase the reliability of power supply in response to intermittent renewable generation, and decrease import/export pressure on the power grid, without sacrificing indoor thermal comfort of building owners. Energy flexibility provided by demand-side response can meet the grid requirement in response to climate condition, stochastic demand, and intermittent renewable integration. Flexible

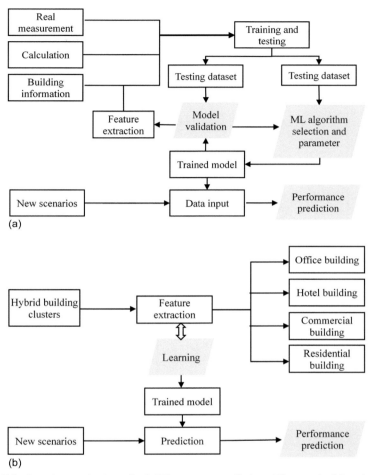

FIG. 8.2 Learning mechanisms for building energy prediction: (A) supervised learning and (B) unsupervised learning (Seyedzadeh & Pour Rahimian, 2021).

demand-side management strategies mainly include smart appliances, HVAC systems, and plug-in loads and storages. The adopted methodologies include demand shifting (Lund, Lindgren, Mikkola, & Salpakari, 2015), peak shaving and valley filling (Vanhoudt et al., 2014), load scheduling (Chellamani & Chandramani, 2019), optimal retail pricing (Yoon, Kim, Zakula, & Moon, 2020), and smart charging (Frendo et al., 2020).

Table 8.3 provides an overview on flexible demand-side management strategies in terms of types, methodology, machine learning approach, and results. In respect of smart appliances, Lund et al. (2015) adopted the demand shifting strategy to improve the penetration of renewable electricity in the multienergy

TABLE 8.2 Classification of machine learning modeling techniques in building load predictions.

Types	Studies	Building types	Algorithms	Results
Supervised learning	Ahmad and Zhang (2020)	Utility company and office building	Gaussian Kernel regression and nonparametric based k-NN	The algorithm is identified with high prediction accuracy and stability.
	Tran, Luong, and Chou (2020)	Residential buildings	Least squares support vector regression and the radial basis function neural network	The evolutionary neural machine inference model is more accurate in energy consumption prediction.
	Zhou and Zheng (2020a)	Office building	Cross-entropy function	The developed data-driven model is accurate with the NMBE less than 10%.
Unsupervised learning	Bourdeau et al. (2021)	Nonresidential buildings	k-means algorithm	Two-cluster classification with different patterns has been identified.
	Somu et al. (2021)		k-means clustering, convolutional neural networks-long short-term memory	Spatiotemporal characteristics in energy consumption can be accurately captured for accurate predictions.
Reinforcement learning (RL)	Liu et al. (2020)	Office building	Deep reinforcement learning	The algorithm can improve the prediction accuracy, with the improvement on mean absolute error by 16%–32%.
	Azuatalam et al. (2020)		Demand response-aware RL controller	RL for normal HVAC control will maximumly reduce energy by 22%, and the demand response-aware RL controller can further reduce the power by 50%.

TABLE 8.3 An overview on flexible demand-side management.

Types	Studies	Methodology	Machine learning	Results
Smart appliances	Chellamani and Chandramani (2019)	Demand shifting	–	Variable renewable electricity can be highly penetrated to provide system energy flexibility.
	Hulst et al. (2015)	Demand response flexibility	–	Smart appliances can improve the economic viability.
HVAC systems	Vijay and Hawkes (2019)	Demand-side flexibility for high-price grid electricity shifting	–	The annual cost can be reduced.
	Vanhoudt et al. (2014)	Peak shaving and renewable penetration	–	Active control will increase electricity consumption.
	Stavrakas and Flamos (2020)	Smart HVAC control	–	The control can increase the renewable penetration.
	Kim (2020)	Demand response of an HVAC system	Supervised learning-based strategy for optimal demand response	The demand response schedule is cost-effective without sacrificing indoor thermal comfort.
	Yoon et al. (2020)	Optimal retail pricing	Data-driven demand response	The pricing optimization can improve the operational flexibility of HAVC systems and ensure grid voltage stability and occupants' thermal comfort.

Continued

TABLE 8.3 An overview on flexible demand-side management—cont'd

Types	Studies	Methodology	Machine learning	Results
Plug-in loads and storages	Chellamani and Chandramani (2019)	Load scheduling	Supervised learning algorithm	Electricity bill and peak load demand can be reduced and consumer comfort can be improved.
	Remani, Jasmin, and Ahamed (2018)	Consumer comfort, stochastic renewable power and tariff	Reinforcement learning to solve the decision-making on residential load scheduling	Price-based demand response program can reduce the economic cost.
	Frendo et al. (2020)	Data-driven smart charging	XGBoost model	Smart charging supported by machine learning models can effectively exploit the infrastructure.

system for energy flexibility enhancement. Hulst et al. (2015) studied demand response flexibility to improve economic performance. In respect of HVAC systems, Stavrakas and Flamos (2020) studied smart HVAC controls for self-consumption enhancement through the reshaping of the load profile. Kim (2020) studied demand response of an HVAC system. Results showed that the demand response schedule is cost-effective without sacrificing indoor thermal comfort. Furthermore, plug-in vehicles, acting as plug-in loads and storages, have been studied in respect of load scheduling (Chellamani & Chandramani, 2019) and data-driven smart charging (Frendo et al., 2020). The obtained benefits include electricity bill savings and reduction in peak load.

8.3.1 Smart appliances

Building appliances and plug-in loads can provide flexibility to local power grids through load scheduling (Chellamani & Chandramani, 2019; Remani et al., 2018) and power shifting (Zhou & Cao, 2019). Chellamani and Chandramani (2019) proposed a supervised learning algorithm on schedulable appliances to autonomously overcome consumers' discomfort. Results show that electricity bills and peak load demand can be reduced, and consumer comfort can be improved. Remani et al. (2018) adopted the reinforcement learning to solve the decision making on residential load scheduling under renewable uncertainty. Postponed electric demand through smart appliances can provide demand shifting (Lund et al., 2015), peak power shaving (Vanhoudt et al., 2014), and reshaping of demand profiles with high matching capability with renewable profiles (Vijay & Hawkes, 2019).

8.3.2 HVAC systems

HVAC (heating, ventilation, and air conditioning) systems consume a large amount of energy in buildings, and the energy flexibility from HVAC systems is of great significance to dynamically balance energy demand and renewable generation. Stavrakas and Flamos (2020) studied energy flexibility provided by buildings (HVAC control setting, occupancy, and activity profiles) and economic benefits to consumers. The flexibility for self-consumption improvement will not change the already existing energy structure. Strategies for energy flexibility improvement from HAVC systems include temperature reset in the thermal zone (Aghniaey & Lawrence, 2018; Sehar, Pipattanasomporn, & Rahman, 2016), thermal mass storage via precooling/preheating (Kevin & Braun, 1997), and renewable energy recharging with controlled chiller water temperature (Zhou & Cao, 2020a). By increasing the temperature reset in cooling conditions and decreasing the temperature reset in heating conditions, building loads can be reduced. An increase of 2°C based on the normal thermostat setting can decrease the peak power by 25% and prolong it by approximately 20 min for continuous operation (Aduda, Labeodan, Zeiler, Boxem, & Zhao, 2016). The

248 Sustainable developments by artificial intelligence & machine learning

precooling/preheating strategies with thermal mass storage can promptly respond to building energy demands with the avoidance of excessive power production. Depending on the storage capacity of thermal inertia and charging/discharging rate, peak load shifting with different magnitudes can be seen (Turner, Walker, & Roux, 2015). Use of PCM (phase change material) wallboards can further reduce electrical energy consumption with enhanced peak load shifting (Qureshi, Nair, & Farid, 2011) due to enhanced energy storage density. Furthermore, the excess renewable-recharging strategy on cooling systems in subtropical regions can improve self-consumption and reduce grid exportation (Zhou & Cao, 2020a).

In terms of the application of artificial intelligence, Kim (2020) developed an optimal demand response for HVAC systems. The demand response schedule is cost-effective without sacrificing indoor thermal comfort. Yoon et al. (2020) studied data-driven demand response for the generation of optimal retail pricing. The pricing optimization can improve the operational flexibility of HAVC systems and ensure grid voltage stability and occupants' thermal comfort.

8.3.3 Plug-in loads and storages

Recently, the integration of plug-in electrical vehicles (PEVs) with buildings has attracted widespread interest worldwide (Barone, Buonomano, Calise, Forzano, & Palombo, 2019; Barone, Buonomano, Forzanoa, Giuzio, & Palombo, 2020; Rehman, Korvola, & Reda, 2020; Zhou, Cao, Hensen, & Lund, 2019). With extended battery storage capacity, PEVs can improve self-consumption with high renewable penetration (Zhou, Cao, et al., 2019), decrease grid reliance (Chen, de Rubens, Noel, Kester, & Sovacool, 2020), and shift both renewable and grid electricity (Zhou & Cao, 2019). Furthermore, batteries and plug loads (such as plug-in electric vehicles) can provide buffer and postponed capability for demand shifting from peak to off-peak periods. Frendo et al. (2020) developed a data-driven smart charging strategy with the XGBoost model. Smart charging supported by machine learning models can effectively exploit the infrastructure.

Fig. 8.3 demonstrates the power-based and hydrogen-based district energy sharing community. The mobility characteristic of PEVs and hydrogen vehicles (HVs) can activate the synergistic functions on building demands of different building types, and be spatiotemporally complementary on solar and wind renewable power supply. Depending on driving schedules, different magnitudes in enhancement of renewable penetration (Zhou, Cao, Kosonen, & Hamdy, 2020) can be realized. However, the battery cycling aging of PEVs (Zhou, Cao, Hensen, & Hasan, 2020) and fuel cell degradation (He et al., 2021) require development of a protective control strategy (Zhou, Cao, Hensen, et al., 2020) for operational and lifetime economic performance enhancement (Zhou & Cao, 2020b).

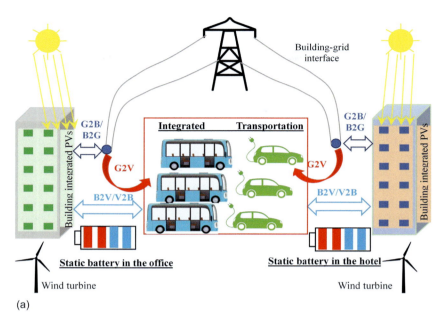

(a)

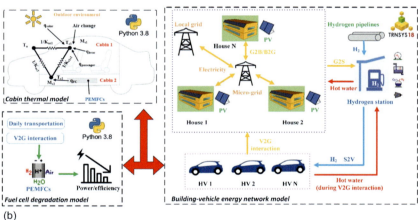

(b)

FIG. 8.3 Schematic diagrams on (A) plug-in electrical vehicles integrated neighborhood energy systems (Zhou, Cao, Kosonen, & Hamdy, 2020); (B) hydrogen-based district energy community (He et al., 2021). *(((A) Reproduced with permission from Zhou, Y., Cao, S., Kosonen, R., & Hamdy, M. (2020). Multi-objective optimisation of an interactive buildings-vehicles energy sharing network with high energy flexibility using the Pareto archive NSGA-II algorithm.* Energy Conversion and Management, *113017. https://doi.org/10.1016/j.enconman.2020.113017. Copyright (2020) with permission from Elsevier. (B) Reproduced with permission from He, Y., Zhou, Y., Wang, Z., Liu, J., Liu, Z., & Zhang, G. (2021). Quantification on fuel cell degradations and techno-economic analysis of a hydrogen-based carbon-neutral interactive residential sharing network with fuel cell powered vehicles.* Applied Energy, *303, 117444. Copyright (2021) with permission from Elsevier.))*

8.4 Machine learning-based advanced controllers

Energy management with artificial intelligence can promote the development of advanced controllers for renewable energy dispatch and flexible energy demands in multienergy systems (Ibrahim, Dong, & Yang, 2020). Fig. 8.4 shows the development procedure of advanced controllers for deterministic and stochastic scenarios. As shown in Fig. 8.4A, by adopting surrogate model training, validation, testing, and transfer learning, well-trained models can be applied to predict building energy demands and renewable generation under deterministic and stochastic uncertainty scenarios. Based on the predicted energy demands of nonlinear systems and renewable generation, both rule-based and model-predictive control strategies can be developed and applied in a multienergy system. Rule-based control strategies mainly include an excess renewable-recharging strategy, dynamic battery charging/discharging behavior, demand-side management, and grid-response strategy. Model-predictive

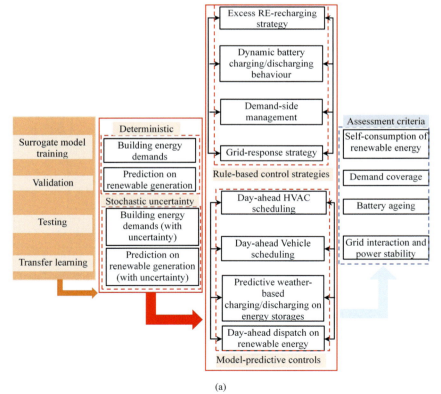

(a)

FIG. 8.4 Development procedure of advanced controllers for deterministic and stochastic scenarios: (A) a simplified diagram;

Machine learning-based hybrid demand-side controller **Chapter | 8** 251

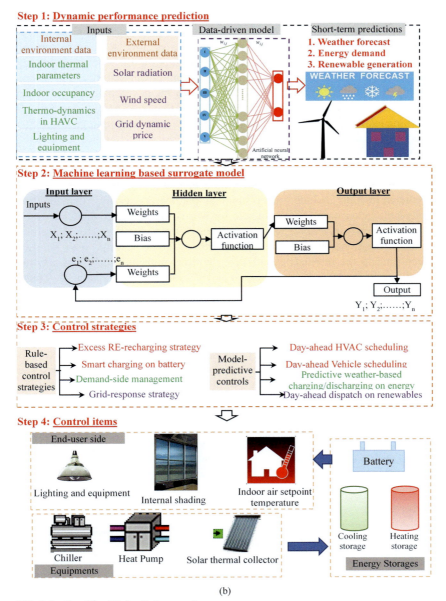

FIG. 8.4—Cont'd (B) detailed approaches.

controls mainly include day-ahead HVAC scheduling, day-ahead vehicle scheduling, predictive weather-based smart charging/discharging of energy storages, and day-ahead dispatch of renewable energy. Assessment criteria for effectiveness analysis include self-consumption of renewable energy, demand coverage, battery aging, grid interaction, power stability, and so on.

Fig. 8.4B demonstrates the detailed steps for the development of advanced controllers, including: Step 1, dynamic performance prediction; Step 2, machine learning-based surrogate model; Step 3, controller strategies; and Step 4, control objectives. In terms of the dynamic performance prediction in Step 1, input parameters include internal (e.g., indoor thermal parameters, indoor occupancy, thermodynamics in HAVC, lighting, and equipment) and external environment data (e.g., solar radiation, wind speed, and grid dynamic price). The predicted results include meteorological parameters, energy demand, and renewable generation. The main role for ML is the surrogate model development, as shown in Step 2, with three layers. Due to the diversity in learning rules, various learning performances with different accuracy can be seen. According to the research experience, the cross-entropy function is accurate for different nonlinear systems (such as PV system (Zhou, Zheng, & Zhang, 2020), high-insulated aerogel glazing system (Zhou & Zheng, 2020b), multienergy systems in the office (Zhou & Zheng, 2020a)), with the error-driven weighting update rule. The control strategies developed in Step 3 include rule-based and model-predictive control strategies. The control items, as shown in Step 4, include equipment (e.g., chillers, heat pumps, solar thermal collectors, and so on), energy storages (e.g., thermal and electrical storages), indoor air set point temperature, internal shading, lighting, and equipment.

Acknowledgment

This research is supported by The Hong Kong University of Science and Technology. The authors also thank the editors for their useful comments and suggestions.

References

Aduda, K. O., Labeodan, T., Zeiler, W., Boxem, G., & Zhao, Y. (2016). Demand side flexibility: Potentials and building performance implications. *Sustainable Cities and Society, 22*, 146–163.

Aghniaey, S., & Lawrence, T. M. (2018). The impact of increased cooling setpoint temperature during demand response events on occupant thermal comfort in commercial buildings: A review. *Energy and Buildings, 173*, 19–27.

Ahmad, T., & Zhang, H. (2020). Novel deep supervised ML models with feature selection approach for large-scale utilities and buildings short and medium-term load requirement forecasts. *Energy.* https://doi.org/10.1016/j.energy.2020.118477.

Alipour, M., Aghaei, J., Norouzi, M., Niknam, T., Hashemi, S., & Lehtonen, M. (2020). A novel electrical net-load forecasting model based on deep neural networks and wavelet transform integration. *Energy.* https://doi.org/10.1016/j.energy.2020.118106.

Aslam, S., Herodotou, H., Mohsin, S. M., Javaid, N., Ashraf, N., & Aslam, S. (2021). A survey on deep learning methods for power load and renewable energy forecasting in smart microgrids. *Renewable and Sustainable Energy Reviews*. https://doi.org/10.1016/j.rser.2021.110992.

Azuatalam, D., Lee, W. L., de Nijs, F., & Liebman, A. (2020). Reinforcement learning for whole-building HVAC control and demand response. *Energy and AI*. https://doi.org/10.1016/j.egyai.2020.100020.

Barone, G., Buonomano, A., Calise, F., Forzano, C., & Palombo, A. (2019). Building to vehicle to building concept toward a novel zero energy paradigm: Modelling and case studies. *Renewable and Sustainable Energy Reviews, 101*, 625–648.

Barone, G., Buonomano, A., Forzanoa, C., Giuzio, G. F., & Palombo, A. (2020). Increasing self-consumption of renewable energy through the building to vehicle to building approach applied to multiple users connected in a virtual micro-grid. *Renewable Energy*. https://doi.org/10.1016/j.renene.2020.05.101.

Bourdeau, M., Basset, P., Beauchêne, S., Silva, D. D., Guiot, T., Werner, D., et al. (2021). Classification of daily electric load profiles of non-residential buildings. *Energy and Buildings*. https://doi.org/10.1016/j.enbuild.2020.110670.

Bünning, F., Heer, P., Smith, R. S., & Lygeros, J. (2020). Improved day ahead heating demand forecasting by online correction methods. *Energy and Buildings*. https://doi.org/10.1016/j.enbuild.2020.109821.

Chellamani, G. K., & Chandramani, P. V. (2019). Demand response management system with discrete time window using supervised learning algorithm. *Cognitive Systems Research, 57*, 131–138.

Chen, C., de Rubens, G. Z., Noel, L., Kester, J., & Sovacool, B. K. (2020). Assessing the socio-demographic, technical, economic and behavioral factors of Nordic electric vehicle adoption and the influence of vehicle-to-grid preferences. *Renewable and Sustainable Energy Reviews*. https://doi.org/10.1016/j.rser.2019.109692.

Chitalia, G., Pipattanasomporn, M., Garg, V., & Rahman, S. (2020). Robust short-term electrical load forecasting framework for commercial buildings using deep recurrent neural networks. *Applied Energy*. https://doi.org/10.1016/j.apenergy.2020.115410.

Feng, Y., Duan, Q., Chen, X., Yakkali, S. S., & Wang, J. (2021). Space cooling energy usage prediction based on utility data for residential buildings using machine learning methods. *Applied Energy*. https://doi.org/10.1016/j.apenergy.2021.116814.

Frendo, O., Graf, J., Gaertner, N., & Stuckenschmidt, H. (2020). Data-driven smart charging for heterogeneous electric vehicle fleets. *Energy and AI*. https://doi.org/10.1016/j.egyai.2020.100007.

Geysen, D., De Somer, O., Johansson, C., Brage, J., & Vanhoudt, D. (2018). Operational thermal load forecasting in district heating networks using machine learning and expert advice. *Energy and Buildings, 162*, 144–153.

He, Y., Zhou, Y., Wang, Z., Liu, J., Liu, Z., & Zhang, G. (2021). Quantification on fuel cell degradations and techno-economic analysis of a hydrogen-based carbon-neutral interactive residential sharing network with fuel cell powered vehicles. *Applied Energy, 303*, 117444.

Hulst, R. D., Labeeuw, W., Beusen, B., Claessens, S., Deconinck, G., & Vanthournout, K. (2015). Demand response flexibility and flexibility potential of residential smart appliances: Experiences from large pilot test in Belgium. *Applied Energy, 155*, 79–90.

Ibrahim, M. S., Dong, W., & Yang, Q. (2020). Machine learning driven smart electric power systems: Current trends and new perspectives. *Applied Energy*. https://doi.org/10.1016/j.apenergy.2020.115237.

254 Sustainable developments by artificial intelligence & machine learning

Idowu, S., Saguna, S., Åhlund, C., & Schelén, O. (2016). Applied machine learning: Forecasting heat load in district heating system. *Energy and Buildings, 133*, 478–488.

Kevin, K., & Braun, J. E. (1997). Application of building precooling to reduce peak cooling requirements. *ASHRAE Transactions, 103*(1), 463–469.

Kim, Y. J. (2020). A supervised-learning-based strategy for optimal demand response of an HVAC system in a multi-zone office building. *IEEE Transactions on Smart Grid, 11*, 4212–4226.

Li, X., & Yao, R. (2021). Modelling heating and cooling energy demand for building stock using a hybrid approach. *Energy and Buildings.* https://doi.org/10.1016/j.enbuild.2021.110740.

Liu, T., Tan, Z., Xu, C., Chen, H., & Li, Z. (2020). Study on deep reinforcement learning techniques for building energy consumption forecasting. *Energy and Buildings.* https://doi.org/10.1016/j.enbuild.2019.109675.

Lund, P. D., Lindgren, J., Mikkola, J., & Salpakari, J. (2015). Review of energy system flexibility measures to enable high levels of variable renewable electricity. *Renewable and Sustainable Energy Reviews, 45*, 785–807.

Luo, X. J. (2020). A novel clustering-enhanced adaptive artificial neural network model for predicting day-ahead building cooling demand. *Journal of Building Engineering.* https://doi.org/10.1016/j.jobe.2020.101504.

Luo, X. J., Oyedele, L. O., Ajayi, A. O., Monyei, C. G., Akinade, O. O., & Akanbi, L. A. (2019). Development of an IoT-based big data platform for day-ahead prediction of building heating and cooling demands. *Advanced Engineering Informatics.* https://doi.org/10.1016/j.aei.2019.100926.

Ngo, N. T. (2019). Early predicting cooling loads for energy-efficient design in office buildings by machine learning. *Energy and Buildings, 182*, 264–273.

Qureshi, W. A., Nair, N. K. C., & Farid, M. M. (2011). Impact of energy storage in buildings on electricity demand side management. *Energy Conversion and Management, 52*, 2110–2120.

Rafati, A., Joorabian, M., & Mashhour, E. (2020). An efficient hour-ahead electrical load forecasting method based on innovative features. *Energy.* https://doi.org/10.1016/j.energy.2020.117511.

Rehman, H., Korvola, T., & Reda, F. (2020). Data analysis of a monitored building using machine learning and optimization of integrated photovoltaic panel, battery and electric vehicles in a Central European climatic condition. *Energy Conversion and Management.* https://doi.org/10.1016/j.enconman.2020.113206.

Remani, T., Jasmin, E. A., & Ahamed, T. P. I. (2018). Residential load scheduling with renewable generation in the smart grid: A reinforcement learning approach. *IEEE Systems Journal*, 3283–3294. https://doi.org/10.1109/JSYST.2018.2855689.

Sehar, F., Pipattanasomporn, M., & Rahman, S. (2016). An energy management model to study energy and peak power savings from PV and storage in demand responsive buildings. *Applied Energy, 173*, 406–417.

Seyedzadeh, S., & Pour Rahimian, F. (2021). Machine learning for building energy forecasting. In *Data-driven modelling of non-domestic buildings energy performance. Green energy and technology.* Cham: Springer. https://doi.org/10.1007/978-3-030-64751-3_4.

Skomski, E., Lee, J. Y., Kim, W., Chandan, V., Katipamula, S., & Hutchinson, B. (2020). Sequence-to-sequence neural networks for short-term electrical load forecasting in commercial office buildings. *Energy and Buildings.* https://doi.org/10.1016/j.enbuild.2020.110350.

Somu, N., Gauthama Raman, M. R., & Ramamritham, K. (2021). A deep learning framework for building energy consumption forecast. *Renewable and Sustainable Energy Reviews.* https://doi.org/10.1016/j.rser.2020.110591.

Stavrakas, V., & Flamos, A. (2020). A modular high-resolution demand-side management model to quantify benefits of demand-flexibility in the residential sector. *Energy Conversion and Management*. https://doi.org/10.1016/j.enconman.2019.112339.

Tan, Z., De, G., Li, M., Lin, H., Yang, S., Huang, L., et al. (2020). Combined electricity-heat-cooling-gas load forecasting model for integrated energy system based on multi-task learning and least square support vector machine. *Journal of Cleaner Production*. https://doi.org/10.1016/j.jclepro.2019.119252.

Tran, D. H., Luong, D. L., & Chou, J. S. (2020). Nature-inspired metaheuristic ensemble model for forecasting energy consumption in residential buildings. *Energy*. https://doi.org/10.1016/j.energy.2019.116552.

Turner, W. J. N., Walker, I. S., & Roux, J. (2015). Peak load reductions: Electric load shifting with mechanical pre-cooling of residential buildings with low thermal mass. *Energy*, *82*, 1057–1067.

Vanhoudt, D., Geysen, D., Claessens, B., Leemans, F., Jespers, L., & Bael, J. V. (2014). An actively controlled residential heat pump: Potential on peak shaving and maximization of self-consumption of renewable energy. *Renewable Energy*, *63*, 531–543.

Vijay, A., & Hawkes, A. (2019). Demand side flexibility from residential heating to absorb surplus renewables in low carbon futures. *Renewable Energy*, *138*, 598–609.

Wang, X., & Ahn, S. H. (2020). Real-time prediction and anomaly detection of electrical load in a residential community. *Applied Energy*. https://doi.org/10.1016/j.apenergy.2019.114145.

Yoon, A. Y., Kim, Y. J., Zakula, T., & Moon, S. I. (2020). Retail electricity pricing via online-learning of data-driven demand response of HVAC systems. *Applied Energy*. https://doi.org/10.1016/j.apenergy.2020.114771.

Zhou, Y., & Cao, S. (2019). Energy flexibility investigation of advanced grid-responsive energy control strategies with the static battery and electric vehicles: A case study of a high-rise office building in Hong Kong. *Energy Conversion and Management*. https://doi.org/10.1016/j.enconman.2019.111888.

Zhou, Y., & Cao, S. (2020a). Quantification of energy flexibility of residential net-zero-energy buildings involved with dynamic operations of hybrid energy storages and diversified energy conversion strategies. *Sustainable Energy, Grids and Networks*. https://doi.org/10.1016/j.segan.2020.100304.

Zhou, Y., & Cao, S. (2020b). Coordinated multi-criteria framework for cycling aging-based battery storage management strategies for positive building–vehicle system with renewable depreciation: Life-cycle based techno-economic feasibility study. *Energy Conversion and Management*. https://doi.org/10.1016/j.enconman.2020.113473.

Zhou, Y., Cao, S., Hensen, J. L. M., & Hasan, A. (2020). Heuristic battery-protective strategy for energy management of an interactive renewables–buildings–vehicles energy sharing network with high energy flexibility. *Energy Conversion and Management*. https://doi.org/10.1016/j.enconman.2020.112891.

Zhou, Y., Cao, S., Hensen, J. L. M., & Lund, P. D. (2019). Energy integration and interaction between buildings and vehicles: A state-of-the-art review. *Renewable and Sustainable Energy Reviews*. https://doi.org/10.1016/j.rser.2019.109337.

Zhou, Y., Cao, S., Kosonen, R., & Hamdy, M. (2020). Multi-objective optimisation of an interactive buildings-vehicles energy sharing network with high energy flexibility using the Pareto archive NSGA-II algorithm. *Energy Conversion and Management*. https://doi.org/10.1016/j.enconman.2020.113017.

Zhou, Y., & Zheng, S. (2020a). Machine-learning based hybrid demand-side controller for high-rise office buildings with high energy flexibilities. *Applied Energy*. https://doi.org/10.1016/j.apenergy.2019.114416.

Zhou, Y., & Zheng, S. (2020b). Stochastic uncertainty-based optimisation on an aerogel glazing building in China using supervised learning surrogate model and a heuristic optimisation algorithm. *Renewable Energy, 155*, 810–826.

Zhou, Y., Zheng, S., & Zhang, G. (2020). Machine-learning based study on the on-site renewable electrical performance of an optimal hybrid PCMs integrated renewable system with high-level parameters' uncertainties. *Renewable Energy, 151*, 403–418.

Zhou, X., Zi, X., Liang, L., Fan, Z., Yan, J., & Pan, D. (2019). Forecasting performance comparison of two hybrid machine learning models for cooling load of a large-scale commercial building. *Journal of Building Engineering, 21*, 64–73.

Chapter 9

A machine learning-based design approach on PCMs-PV systems with multilevel scenario uncertainty

Yuekuan Zhou[a,b]

[a]*Sustainable Energy and Environment Thrust, Function Hub, The Hong Kong University of Science and Technology, Guangzhou, China,* [b]*Department of Mechanical and Aerospace Engineering, The Hong Kong University of Science and Technology, Clear Water Bay, Hong Kong SAR, China*

9.1 Introduction

Due to the energy shortage crisis and daily intensifying global warming, resulting from reliance on traditional fossil fuels, resorting to renewable energy systems for cleaner power production is one of most promising solutions. Solar energy—coming from a nuclear fusion reaction with the lighter nuclei polymerized into the heavier nuclei with a huge amount of released energy—becomes one of the most attractive renewable energy sources, due to the large abundance in quantity without environmentally polluting byproducts. As the main system for solar energy utilization, solar cells have been rapidly developed, with the solar-to-power conversion efficiency at around 15% under standard testing conditions. The relatively low power efficiency somewhat hinders the techno-economic attractiveness of solar cells for renewable energy utilization.

Solar cell cooling to mitigate the extreme temperatures at midday has attracted widespread interest, as the power efficiency is inversely correlated with the solar cell temperature. Researchers have concentrated on passive, active, and combined passive/active strategies for solar cell cooling (Zhou, Zheng, Liu, et al., 2020). Passive cooling is relatively simple and cost-saving without requirement for additional energy inputs; however, the cooling potential is relatively limited. By contrast, the cooling potential of active cooling strategies is much greater, but initial investment cost, associated facility

---

Sustainable Developments by Artificial Intelligence and Machine Learning for Renewable Energies.
https://doi.org/10.1016/B978-0-323-91228-0.00010-0
Copyright © 2022 Elsevier Inc. All rights reserved.

maintenance cost, and additional energy inputs are required (Tang, Zhou, Zheng, & Zhang, 2020). In recent years, combined strategies have been investigated, and smart operation has been found to be highly dependent on the meteorological parameters (Zhou, Zheng, & Zhang, 2019a, 2019b), such as ambient temperature, solar radiation, and so on.

Recently, artificial intelligence has experienced rapid development, and the applications of advanced machine learning techniques in solar PV systems have been studied (Zhou, Zheng, & Zhang, 2019c) in terms of thermodynamic behavior prediction and power performance estimation, with straightforward mathematical associations between multivariant and outputs. Compared to physics-based models, the machine learning (ML)-based data-driven models are much simpler in modeling and much more computationally efficient for the capture of thermodynamics of nonlinear systems (Zhou, Zheng, & Zhang, 2020a). After being well trained, tested, and validated, the surrogate model is integrated in optimization algorithms to address techno-economic contradictions, with optimal design and operating parameters. Compared to traditional optimization approaches with the calling back to mathematical models for each thread, the ML-based optimization approach is more computationally efficient, whereas the reliability of the optimal results might be lower, due to the ignorance of heat transfer principles and mechanisms in ML models (Zhou & Zheng, 2020a).

Considering the uncertainties of multivariants in the real operational conditions, uncertainty analysis on multivariants is necessary. The uncertainty-based optimization can improve system reliability and robustness to address performance failure (Zhou & Zheng, 2020b) resulting from the lack of inherent randomness of the performance behavior and lack of knowledge for the appropriate parameter values. Compared to deterministic optimization, the most challenging issue of the uncertainty-based optimization is the considerable computational load resulting from the stochastic nature of uncertainty for each variable. The ML-based surrogate models provide the possibility for uncertainty-based optimization with computationally efficient calculations for uncertainty-based performance prediction.

In this chapter, an overview on solar cell cooling techniques has been reviewed, including passive, active, and combined passive/active strategies. The mechanism of machine learning for performance prediction of nonlinear systems is studied in terms of training, validation, and testing processes. Following this, the prospects of machine learning applications in PCMs-PV systems have been presented, including surrogate model development for thermal and electrical performance prediction, single-objective, multiobjective, and robust optimizations with multilevel scenario uncertainties. Last but not least, challenges and outlooks are presented, including the uncertainty quantification and probability density function, stochastic sampling size and uncertainty-based optimization function, hybrid learning and advanced optimization algorithms, and multicriteria decision-marking for trade-off solutions.

9.2 Overview on PCMs-PV systems and operations

Due to the negative relationship between solar cell temperature and PV efficiency, PV cooling techniques have attracted widespread interest. Phase change materials (PCMs) with considerable latent heat within a small fusion temperature range are a promising option to absorb the heat from solar cells and mitigate extreme temperatures. Convective, conductive, and radiative heat transfer mechanisms are the most common techniques for PV cooling (Kandeal et al., 2020). Depending on structural design and the adopted cooling techniques, PCMs-PV cooling systems can be mainly classified into passive, active and hybrid cooling systems. Table 9.1 lists an overview on passive, active, and hybrid PCMs-PV systems, including both rooftop PVs and façade-integrated PVs. Passive PCMs-PV cooling systems mainly include PCM copper with foam matrix, heat sinks, nanoparticles, and natural ventilation. Active PCMs-PV cooling systems mainly include forced ventilation-based BIPV/PCM (building-integrated photovoltaics/phase change materials), water-PCMs-PVT (photovoltaic thermal), and nanofluid-PCMs-PVT systems. In addition, integrated passive/active systems have been proposed, designed, constructed, and studied in the literature.

9.2.1 Passive PCMs-PV systems

The temperature regulation of solar cells is the most effective strategy for PV efficiency improvement. PCMs are promising candidates due to high energy density and thermal and chemical stability (Browne, Norton, & McCormack, 2015). In the literature, passive cooling techniques mainly include PCM copper with foam matrix (Abdulmunem et al., 2020), heat sinks (Rabie et al., 2019), nanoparticles (Nada, El-Nagar, & Hussein, 2018), and natural ventilation (Gan & Xiang, 2020). Fig. 9.1 shows a schematic diagram of a finned double-pass double glass-covered concentrated photovoltaic-thermal (PVTC) (Elsafi & Gandhidasan, 2015) and a ventilated PCMs-PV system (Gan and Xiang, 2020). Abdulmunem et al. (2020) experimentally studied passive PV cooling with PCM copper and multiwalled carbon nanotubes. Results showed that the proposed passive cooling strategy can reduce the cell's temperature by 13.29% and improve electricity generation by 5.68%. Arıcı et al. (2018) conducted a case study on the integrated PCM location, achieving a peak cell temperature of 10.26°C, together with efficiency improvement of up to 3.73%. Hernandez-Perez, Carrillo, Bassam, Flota-Banuelos, and Patino-Lopez (2020) designed a heat sink to dissipate the solar cell temperature. A maximum drop in cell temperature of 10°C was seen during the period with peak irradiance. Rabie et al. (2019) designed phase change material heat sinks to improve the power performance of a PCMs-PV system. Nada et al. (2018) added nanoparticles to PCM-Integrated PV modules for thermal regulation and efficiency enhancement. The PCM and PCM/nanoparticles can improve the efficiency by

TABLE 9.1 A holistic overview on passive, active, and hybrid PCMs-PV systems.

Systems	Studies	Integrated forms	Design parameters	Applications	Conclusions
Passive systems	Abdulmunem, Samin, Rahman, Hussien, and Mazali (2020)	PCM copper with foam matrix and multiwalled carbon nanotubes.	Phase change material/copper foam matrix without multiwalled carbon nanotubes	Building integrated PVs	The proposed passive cooling strategy can reduce the cell temperature by 13.29% and improve electricity generation by 5.68%.
	Arıcı, Bilgin, Nižetić, and Papadopoulos (2018)	PCMs-PV	Integrated PCM location	Rooftop	Peak cell temperature can be reduced by 10.26°C, together with efficiency improvement of up to 3.73%.
	Rabie, Emam, Ookawara, and Ahmed (2019)	PCMs-PV	Inclination angle and concentration ratio	Rooftop	The peak cell temperature drops from 92°C to 74°C.
Active systems	Zhou et al. (2019a)	Water-based PCMs-PV system	Inlet water temperature and mass flow rate	Rooftop	PV efficiency can be improved with the active cooling.
	Hosseinzadeh, Sardarabadi, and Passandideh-Fard (2018)	Nanofluid-based PCMs-PV/T system	Conventional PV module and nanofluid PVT/PCM system	Rooftop	Nanofluid-based PVT/PCM system can improve the output overall exergy and exergy efficiency.
	Kant, Pitchumani, Shukla, and Sharma (2019)	Air ventilation-based BIPV/PCM system	Geometrical and operating parameters	Building façade	Optimal parameters can be identified to maximize the PV generation.
Integrated passive/active systems	Zhou, Zheng, and Zhang (2019c)	Hybrid ventilations + radiative cooling wall + ventilated roof	PCM melting temperature, PCM thickness, and cooling water	Building façade and roof	The photovoltaic efficiency was increased from 11.5% to 14.99%.
	Nasef, Nada, and Hassan (2019)	Interactive passive and active cooling	Concentration ratio and heat transfer velocity	Concentrated photovoltaic solar cells	Cell temperature is reduced and system efficiency is improved.

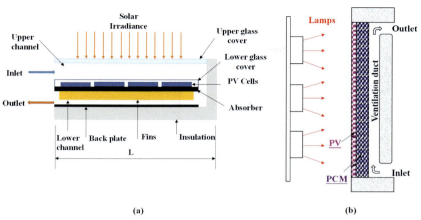

FIG. 9.1 Schematic diagram of: (A) finned double-pass double glass-covered PVTC (Elsafi and Gandhidasan, 2015); (B) ventilated PCMs-PV system (Gan and Xiang, 2020). *((A)*Reproduced with permission from *Elsafi, A. M., & Gandhidasan, P. (2015). Comparative study of double-pass flat and compound parabolic concentrated photovoltaic–thermal systems with and without fins.* Energy Conversion and Management, *98, 59–68. Copyright (2015) with permission from Elsevier. (B)* Reproduced with permission from *Gan, G., & Xiang, Y. (2020). Experimental investigation of a photovoltaic thermal collector with energy storage for power generation, building heating and natural ventilation.* Renewable Energy, *150, 12–22. Copyright (2020) with permission from Elsevier.)*

5.7% and 13.2%, respectively. Gan and Xiang (2020) designed a naturally ventilated PCMs-PV system for power generation and building heating applications. Results showed that the photovoltaic efficiency can be increased by 10%.

9.2.2 Active PCMs-PV systems

In the literature, depending on the media of coolants, active PCMs-PV cooling techniques include air-based (Kant et al., 2019), water-based (Hassan et al., 2020), and nanofluid-based cooling (Hosseinzadeh et al., 2018). Fig. 9.2 shows active PV cooling solutions for hydro water-cooled PV (Kabeel, Abdelgaied, & Sathyamurthy, 2019) and nanofluid (Abdallah et al., 2018; Hosseinzadeh et al., 2018). Kant et al. (2019) parametrically studied an air ventilative BIPV/PCM system. Parametrical analysis indicates that the optimum values of PCM thickness, BIPV height, air gap between BIPV/PCM and wall, and air mass flow rate are 0.04m, 3m, 0.02m, and 0.18 kg/s, respectively. Hassan et al. (2020) conducted an experimental study on traditional PV/T, water-PCM PVT, and nanofluid-PCM PVT systems. The nanofluid-PCM PVT system showed the highest overall efficiency, which was 12% higher than the water-PCM PVT system, and 23.9% higher than conventional PV systems. Hosseinzadeh et al. (2018) studied energy and exergy performances of a nanofluid-based PCMs-PV/T system. The comparison with PVT and nanofluid-PV systems indicated

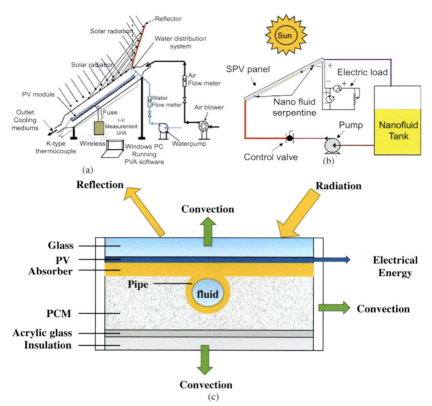

FIG. 9.2 Active PV cooling solutions: (A) hydro water-cooled PV (Kabeel et al., 2019), (B) nanofluid (Abdallah et al., 2018), (C) heat transfer mechanism. *((A) Reproduced with permission from Kabeel, A. E., Abdelgaied, M., & Sathyamurthy, R. (2019). A comprehensive investigation of the optimization cooling technique for improving the performance of PV module with reflectors under Egyptian conditions. Solar Energy, 186, 257–263. Copyright (2019) with permission from Elsevier. (B) Reproduced with permission from Abdallah, S. R., Elsemary, I. M., Altohamy, A. A., Abdelrahman, M., Attia, A. A., & Abdellatif, O. E. (2018). Experimental investigation on the effect of using nano fluid (Al_2O_3-water) on the performance of PV/T system. Thermal Science and Engineering Progress, 7, 1–7. Copyright (2018) with permission from Elsevier.)*

that the nanofluid-based PVT/PCM system showed the maximum output overall exergy and efficiency of 114.99 W/m^2 and 13.61%.

9.2.3 Combined passive/active PCMs-PV systems

Fig. 9.3 demonstrates a hybrid PCMs-PV system integrated with active cooling and hybrid ventilation. Passive strategies include natural ventilation for heat dissipation, conduction between PCM and solar cell, and high-reflective/high-absorptive coating in the curtain to reflect/absorb solar radiation in summer/winter. Active strategies include active water-based PV cooling, mechanical

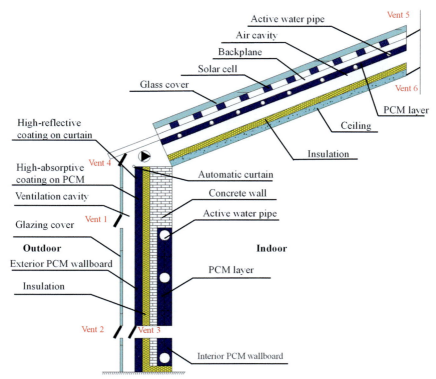

FIG. 9.3 The structural constitution of a ventilative system (Zhou, Zheng, Liu, et al., 2020). *(Reproduced with permission from Zhou, Y., Zheng, S., Liu, Z., Wen, T., Ding, Z., Yan, J., et al. (2020). Passive and active phase change materials integrated building energy systems with advanced machine-learning based climate-adaptive designs, intelligent operations, uncertainty-based analysis and optimisations: A state-of-the-art review.* Renewable and Sustainable Energy Reviews, *109889. https://doi.org/10.1016/j.rser.2020.109889. Copyright (2020) with permission from Elsevier.)*

ventilation, and indoor radiative cooling. Operational strategies include natural ventilation, mechanical ventilation, and insulation modes (Zhou et al., 2019c).

9.3 Mechanism for machine learning on performance prediction of nonlinear systems

The underlying mechanism of supervised machine learning method for performance prediction of nonlinear systems is to set up the mathematical relationship between multiple inputs and outputs, without professional knowledge in specific fields. By training with different learning algorithms, the dynamic update of weight factors aims to minimize the predicted errors.

Fig. 9.4 demonstrates the development of a building energy consumption forecast model, including original data source and preparation, feature

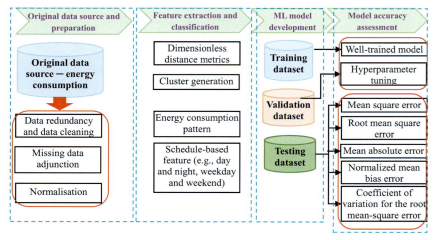

FIG. 9.4 Building energy consumption forecast model.

extraction and classification, ML model development, and model accuracy assessment. The stage of original data source and preparation includes data redundancy and data cleaning, missing data adjunction, and data normalization. Feature extraction and classification include dimensionless distance metrics, cluster generation, energy consumption pattern, and schedule-based feature. The original dataset is classified into training, validation, and testing datasets for trained model, hyperparameter tuning, and model accuracy assessment, respectively. The accuracy assessment criteria with statistic indicators are included.

The underlying mechanism for mathematical training of ML is shown in Fig. 9.5. The input layer consists of inputs and difference (i.e., error) between the actual data and the model's output. Weighting factors are updated in the hidden layer, in accordance with the quantified errors, to minimize the difference between the actual data and predicted results. Through the iterative calculation and dynamic update on weighting factors, a well-trained data-driven model can be developed. The training accuracy and efficiency are highly dependent on the adopted training algorithms. As shown in Fig. 9.5, there are four training algorithms to quantify the error and to update the weights in the hidden layer.

9.4 Application of machine learning in PCMs-PV systems

9.4.1 Surrogate model for performance prediction

Compared to physics-based models, the ML model is much simpler for modeling development and more accurate and efficient for dynamic performance prediction of nonlinear systems. Al-Waeli, Kazem, Yousif, Chaichan, and Sopian (2020) trained an ANN (artificial neural network) model to predict the thermal

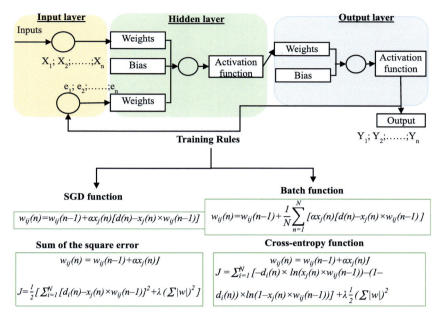

FIG. 9.5 Principle and mechanism for ML training.

and power performance of a nano-PCM PVT system. The model shows accurate prediction performance with high prediction efficiency. In addition, hybrid models with integrated optimization algorithms can improve the prediction accuracy. Alnaqi, Moayedi, Shahsavar, & Nguyen (2019) concluded that the hybrid model shows more accurate prediction performance than conventional models.

Fig. 9.6 demonstrates the process of surrogate model development. The input parameters include meteorological, thermo-physical, design, and operating parameters. With the multiplication of intermediate weight matrix, the prediction results can be generated. Afterwards, the intermediate weight matrix will be updated, following the principle to decrease the error between predicted and real results. The dynamic update of intermediate weight matrix does not end up until the tracked error within the requirement. Through the data-driven model for dynamic performance prediction, both prediction accuracy and efficiency can be ensured (Zhou & Zheng, 2020c).

9.4.2 System optimization

Due to the straightforward mathematical association between inputs and outputs with updated weighting factors, machine learning can accurately and

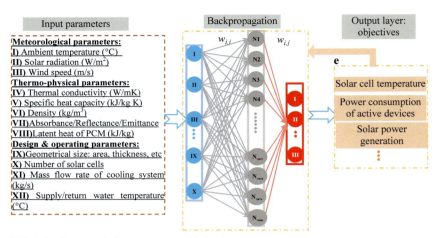

FIG. 9.6 An example for surrogate model development process.

efficiently predict system performance through back-propagation for minimizing errors between prediction and real data. In the literature, researchers are mainly focused on machine learning-based single- and multiobjective optimizations.

9.4.2.1 Single-objective optimization

Due to accurate and efficient performance prediction, machine learning-trained models were developed for parametrical analysis, so as to assist the geometrical design of PV systems. Shahsavar, Moayedi, Al-Waeli, Sopian, and Chelvanathan (2020) developed machine learning predictive models for parametrical analysis. Comparative analysis between different training approaches indicates that the random forest outperforms other algorithms in terms of performance evaluation criteria. Furthermore, the integration of machine learning-based surrogate models in advanced optimization algorithms has great prospects, due to the full utilization of prediction efficiency and accuracy of machine learning models, and the excellent identification of the optimal solution from optimization algorithms. Zhou, Zheng, and Zhang (2020b) optimized the design and operating parameters of an active PCMs-PV system to maximize the annual equivalent overall output energy in five different climate zones in China. The proposed approach is superior to the Taguchi standard orthogonal array, and the machine learning approach is worthy of being popularized. Considering the power consumption of active devices, Tang et al. (2020) conducted an exergy-based optimization on an active PCMs-PV system and improved the overall exergy by 2.6%, to 872.06 kWh.

9.4.2.2 Multiobjective optimization of the hybrid renewable system using the Pareto NSGA-II

In addition to single-objective optimization, multiobjective optimization with machine learning techniques' integration is also promising. The advantages of an ML-based multiobjective optimization approach include multiobjective performance prediction with high efficiency and accuracy and excellent identification of Pareto optimal solutions from the Pareto archive (nondominated sorting genetic algorithm) NSGA-II algorithm. Perera, Wickramasinghe, Nik, and Scartezzini (2019) adopted supervised learning to develop the surrogate model to be integrated into a hybrid optimization algorithm for multiobjective optimization. The proposed technique can reduce computational time by 84%. Zhou and Zheng (2020a) developed a novel multiobjective optimization approach, integrating multiobjective surrogate models and NSGA-II. The proposed approach can be an effective tool for trade-off solutions along the Pareto front, with high efficiency.

9.4.3 Robust optimization with multilevel scenario uncertainty

Considering the uncertainties of multivariables in the real operational conditions, uncertainty analysis on multivariables is necessary. Uncertainty-based optimization can improve system reliability and robustness to address performance overestimation or underestimation. Zhou, Zheng, & Zhang (2020c) studied the uncertainty analysis on a deterministic-based optimal PCMs-PV system with a supervised learning trained surrogate model. Results showed that the scenario uncertainty can improve the peak power of the PCMs-PV system from 11.5 to 20 kW. However, due to thousands of scenario cases with uncertainty from the stochastic sampling, the uncertainty-based optimization is challenging with the traditional mathematical model, especially considering the sophisticated heat transfer, dynamic iteration, and convergence on latent PCM storage. The ML techniques make it possible to address the computational complexity through the straightforward mathematical association between inputs and outputs. Zhou and Zheng (2020d) developed a novel uncertainty-based optimization method with supervised learning surrogate model, which can improve the peak power and renewable generation to 25 kW and 2340 kWh.

Fig. 9.7A demonstrates the general steps for multilevel uncertainty-based optimization, consisting of data-driven models for performance prediction, optimization function, optimization algorithms, and results analysis. The cutting-edge techniques include the ML-based performance prediction with multilevel uncertainties (as shown in Step 1, Fig. 9.7B), and flexible integration of optimization function (as shown in Step 2, Fig. 9.7B), in a metaheuristic optimization algorithm. The novel approach can be applied to other systems for the highly efficient utilization of renewable energy in a reliable and robust way.

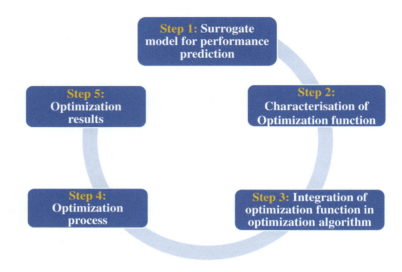

FIG. 9.7 (A) Roadmap for multi-level uncertainty-based optimization (Zhou and Zheng, 2020d); *(Continued)*

9.5 Challenges and outlooks

9.5.1 Uncertainty quantification and probability density function

Due to the complexity of uncertainty sources, especially for different types on different parameters, achieving accurate models for uncertainty quantifications of each parameter will be challenging. The combination of different probability density functions might be an effective solution for quantification of frequency and uncertainty magnitudes (Zhou, Zheng, & Zhang, 2020a). Furthermore, forward uncertainty quantification (such as Monte Carlo sampling-based simulation, variations of Monte-Carlo methods, nonsampling, and nonprobabilistic uncertainty propagation) and inverse uncertainty quantification (such as frequentist and Bayesian techniques) are worthy of being investigated (Tian et al., 2018).

9.5.2 Stochastic sampling size and uncertainty-based optimization function

In the ML-based uncertainty-based optimization approach, one of the critical steps is the optimization function. The optimization function is dependent on the stochastic uncertainty sampling size and the ML model (such as structural

PCMs-PV systems with multilevel scenario uncertainty Chapter | 9 **269**

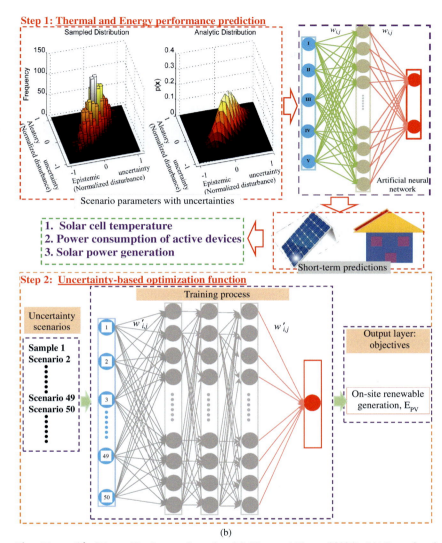

Fig. 17, cont'd (B) specifications on Steps 1 and 2 (Zhou and Zheng, 2020d). *((A) Reproduced with permission from Zhou, Y., & Zheng, S. (2020). Multi-level uncertainty optimisation on phase change materials integrated renewable systems with hybrid ventilations and active cooling. Energy, 117747. https://doi.org/10.1016/j.energy.2020.117747. Copyright (2020) with permission from Elsevier. (B) Partially reproduced with permission from Zhou, Y., & Zheng, S. (2020). Multi-level uncertainty optimisation on phase change materials integrated renewable systems with hybrid ventilations and active cooling.* Energy, *117747. https://doi.org/10.1016/j.energy.2020.117747. Copyright (2020) with permission from Elsevier.)*

configuration of neural network, learning algorithm, learning rate, and so on). Through the parameter adjustment and algorithm comparison processes, the prediction accuracy can be guaranteed. However, the stochastic sampling size is a big challenge for database preparation. Comparative and sensitivity analysis on stochastic sampling size is necessary to give guidance to technicians for uncertainty-based optimization.

9.5.3 Hybrid learning and advanced optimization algorithms

Due to the specific application limitation of each ML algorithm, the complicated prediction problem might need a hybrid combination of learning algorithms, with complementary functions of each algorithm, to achieve accurate predictions. Furthermore, the development of advanced optimization algorithms is necessary to generate optimal solutions.

9.5.4 Multicriteria decision-marking for trade-off solutions

In respect to Pareto front solutions, the multiobjective optimization, multicriteria decision-making approaches need to be explored to identify the "best of the best" optimal solution from all nondominated optimal solutions. The challenge is that the most optimal solution is dependent on objective-oriented preference and adopted approach. In the literature, multicriteria decision-making approaches mainly include the Shannon entropy, Euclidean distance-based, and fuzzy membership function methods. Future studies need to select the "best of the best" optimal solution with the incorporation of preferences of stakeholders on each objective.

Acknowledgment

This research is supported by The Hong Kong University of Science and Technology. The authors also thank the editors for their useful comments and suggestions.

References

Abdallah, S. R., Elsemary, I. M., Altohamy, A. A., Abdelrahman, M., Attia, A. A., & Abdellatif, O. E. (2018). Experimental investigation on the effect of using nano fluid (Al_2O_3-water) on the performance of PV/T system. *Thermal Science and Engineering Progress, 7*, 1–7.

Abdulmunem, A. R., Samin, P. M., Rahman, H. A., Hussien, H. A., & Mazali, I. I. (2020). Enhancing PV cell's electrical efficiency using phase change material with copper foam matrix and multi-walled carbon nanotubes as passive cooling method. *Renewable Energy, 160*, 663–675.

Alnaqi, A. A., Moayedi, H., Shahsavar, A., & Nguyen, T. K. (2019). Prediction of energetic performance of a building integrated photovoltaic/thermal system thorough artificial neural network and hybrid particle swarm optimization models. *Energy Conversion and Management, 183*, 137–148.

Al-Waeli, A. H. A., Kazem, H. A., Yousif, J. H., Chaichan, M. T., & Sopian, K. (2020). Mathematical and neural network modeling for predicting and analyzing of nanofluid-nano PCM photovoltaic thermal systems performance. *Renewable Energy, 145*, 963–980.

Arıcı, M., Bilgin, F., Nižetić, S., & Papadopoulos, A. M. (2018). Phase change material based cooling of photovoltaic panel: A simplified numerical model for the optimization of the phase change material layer and general economic evaluation. *Journal of Cleaner Production, 189*, 738–745.

Browne, M. C., Norton, B., & McCormack, S. J. (2015). Phase change materials for photovoltaic thermal management. *Renewable and Sustainable Energy Reviews, 47*, 762–782.

Elsafi, A. M., & Gandhidasan, P. (2015). Comparative study of double-pass flat and compound parabolic concentrated photovoltaic–thermal systems with and without fins. *Energy Conversion and Management, 98*, 59–68.

Gan, G., & Xiang, Y. (2020). Experimental investigation of a photovoltaic thermal collector with energy storage for power generation, building heating and natural ventilation. *Renewable Energy, 150*, 12–22.

Hassan, A., Wahab, A., Qasim, M. A., Janjua, M. M., Ali, M. A., Ali, H. M., et al. (2020). Thermal management and uniform temperature regulation of photovoltaic modules using hybrid phase change materials-nanofluids system. *Renewable Energy, 145*, 282–293.

Hernandez-Perez, J. G., Carrillo, J. G., Bassam, A., Flota-Banuelos, M., & Patino-Lopez, L. D. (2020). A new passive PV heatsink design to reduce efficiency losses: A computational and experimental evaluation. *Renewable Energy, 147*, 1209–1220.

Hosseinzadeh, M., Sardarabadi, M., & Passandideh-Fard, M. (2018). Energy and exergy analysis of nanofluid based photovoltaic thermal system integrated with phase change material. *Energy, 147*, 636–647.

Kabeel, A. E., Abdelgaied, M., & Sathyamurthy, R. (2019). A comprehensive investigation of the optimization cooling technique for improving the performance of PV module with reflectors under Egyptian conditions. *Solar Energy, 186*, 257–263.

Kandeal, A. W., Thakur, A. K., Elkadeem, M. R., Elmorshedy, M. F., Ullah, Z., Sathyamurthy, R., et al. (2020). Photovoltaics performance improvement using different cooling methodologies: A state-of-art review. *Journal of Cleaner Production.* https://doi.org/10.1016/j.jclepro.2020.122772.

Kant, K., Pitchumani, R., Shukla, A., & Sharma, A. (2019). Analysis and design of air ventilated building integrated photovoltaic (BIPV) system incorporating phase change materials. *Energy Conversion and Management, 196*, 149–164.

Nada, S. A., El-Nagar, D. H., & Hussein, H. M. S. (2018). Improving the thermal regulation and efficiency enhancement of PCM-integrated PV modules using nano particles. *Energy Conversion and Management, 166*, 735–743.

Nasef, H. A., Nada, S. A., & Hassan, H. (2019). Integrative passive and active cooling system using PCM and nanofluid for thermal regulation of concentrated photovoltaic solar cells. *Energy Conversion and Management.* https://doi.org/10.1016/j.enconman.2019.112065.

Perera, A. T. D., Wickramasinghe, P. U., Nik, V. M., & Scartezzini, J. L. (2019). Machine learning methods to assist energy system optimization. *Applied Energy, 243*, 191–205.

Rabie, R., Emam, M., Ookawara, S., & Ahmed, M. (2019). Thermal management of concentrator photovoltaic systems using new configurations of phase change material heat sinks. *Solar Energy, 183*, 632–652.

Shahsavar, A., Moayedi, H., Al-Waeli, A. H., Sopian, K., & Chelvanathan, P. (2020). Machine learning predictive models for optimal design of building-integrated photovoltaic-thermal collectors. *International Journal of Energy Research.* https://doi.org/10.1002/er.5323.

Tang, L., Zhou, Y., Zheng, S., & Zhang, G. (2020). Exergy-based optimisation of a phase change materials integrated hybrid renewable system for active cooling applications using supervised machine learning method. *Solar Energy*, *195*, 514–526.

Tian, W., Heo, Y., Wilde, P., Li, Z., Yan, D., Park, C. S., et al. (2018). A review of uncertainty analysis in building energy assessment. *Renewable and Sustainable Energy Reviews*, *93*, 285–301.

Zhou, Y., & Zheng, S. (2020a). Machine learning-based multi-objective optimisation of an aerogel glazing system using NSGA-II—Study of modelling and application in the subtropical climate Hong Kong. *Journal of Cleaner Production*. https://doi.org/10.1016/j.jclepro.2020.119964.

Zhou, Y., & Zheng, S. (2020b). Stochastic uncertainty-based optimisation on an aerogel glazing building in China using supervised learning surrogate model and a heuristic optimisation algorithm. *Renewable Energy*. https://doi.org/10.1016/j.renene.2020.03.122.

Zhou, Y., & Zheng, S. (2020c). Machine-learning based hybrid demand-side controller for high-rise office buildings with high energy flexibilities. *Applied Energy*. https://doi.org/10.1016/j.apenergy.2019.114416.

Zhou, Y., & Zheng, S. (2020d). Multi-level uncertainty optimisation on phase change materials integrated renewable systems with hybrid ventilations and active cooling. *Energy*. https://doi.org/10.1016/j.energy.2020.117747.

Zhou, Y., Zheng, S., Liu, Z., Wen, T., Ding, Z., Yan, J., et al. (2020). Passive and active phase change materials integrated building energy systems with advanced machine-learning based climate-adaptive designs, intelligent operations, uncertainty-based analysis and optimisations: A state-of-the-art review. *Renewable and Sustainable Energy Reviews*. https://doi.org/10.1016/j.rser.2020.109889.

Zhou, Y., Zheng, S., & Zhang, G. (2019a). Study on the energy performance enhancement of a new PCMs integrated hybrid system with the active cooling and hybrid ventilations. *Energy*, *179*, 111–128.

Zhou, Y., Zheng, S., & Zhang, G. (2019b). Multivariable optimisation of a new PCMs integrated hybrid renewable system with active cooling and hybrid ventilations. *Journal of Building Engineering*, *26*, 100845. https://doi.org/10.1016/j.jobe.2019.100845.

Zhou, Y., Zheng, S., & Zhang, G. (2019c). Artificial neural network based multivariable optimisation of a hybrid system integrated with phase change materials, active cooling and hybrid ventilations. *Energy Conversion and Management*. https://doi.org/10.1016/j.enconman.2019.111859.

Zhou, Y., Zheng, S., & Zhang, G. (2020a). A state-of-the-art-review on phase change materials integrated cooling systems for deterministic parametrical analysis, stochastic uncertainty-based design, single and multi-objective optimisations with machine learning applications. *Energy and Buildings*. https://doi.org/10.1016/j.enbuild.2020.110013.

Zhou, Y., Zheng, S., & Zhang, G. (2020b). Machine learning-based optimal design of a phase change material integrated renewable system with on-site PV, radiative cooling and hybrid ventilations—study of modelling and application in five climatic regions. *Energy*. https://doi.org/10.1016/j.energy.2019.116608.

Zhou, Y., Zheng, S., & Zhang, G. (2020c). Machine-learning based study on the on-site renewable electrical performance of an optimal hybrid PCMs integrated renewable system with high-level parameters' uncertainties. *Renewable Energy*, *151*, 403–418.

Chapter 10

Agent-based peer-to-peer energy trading between prosumers and consumers with cost-benefit business models

Yuekuan Zhou[a,b] and Jia Liu[c]

[a]*Sustainable Energy and Environment Thrust, Function Hub, The Hong Kong University of Science and Technology, Guangzhou, China,* [b]*Department of Mechanical and Aerospace Engineering, The Hong Kong University of Science and Technology, Clear Water Bay, Hong Kong SAR, China,* [c]*Department of Building Environment and Energy Engineering, Faculty of Construction and Environment, Hong Kong Polytechnic University, Kowloon, Hong Kong, China*

10.1 Introduction

Due to the uneven spatiotemporal distribution of solar and wind energy resources, the intermittence in renewable energy will lead to power outages and frequent fluctuations of local power grids. Effective strategies to mitigate the grid power pressure or burden include demand-side management, power-to-X conversions, energy storages, etc. The underlying principle for demand-side management is reshaping the demand power profile with high correspondence with the renewable power curve. The underlying mechanisms for power-to-X conversions and energy storages are based on surplus energy management. Due to the nonlinearity of energy demands with multiple integrated units (such as building appliances, lighting and HVAC systems, stochastic occupants' behaviors, and so on), demand-side management strategies can improve the renewable penetration with energy flexibility provided by end-users. Meanwhile, the power-to-X conversions and energy storages will lead to power loss, the decrease in energy quality, and the increase in associated investment cost. Peer-to-peer (P2P) energy sharing and trading provide new approaches to improve renewable energy penetration, with considerations on diversity of demand and renewable generation for each building.

P2P energy trading can be achieved with different energy forms, such as thermal, electrical, and hydrogen energy forms. The underlying mechanism for P2P

Sustainable Developments by Artificial Intelligence and Machine Learning for Renewable Energies.
https://doi.org/10.1016/B978-0-323-91228-0.00011-2
Copyright © 2022 Elsevier Inc. All rights reserved.

273

energy trading is that, due to the spatiotemporal diversity, the demand shortage of neighborhood buildings is covered with onsite renewable energy in a geographically close building. Therefore, the import/export pressure on energy grids will be significantly reduced. Internal pricing, as the most critical driving factor on dynamic energy trading behavior, can be determined by the supply to demand ratio (SDR), midmarket rate (MMR) and bill sharing (BS) methods. However, the impact of multivariants on internal pricing has rarely been studied, such as diversity of surplus renewable energy and demand shortage of each building, energy trading priority, or preference of each building owner, and the internal price-driven energy trading behavior in the electricity market.

The advances in artificial intelligence and internet of things (IoT) lead to the applications of blockchain and machine learning technologies in P2P energy trading. Artificial intelligence, as an efficient technology for market prediction, has been applied by researchers for trading price prediction (Chen, Lin, & Song, 2019), energy sharing decision making (Chen & Su, 2018; Zhou, Hu, Gu, Jiang, & Zhang, 2019), near-optimal pricing strategy searching (Chen & Bu, 2019; Xu, Xu, et al., 2020), optimal energy trading policy making (Xu, Yu, Bi, & Zhang, 2020), and so on.

In this chapter, agent-based P2P energy trading with dynamic internal pricing has been presented, in terms of various energy trading forms, underlying mechanisms, and mathematical models for dynamic internal pricing. Applications and prospects of blockchain and machine learning technologies in P2P energy trading have been reviewed, to show the recent progresses and advances. In order to promote the end-users' participation, decentralized electricity market design and techno-economic incentives are reviewed, so as to improve the social acceptance and widespread popularity. Last but not least, challenges and outlooks are presented, including the identification on individual trading price and dynamic price-based trading strategy, initiatives of internal pricing with considerations on diversity of surplus renewable energy and demand shortage of each building, energy trading priority or preference of each building owner, the internal price-driven energy trading behavior in the electricity market, and cost-benefit allocation schemes with fair distributions, to enable the participation motivations of each building owner.

10.2 Agent-based peer-to-peer energy trading with dynamic internal pricing

10.2.1 P2P energy trading modes with different energy forms

Depending on the energy trading forms, P2P energy trading mainly includes thermal, electrical, and hydrogen energy systems. Table 10.1 summarizes the P2P energy trading in thermal and electrical energy forms. Most studies are focused on electricity in the trading market. Davoudi et al. (2021) proposed a P2P thermal energy transaction framework. By determining the optimal

Agent-based peer-to-peer energy trading **Chapter | 10 275**

TABLE 10.1 P2P energy trading with different energy forms.

Energy forms	Studies	Approaches	Results
Thermal energy	Davoudi, Moeini-Aghtaie, and Ghorani (2021)	A P2P platform for thermal energy transaction	By determining the optimal participation strategy, benefits and applicability can be obtained from the proposed framework
Electrical energy	Sousa et al. (2019)	Consumer-centric markets	Strategies to address the prosumers' imbalance
	Wang, Taha, Wang, Kvaternik, and Hahn (2019)	Blockchain-enabled energy trade in smart grids	Seamless energy trading can be realized
	Cui, Wang, and Xiao (2019)	Total social energy cost minimization in the first stage and a noncooperative game for mutual energy sharing	The approach can promote energy-efficient buildings in a computationally efficient and sustainable regional building cluster
	Zhang, Li, and Li (2020)	Energy trading and uncertainty trading	Compared to separate trading, joint trading on energy and uncertainty can improve the balanced PV forecast error from 43.6% to 55.3%
	Tushar, Saha, Yuen, Smith, and Poor (2020)	Key features of P2P transactions and benefit analysis for grid and prosumers	P2P trading is promising, but challenges need to be addressed
	Morstyn, Farrell, Darby, and McCulloch (2018)	Federated power plant with P2P transactions	Prosumers are incentivized to selforganize into coalitions with grid services
Hydrogen energy	Liu et al. (2021)	Individual trading price and dynamic price-based trading strategy	The dynamic price-based trading strategy is cost-saving

participation strategy, benefits and applicability can be obtained from the proposed framework. In addition to thermal energy, researchers are mainly focused on P2P energy trading on distributed renewable power from solar PVs. Zhang et al. (2020) developed combined joint trading on energy and uncertainty to improve the balanced PV forecast error from 43.6% to 55.3%. Morstyn et al. (2018) designed a federated power plant with P2P transactions to motivate prosumers to selforganize into coalitions with grid services. Furthermore, with the wide deployment of hydrogen-fueled vehicles, Liu, Yang, and Zhou (2021) studied P2P energy trading with identification on individual trading price and dynamic price-based trading strategy.

Fig. 10.1 demonstrates an overview framework of a renewable-hydrogen-based energy community with P2P energy trading. Liu et al. (2021) proposed an individual trading price, a time-of-use grid penalty cost model, and a dynamic price-based trading strategy to study the economic potential. By adopting the proposed techniques, the net grid import energy, annual electricity cost, and grid penalty cost can be reduced by 8.93%, 14.54%, and 142.87%, respectively.

10.2.2 Mechanisms and mathematical models for dynamic internal pricing

10.2.2.1 Theoretical mechanisms

In the literature, popular mechanisms for the identification of trading price mainly include supply-to-demand ratio mechanism, MMR mechanism, and BS mechanism. Long et al. (2017a) quantitatively studied three market paradigms, i.e., the BS, MMR, and an auction-based pricing strategy, to disclose P2P trading mechanisms in terms of local energy exchange prices and individual customers' energy costs. Zhou, Wu, and Long (2018) concluded that the SDR method outperformed other approaches.

P2P trading mechanisms have been applied for different purposes. In order to improve the renewable penetration in neighboring PV prosumers, a dynamic internal pricing model was proposed by Liu et al. (2017). Meinke, Sun, and Jiang (2020) proposed a P2P trading method based on an internal pricing model with supply-demand ratio, which can improve selfconsumption and reduce peak demand hours. In order to guarantee that all P2P participants can gain economic benefits, P2P energy sharing (Long, Wu, Zhou, & Jenkins, 2018) can decrease the energy cost by 30% and the electricity bill by 12.4% for prosumers.

In order to realize fair benefits for both producers and consumers, the midmarket mechanism was adopted. Tushar et al. (2018) adopted the MMR for P2P trading. Results showed that the stability of the coalition and the benefit to the prosumers can be ensured under the MMR.

In addition, Long et al. (2017b) adopted the BS mechanism for P2P energy sharing among district consumers. Long et al. (2017a) validated the effectiveness of BS strategy on a P2P energy sharing scheme of a residential community

Agent-based peer-to-peer energy trading **Chapter | 10 277**

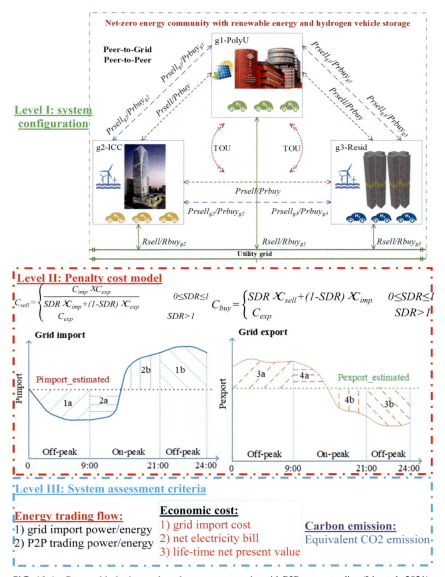

FIG. 10.1 Renewable-hydrogen-based energy community with P2P energy trading (Liu et al., 2021).

microgrid, in terms of local energy exchange prices and individual customers' energy costs. Zhou et al. (2018) comparatively studied the BS mechanism with other energy sharing mechanisms, and concluded that the BS mechanism is almost the same as the conventional paradigm, whereas the SDR mechanism outperforms others.

278 Sustainable developments by artificial intelligence & machine learning

10.2.2.2 Mathematical models

In the literature, the annual SDR was widely adopted, whereas deficiencies of the approach include the lack of diversity of surplus renewable energy and demand shortage of each building, energy trading priority or preference of each building owner, and the internal price-driven energy trading behavior in the electricity market. In order to overcome these deficiencies, several mathematical models are listed below.

After directly covering the associated building energy demand, the onsite renewable energy was managed for P2P energy sharing. From the perspective of the entire community, the annual SDR was calculated (Liu et al., 2017) by Eq. (10.1):

$$SDR = \frac{\sum_{i=1}^{n} \int_{0}^{t_{end}} (P_{RE,i,t} - P_{demand,i,t}) dt}{\sum_{i=1}^{n} \int_{0}^{t_{end}} (P_{demand,i,t} - P_{RE,i,t}) dt} \tag{10.1}$$

Furthermore, considering the diversity of surplus renewable energy and demand shortage of each building, the annual supply to demand ratio of the ith building (SDR_i) can be calculated by Eq. (10.2):

$$SDR_i = \frac{\int_{0}^{t_{end}} (P_{RE,i,t} - P_{demand,i,t} - P_{P2P,sell,i,t}) dt}{\int_{0}^{t_{end}} (P_{demand,i,t} - P_{RE,i,t} - P_{P2P,buy,i,t}) dt} \tag{10.2}$$

In the electricity market with internal price-driven energy trading behavior, the onsite renewable generation might not be directly used to cover associated building demand. In this situation, the dynamic supply to demand ratio ($SDR_{i,t}$) is calculated by the dynamic supply power of the ith building ($P_{RE,i,t}$) and the dynamic demand of the ith building ($P_{demand,i,t}$). The mathematical Eq. (10.3) can be adopted:

$$SDR_{i,t} = \frac{P_{RE,i,t}}{P_{demand,i,t}} \tag{10.3}$$

With consideration of the dynamic trading power, the dynamic supply to demand ratio ($SDR_{i,t}$) can be updated by Eq. (10.4):

$$SDR_{i,t} = \frac{P_{RE,i,t} - P_{P2P,sell,i,t}}{P_{demand,i,t} - P_{P2P,buy,i,t}} \tag{10.4}$$

The sell and buy prices for each peer are dynamically dependent on the SDR (Liu et al., 2017), as shown in Eqs. (10.5) and (10.6):

$$C_{sell} = \begin{cases} \dfrac{C_{imp} \times C_{exp}}{SDR \times C_{imp} + (1 - SDR) \times C_{exp}} & 0 \leq SDR \leq 1 \\ C_{exp} & SDR > 1 \end{cases} \tag{10.5}$$

$$C_{buy} = \begin{cases} SDR \times C_{sell} + (1 - SDR) \times C_{imp} & 0 \leq SDR \leq 1 \\ C_{exp} & SDR > 1 \end{cases} \tag{10.6}$$

where C_{exp} and C_{imp} refer to grid export cost (0.2 HK\$/kWh) and import cost from the retailer (Parag & Sovacool, 2016).

As the price is dynamically dependent on the SDR, the selling and purchasing prices are completely different between different peers. According to the auction theory, in which the sellers raise higher revenues and enable buyers to procure at a lower cost, the success of energy trading and dynamic power trading should be based on the following constraints:

$$\begin{cases} \text{Constraint } 1 : C_{sell,A} \leq C_{buy,B} \\ \text{Constraint } 2 : C_{buy,B} \geq C_{buy,C} \geq C_{buy,D} \end{cases} \tag{10.7}$$

where A, B, C, and D refer to different participators in the P2P sharing platform.

Based on the sell price for the peer A, $C_{sell,A}$, and the buy price for the peer B, $C_{buy,B}$, the transaction price can be reached based on negotiation after adopting the proposed aggregate auction mechanism. In this study, the transaction price (C_{A2B}) after the proposed aggregate auction mechanism for market equilibrium is determined based on the SDR:

$$C_{A2B} = \begin{cases} \dfrac{C_{sell,A} \times C_{buy,B}}{SDR_A \times \left(C_{buy,B} - C_{sell,A} \right) + C_{sell,A}} & 0 \leq SDR \leq 1 \\ C_{sell,A} & SDR > 1 \end{cases} \tag{10.8}$$

It is noteworthy that in the scenario with dynamic supply to demand ratio ($SDR_{i,t}$), the internal sell/buy price is dependent on dynamic power trading, which is contrarily dependent on the internal sell/buy price, as shown in Eq. (10.8). The mutual influence between the dynamic power trading and the internal sell/buy price poses challenges for the dynamic iteration to reach equilibrium. However, the current literature provides limited progress toward advanced techniques for dynamic equilibrium and dynamic sell/buy price. The existence of Nash equilibriums is highly dependent on the P2P energy sharing mechanism, in that, each participator follows the bidding trading strategy when simulation converges, to maximize the benefit. Otherwise, the trading strategy follows the last-defense mechanism. Paudel, Chaudhari, Long, and Gooi (2019) adopted an M-leader and N-follower Stackelberg game approach for modeling the dynamic interaction between buyers and sellers to reach the equilibrium state. The evolutionary game theory was effective for providing benefits to the community. Kang et al. (2017) adopted the iterative double auction mechanism to address electricity pricing and the amount of traded electricity. Results showed that the double auction mechanism can maximize social welfare with privacy protection.

Based on the research experience of the authors, the uniform peer energy trading price model based on the total SDR of the community is shown in

280 Sustainable developments by artificial intelligence & machine learning

Fig. 10.2A, with the inverse-proportional relationship between price and SDR (Liu et al., 2017). The *SDR* is formulated by Eq. (10.9) as shown below.

$$SDR = \frac{\sum P_{REgi\_sur}}{\sum P_{Loadgi\_shor}} \quad (10.9)$$

The P2P energy selling price (*Prsell*) in the net-zero energy community can be formulated as the piecewise function of *SDR* as shown in Eq. (10.10) (Liu et al., 2017):

$$Prsell = f(SDR) = \begin{cases} \dfrac{Rsell \cdot Rbuy_{g3}}{\left(Rbuy_{g3} - Rsell\right) \cdot SDR + Rsell}, & 0 \leq SDR \leq 1 \\ Rsell, & 1 < SDR \end{cases} \quad (10.10)$$

The buy-in price (*Prbuy*) is also the piecewise function of *SDR* as per Eq. (10.11) (Liu et al., 2017):

$$Prbuy = f(SDR) = \begin{cases} Prsell \cdot SDR + Rbuy_{g3} \cdot (1 - SDR), & 0 \leq SDR \leq 1 \\ Rsell, & 1 < SDR \end{cases}$$
$$(10.11)$$

The surplus ratio (*SR*) for each user is quantified by Eq. (10.12):

$$SR_{gi\_sur} = \frac{P_{REgi\_sur}}{P_{REgi}} \quad (10.12)$$

The demand ratio (*DR*) for each user is quantified by Eq. (10.13):

$$DR_{gi\_shor} = \frac{P_{Loadgi\_shor}}{P_{Loadgi}} \quad (10.13)$$

The selling price and buy-in price of each user can be quantified by Eqs. (10.14) and (10.15).

$$Prsell_{gi} = f\left(SR_{gi\_sur}\right) = \frac{Rsell \cdot Rbuy_{gi}}{\left(Rbuy_{gi} - Rsell\right) \cdot SR_{gi\_sur} + Rsell} \quad (10.14)$$

$$Prbuy_{gi} = f\left(DR_{gi\_shor}\right) = \left(Rbuy_{gi} - Rsell\right) \cdot DR_{gi\_shor} + Rsell \quad (10.15)$$

For the definition of terminologies in equations, the readers are recommended to refer for our previous publication (Liu et al., 2021) for more details. Several critical challenges need to be studied to promote energy trading between multiple peers:

(1) Instead of using constant internal trading price based on annual SDR, the time-variant internal trading price, calculated by dynamic supply to demand ratio ($SDR_{i,t}$) of each peer, needs to be implemented in the P2P

Agent-based peer-to-peer energy trading Chapter | 10 **281**

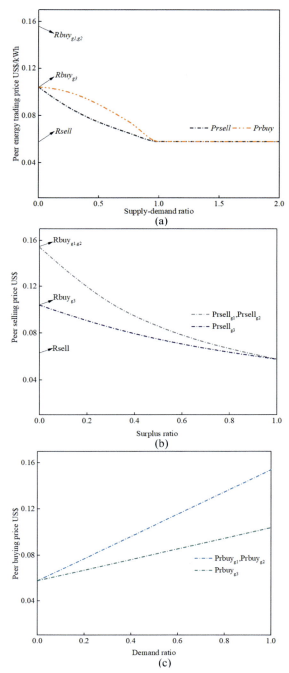

FIG. 10.2 P2P trading price. (A) A uniform model; (B) a user-dependent model (Liu et al., 2021).

energy trading platform, which can provide more flexibility and resilience to both prosumers and consumers.

(2) Referring to the quantity theory of money (*Quantity theory of money*, n.d.), dynamic power trading is inversely proportional to the internal sell/buy price. In the microgrid-based P2P market with a diversity of selling prices among prosumers and buying prices among consumers, the energy trading priority between prosumers and consumers is full of challenge, especially considering the power surplus/shortage level, different types of prosumer, the diverse preferences, and so on. From the price perspective, the mechanism of aggregate auction in the stock market can be applied to guide the P2P energy trading behavior with transacted trading price and maximum trading volume.

(3) An automatic P2P energy trading platform needs to be developed, following the aggregate auction principle, to promote the P2P energy trading among participators. However, difficulties include the capture of prosumers' diverse preferences within a platform.

10.3 Blockchain and machine learning technologies in P2P energy trading

10.3.1 Blockchain in P2P energy trading

As an open and distributed ledger that can efficiently record transactions between two peers (Iansiti & Lakhani, 2017), blockchain technology has been widely applied in energy trading. Esmat, de Vos, Ghiassi-Farrokhfal, Palensky, and Epema (2020) proposed a novel decentralized platform, as shown in Fig. 10.3, to keep the balance between economy and information privacy. In the market layer, prosumers can trade various market products. In the blockchain layer, market results are securely stored, and cash trading is realized only when the energy is delivered following a smart contract.

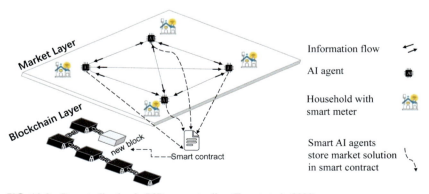

FIG. 10.3 Decentralization in P2P energy trading (Esmat et al., 2020).

Diestelmeier (2019) studied the policy implications in the EU electricity market for prosumers' integration with blockchain technology. Studies on blockchain technology in P2P energy trading mainly focus on the blockchain-based microgrid energy market (Brousmiche, Menegazzi, Boudeville, & Fantino, 2020; Mengelkamp et al., 2018), blockchain-based energy trading platform (Hua, Jiang, Sun, & Wu, 2020), replacement of central intermediaries for information transmission, and undesired inherent delay on energy transactions (Okoye et al., 2020). In terms of the blockchain-based microgrid energy market, Mengelkamp et al. (2018) designed a blockchain-based market for automatic information transmission. Socioeconomic incentives require market design to effectively allocate local energy generation. Brousmiche et al. (2020) applied blockchain and vehicle-to-grid interaction to support the P2P energy market. The adopted technologies can promote energy trading between households, improve flexibility, and decrease energy dependence. However, several challenges can be seen. For application in P2P energy trading platforms, Peck and Wagman (2017) developed an energy trading platform for energy sharing and trading of neighbors' rooftop solar power based on blockchain. Results indicated that net metering and feed-in tariff programs are critical elements in determining trading behaviors of peers. Hua et al. (2020) proposed a blockchain-based P2P trading framework to promote the regional energy balance and carbon saving. By reshaping the energy trading behaviors of prosumers through the adjustment of bidding/selling prices, the proposed technique in the decentralized framework can reduce daily carbon emissions. Okoye et al. (2020) developed a blockchain-enhanced transaction model to address the undesired inherent delay on energy transactions. Results showed that the cyber-enhanced transactive microgrid model can improve the transaction speed and provide great convenience. Andoni et al. (2019) reviewed challenges and opportunities of blockchain technology in distributed P2P energy trading systems. Through the case study from P2P energy trading and IoT applications to decentralized marketplaces, the main functions of blockchain includes transparent, tamper-proof, and secure systems. However, the undesired inherent delay will directly lead to failure in coverage of urgent demand.

10.3.2 Machine learning technologies in P2P energy trading

In the automatic P2P energy trading platform, market prediction is one of the most effective economic incentives for dynamic operations and trading, with maximum economic benefits for all participants. Artificial intelligence, as an efficient technology for market prediction, has been applied by researchers for trading price prediction (Chen et al., 2019), energy sharing decision making (Chen & Su, 2018; Zhou et al., 2019), near-optimal pricing strategy searching (Chen & Bu, 2019; Xu, Xu, et al., 2020), optimal energy trading policy making (Xu, Yu, et al., 2020), and so on.

In order to predict trading price in the electricity market, Chen et al. (2019) developed a prediction-integration strategy optimization (PISO) model to optimize prosumers' operations and trading strategies. Results indicated that, compared to continuous double auction (CDA) market, the PISO can further improve economic profit, with improved integration of flexible resources into electricity markets. Chen and Su (2018) adopted deep reinforcement learning (DRL) technology (a deep Q-learning) to address the decision-making process of local market participation. With the adoption of deep Q-learning, prosumers' willingness for participation in the localized energy ecosystem can be improved with fast-responsive decision-making on energy trading. In terms of near-optimal pricing strategy searching, Xu, Xu, et al. (2020) developed data-driven game-based pricing based on stochastic rooftop PV power prediction. By following the leader-followers Stackelberg game, Q-learning-based decision-making can dynamically search for near-optimal pricing strategies. Chen and Bu (2019) adopted a DRL algorithm to identify the optimal strategy for distributed energy trading. Results showed that the P2P energy trading model was applicable for real applications. From the perspective of optimal energy trading policy making, Xu, Yu, et al. (2020) concluded that the proposed trading policy can effectively minimize the energy cost of each microgrid and ensure data transaction security.

Fig. 10.4 demonstrates the structural configuration of an extreme learning machine (ELM) for market prediction. The inputs include the bidding price (λ_t) and bidding quantity in trading cycle t (P_t^{CDA}), respectively. The market prediction model (F_t) can predict the participant's selling profit or purchasing cost. The ELM-based market prediction model can assist participants to intelligently join the P2P energy market with quantified profits or costs.

10.4 Electricity market and techno-economic incentives for P2P energy market

In order to improve the widespread acceptance of P2P energy trading among building owners and promote P2P energy sharing in the market, development

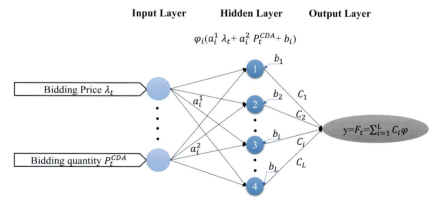

FIG. 10.4 Structural configuration of extreme learning machine (ELM)-based market prediction models (Chen et al., 2019).

and construction of the electricity market and techno-economic incentives are essential. Hahnel, Herberz, Pena-Bello, Parra, and Brosch (2020) studied the participation willingness of homeowners in the P2P energy community, and concluded that electricity prices and state of charge of battery are two critical factors to determine the homeowners' energy trading behavior.

10.4.1 Decentralized electricity market design

Compared to the incentives for local trading, the electricity market design is more critical. Guerrero, Gebbran, Mhanna, Chapman, and Verbič (2020) systematically reviewed transactive energy systems with P2P energy trading in the decentralized energy markets. The increase in distributed renewable integration will necessitate the consideration of network constraints. Studies on decentralized electricity market design are mainly focused on layer constitution of prosumer markets (Parag & Sovacool, 2016), flexibility provided by combined trade and storage (Lüth, Zepter, del Granado, & Egging, 2018), advanced algorithms (Khorasany, Mishra, & Ledwich, 2020), and social welfare-based on a multibilateral economic dispatch and consumers' preferences (Sorin, Bobo, & Pinson, 2019). Parag and Sovacool (2016) proposed three prosumer markets for the transition toward sustainability and carbon-neutrality. Advanced market design calls for combined efforts. Lüth et al. (2018) studied the optimal local electricity market design with flexibility provided by central battery storages. They concluded that the approach can provide cost savings of 31%. Khorasany et al. (2020) designed a decentralized bilateral energy trading system. Results showed that the market players can maximize their welfare via energy trading following line flow constraints. Furthermore, the proposed technique can effectively reduce the data exchange, and shows a faster convergence. Sorin et al. (2019) designed a distributed P2P market structure based on a multibilateral economic dispatch. Different from centralized market structures, the decentralized market can generate more social welfare with considerations of consumers' preferences.

10.4.2 Techno-economic incentives

With the decline in regulatory tariff-based incentives, P2P energy sharing is essential to promote the deployment of renewable systems. Nguyen, Peng, Sokolowski, Alahakoon, and Yu (2018) studied the energy saving of each household in a local community under P2P energy sharing. Results showed that the maximal cost saving at 28% can be realized on weekdays. Tushar et al. (2019) developed a game-theoretic P2P energy trading scheme to encourage homeowners to form a collaborative coalition. Compared to the traditional feed-in-tariff scheme, the scheme can reduce carbon emissions and economic cost.

Furthermore, techno-economic incentives can further improve the popularity and economic competitiveness of P2P energy sharing. Cali and Çakir (2019) adopted an alternative incentive mechanism (i.e., the fixed stipend and the decaying stipend incentive) with distributed ledger technology to promote the P2P local energy markets. Results showed that the proposed energy policy instruction can effectively combine feed-in tariff and flexibility grid access to accelerate the widespread acceptance of decentralized local market structures. In addition to subsidized flat feed-in tariffs, incentives can include economic and behavioral motivations. Zepter, Lüth, Del Granado, and Egging (2019) studied the synergistic operation of P2P energy trading and residential storage to reduce electricity bills. Results showed that each strategy can reduce electricity bills by 20%–30%.

10.5 Challenges and outlook

Future studies can be focused on:

(1) development of individual energy trading price models and time-of-use trading management strategy;
(2) initiatives of internal pricing with considerations on diversity of surplus renewable energy and demand shortage of each building, energy trading priority or preference of each building owner, and the internal price-driven energy trading behavior in the electricity market; and.
(3) cost-benefit allocation schemes with fair distributions, to enable the participation motivations of each building owner.

Acknowledgment

This research is supported by The Hong Kong University of Science and Technology and The Hong Kong Polytechnic University. The authors also thank the editors for their useful comments and suggestions. All copyright licenses have been successfully applied for all cited graphics, images, tables, and/or figures.

References

Andoni, M., Robu, V., Flynn, D., Abram, S., Geach, D., Jenkins, D., et al. (2019). Blockchain technology in the energy sector: A systematic review of challenges and opportunities. *Renewable and Sustainable Energy Reviews*, *100*, 143–174.

Brousmiche, K., Menegazzi, P., Boudeville, O., & Fantino, E. (2020). Peer-to-peer energy market place powered by blockchain and vehicle-to-grid technology. In *2nd Conference on blockchain research & applications for innovative networks and services (BRAINS)*. https://doi.org/10.1109/BRAINS49436.2020.9223276.

Cali, U., & Çakir, O. (2019). Energy policy instruments for distributed ledger technology empowered peer-to-peer local energy markets. *IEEE Access*, 82888–82900.

Chen, T., & Bu, S. (2019). Realistic peer-to-peer energy trading model for microgrids using deep reinforcement learning. In *2019 IEEE PES innovative smart grid technologies Europe (ISGT-Europe)*. https://doi.org/10.1109/ISGTEurope.2019.8905731.

Chen, K., Lin, J., & Song, Y. (2019). Trading strategy optimization for a prosumer in continuous double auction-based peer-to-peer market: A prediction-integration model. *Applied Energy*, *242*, 1121–1133.

Chen, T., & Su, W. (2018). Local energy trading behavior modeling with deep reinforcement learning. *IEEE Access*, *6*, 62806–62814.

Cui, S., Wang, Y. W., & Xiao, J. W. (2019). Peer-to-peer energy sharing among smart energy buildings by distributed transaction. *IEEE Transactions on Smart Grid*, *10*(6), 6491–6501.

Davoudi, M., Moeini-Aghtaie, M., & Ghorani, R. (2021). Developing a new framework for transactive peer-to-peer thermal energy market. *IET Generation, Transmission and Distribution*. https://doi.org/10.1049/gtd2.12150.

Diestelmeier, L. (2019). Changing power: Shifting the role of electricity consumers with blockchain technology–Policy implications for EU electricity law. *Energy Policy*, *128*, 189–196.

Esmat, A., de Vos, M., Ghiassi-Farrokhfal, Y., Palensky, P., & Epema, D. (2020). A novel decentralized platform for peer-to-peer energy trading market with blockchain technology. *Applied Energy*. https://doi.org/10.1016/j.apenergy.2020.116123.

Guerrero, J., Gebbran, D., Mhanna, S., Chapman, A. C., & Verbič, G. (2020). Towards a transactive energy system for integration of distributed energy resources: Home energy management, distributed optimal power flow, and peer-to-peer energy trading. *Renewable and Sustainable Energy Reviews*. https://doi.org/10.1016/j.rser.2020.110000.

Hahnel, U. J. J., Herberz, M., Pena-Bello, A., Parra, D., & Brosch, T. (2020). Becoming prosumer: Revealing trading preferences and decision-making strategies in peer-to-peer energy communities. *Energy Policy*. https://doi.org/10.1016/j.enpol.2019.111098.

Hua, W., Jiang, J., Sun, H., & Wu, J. (2020). A blockchain based peer-to-peer trading framework integrating energy and carbon markets. *Applied Energy*. https://doi.org/10.1016/j.apenergy.2020.115539.

Iansiti, M., & Lakhani, K. R. (2017). The truth about blockchain. *Harvard Business Review*. Harvard University. Archived from the original on 18 January 2017. Retrieved 17 January 2017. The technology at the heart of bitcoin and other virtual currencies, blockchain is an open, distributed ledger that can record transactions between two parties efficiently and in a verifiable and permanent way.

Kang, J., Yu, R., Huang, X., Maharjan, S., Zhang, Y., & Hossain, E. (2017). Enabling localized peer-to-peer electricity trading among plug-in hybrid electric vehicles using consortium blockchains. *IEEE Transactions on Industrial Informatics*, *13*(6), 3154–3164.

Khorasany, M., Mishra, Y., & Ledwich, G. (2020). A decentralized bilateral energy trading system for peer-to-peer electricity markets. *IEEE Transactions on Industrial Electronics*, *67*(6), 4646–4657.

Liu, J., Yang, H., & Zhou, Y. (2021). Peer-to-peer energy trading of net-zero energy communities with renewable energy systems integrating hydrogen vehicle storage. *Applied Energy*, *298*. https://doi.org/10.1016/j.apenergy.2021.117206, 117206.

Liu, N., Yu, X., Wang, C., Li, C., Ma, L., & Lei, J. (2017). Energy-sharing model with price-based demand response for microgrids of peer-to-peer prosumers. *IEEE Transactions on Power Systems*, *32*(5), 3569–3583. https://doi.org/10.1109/TPWRS.2017.2649558.

Long, C., Wu, J., Zhang, C., Thomas, L., Cheng, M., & Jenkins, N. (2017a). Peer-to-peer energy trading in a community microgrid 2017. *IEEE Power & Energy Society General Meeting*. https://doi.org/10.1109/PESGM.2017.8274546.

Long, C., Wu, J., Zhang, C., Thomas, L., Cheng, M., & Jenkins, N. (2017b). Peer-to-peer energy trading in a community microgrd. In *Proc IEEE PES general meeting, Chicago, IL, USA* (pp. 1–5).

Long, C., Wu, J., Zhou, Y., & Jenkins, N. (2018). Peer-to-peer energy sharing through a two-stage aggregated battery control in a community microgrid. *Applied Energy, 226,* 261–276.

Lüth, A., Zepter, J. M., del Granado, P. C., & Egging, R. (2018). Local electricity market designs for peer-to-peer trading: The role of battery flexibility. *Applied Energy, 229,* 1233–1243.

Meinke, R. J., Sun, H., & Jiang, J. (2020). Optimising demand and bid matching in a peer-to-peer energy trading model. In *ICC 2020—2020 IEEE international conference on communications (ICC).* https://doi.org/10.1109/ICC40277.2020.9148652.

Mengelkamp, E., Gärttner, J., Rock, K., Kessler, S., Orsini, L., & Weinhardt, C. (2018). Designing microgrid energy markets: A case study: The Brooklyn Microgrid. *Applied Energy, 210,* 870–880.

Morstyn, T., Farrell, N., Darby, S. J., & McCulloch, M. D. (2018). Using peer-to-peer energy-trading platforms to incentivize prosumers to form federated power plants. *Nature Energy, 3,* 94–101.

Nguyen, S., Peng, W., Sokolowski, P., Alahakoon, D., & Yu, X. (2018). Optimizing rooftop photovoltaic distributed generation with battery storage for peer-to-peer energy trading. *Applied Energy, 228,* 2567–2580.

Okoye, M. O., Yang, J., Cui, J., Lei, Z., Yuan, J., Wang, H., et al. (2020). A blockchain-enhanced transaction model for microgrid energy trading. *IEEE Access.* https://doi.org/10.1109/ACCESS.2020.3012389.

Parag, Y., & Sovacool, B. K. (2016). Electricity market design for the prosumer era. *Nature Energy, 1,* 16032.

Paudel, A., Chaudhari, K., Long, C., & Gooi, H. B. (2019). Peer-to-peer energy trading in a prosumer-based community microgrid: A game-theoretic model. *IEEE Transactions on Industrial Electronics, 66*(8), 6087–6097.

Peck, M. E., & Wagman, D. (2017). Energy trading for fun and profit buy your neighbor's rooftop solar power or sell your own-it'll all be on a blockchain. *IEEE Spectrum.* https://doi.org/10.1109/MSPEC.2017.8048842.

Quantity theory of money. Website https://en.wikipedia.org/wiki/Quantity_theory_of_money.

Sorin, E., Bobo, L., & Pinson, P. (2019). Consensus-based approach to peer-to-peer electricity markets with product differentiation. *IEEE Transactions on Power Systems, 34*(2), 994–1004.

Sousa, T., Soares, T., Pinson, P., Moret, F., Baroche, T., & Sorin, E. (2019). Peer-to-peer and community-based markets: A comprehensive review. *Renewable and Sustainable Energy Reviews, 104,* 367–378.

Tushar, W., Saha, T. K., Yuen, C., Liddell, P., Bean, R., & Vincent, H. (2018). Poor peer-to-peer energy trading with sustainable user participation: A game theoretic approach. *IEEE Access.* https://doi.org/10.1109/ACCESS.2018.2875405.

Tushar, W., Saha, T. K., Yuen, C., Morstyn, T., McCulloch, M. D., Poor, H. V., et al. (2019). A motivational game-theoretic approach for peer-to-peer energy trading in the smart grid. *Applied Energy, 243,* 10–20.

Tushar, W., Saha, T. K., Yuen, C., Smith, D., & Poor, H. V. (2020). Peer-to-peer trading in electricity networks: An overview. *IEEE Transactions on Smart Grid, 11*(4), 3185–3200.

Wang, S., Taha, A. F., Wang, J., Kvaternik, K., & Hahn, A. (2019). Energy crowdsourcing and peer-to-peer energy trading in blockchain-enabled smart grids. *IEEE Transactions on Systems, Man, and Cybernetics: Systems, 49*(8), 1612–1623.

Xu, X., Xu, Y., Wang, M. H., Li, J., Xu, Z., Chai, S., et al. (2020). Data-driven game-based pricing for sharing rooftop photovoltaic generation and energy storage in the residential building cluster under uncertainties. *IEEE Transactions on Industrial Informatics.* https://doi.org/10.1109/TII.2020.3016336.

Xu, Y., Yu, L., Bi, G., & Zhang, M. (2020). Deep reinforcement learning and blockchain for peer-to-peer energy trading among microgrids. In *International conferences on internet of things (iThings) and IEEE green computing and communications (GreenCom) and IEEE cyber, physical and social computing (CPSCom) and IEEE smart data (SmartData) and IEEE congress on cybermatics (cybermatics)*. https://doi.org/10.1109/iThings-GreenCom-CPSCom-SmartData-Cybermatics50389.2020.00071.

Zepter, J. M., Lüth, A., Del Granado, P. C., & Egging, R. (2019). Prosumer integration in wholesale electricity markets: Synergies of peer-to-peer trade and residential storage. *Energy and Buildings, 184*, 163–176.

Zhang, Z., Li, R., & Li, F. (2020). A novel peer-to-peer local electricity market for joint trading of energy and uncertainty. *IEEE Transactions on Smart Grid, 11*(2), 1205–1215.

Zhou, S., Hu, Z., Gu, W., Jiang, M., & Zhang, X. P. (2019). Artificial intelligence based smart energy community management: A reinforcement learning approach. *CSEE Journal of Power and Energy Systems, 5*(1), 1–10.

Zhou, Y., Wu, J., & Long, C. (2018). Evaluation of peer-to-peer energy sharing mechanisms based on a multiagent simulation framework. *Applied Energy, 222*, 993–1022.

Chapter 11

Machine learning-based hybrid demand-side controller for renewable energy management

Padmanabhan Sanjeevikumar[a], Tina Samavat[b], Morteza Azimi Nasab[a], Mohammad Zand[a], and Mohammad Khoobani[b]

[a]*Department of Business Development and Technology, CTIF Global Capsule, Aarhus University, Herning, Denmark,* [b]*Department of Mechanical and Mechatronics Engineering, Shahrood University of Technology, Shahroud, Iran*

11.1 Introduction

Traditionally, power systems are designed for manufacturing electricity. In the early stages, this was a simple matter, but with the increasing number of consumers in industrial and residential areas and the variety of time and amount of energy they need, providing sustainable power has become more complex while maintaining the equivalency of supply and demand. This rapid growth in demand for electricity made the power generator centers think of increasing their production. However, increasing the production capacity or building a new power plant is very costly and time-consuming (Tian, Zhang, Yuan, Che, & Zafetti, 2020). Furthermore, due to the reduction of consumption during specific periods, such as after midnight, power production must be decreased, which would be very difficult or, sometimes, impossible.

Consumption by specific consumers, especially in the industrial sector, was one of the proposed solutions. However, the grids faced another problem: the unbalanced demand of the residential sector at different times during the day (Azimi Nasab, Zand, Eskandari, Sanjeevikumar, & Siano, 2021). To protect the grid from fluctuations in demand, a discussion called demand-side management (DSM) was raised and progressed up to the present moment to become one of the most critical topics in the electricity generation industry (McIlwaine et al., 2021).

Sustainable Developments by Artificial Intelligence and Machine Learning for Renewable Energies.
https://doi.org/10.1016/B978-0-323-91228-0.00003-3
Copyright © 2022 Elsevier Inc. All rights reserved.

292 Sustainable developments by artificial intelligence & machine learning

Furthermore, the increasing scarcity of fossil fuels soon and the harmful effects of using fossil fuels on the environment, such as CO_2 emissions resulting from burning fuels to extract energy, has led to a transition from old destructive methods to environmentally friendly ones, known as green energy sources (Hosseini-Fashami, Motevali, Nabavi-Pelesaraei, Hashemi, & Chau, 2019). Being inexpensive, available, and easily exploitable, renewable energy sources (RESs) are regarded by experts as an alternative way. On the other hand, the unreliability of green energy because of its high dependency on environmental status for prolonged and continuous use is a significant drawback (Jurasz, Ceran, & Orłowska, 2020). Using storage sections and combinations of different sources and making a connection to the grid can solve this problem, which results in the formation of hybrid sources. For this purpose, it is crucial to analyze possible sources and a variety of consumers and their consumption patterns, to sum up with a program for each zone that balances the combination of sources at various times based on complex parameters such as season or the type of consumer. This analysis must also be considered in DSM (Sarker et al., 2021).

With the growing importance of DSM, researchers have made great efforts to design the best controller for networks and consumers by using new and creative methods to make the best decisions and provide intelligent solutions, in addition to guaranteeing system stability and increasing plant productivity (Rohani & Joorabian, 2019). So far, several methods and various controlling techniques have been introduced to achieve this goal. As a result of trying to make controllers more intelligent and increase accuracy to act like humans in facing problems and different conditions that may happen while operating a system, a data-science technology is developing. This technique, known as machine leading (ML), is a subdiscipline of artificial intelligence (AI) and tries to teach machines to pick the best reaction and erase the need for overseer presence in all stages (Tightiz et al., 2020).

Machine learning is an AI subset that allows a system to learn from data rather than programming. The learning process focuses on developing computer programs to access data and use it for their learning. At another level, the learning machine can be thought of as the process of training a computer system to make accurate predictions when given the correct information (Iqbal, Maniak, Doctor, & Karyotis, 2019). These predictions may identify, for example, whether the fruit in a photo is an apple, an orange, or a banana. How to communicate AI or machine learning factors with data is determined based on different models and algorithms.

Meanwhile, machine learning has attracted much attention, even regarding complex issues, due to the minimization of a person's need to control performance, speed, and high accuracy. These powerful methods alone have met the requirements of various fields. To increase the power of these methods, a combination can be used to coordinate to learning faster and taking action more quickly while facing a problem (Antonopoulos et al., 2020).

Numerous articles have been published in balancing and consumption, which can be divided into two general categories. Another way is to connect the network so that by exchanging energy and supplying the required energy at times when the system cannot provide it, the excess energy produced is transferred to the network (Canale, Di Fazio, Russo, Frattolillo, & Dell'Isola, 2021; Jyoti Saharia, Brahma, & Sarmah, 2018). In Kiptoo, Adewuyi, Lotfy, Ibrahimi, and Senjyu (2020), a network consisting of renewable sources (wind turbine and solar panel) has been considered, which has been managed and forecasted by the production of each section with different DSM techniques. This is done based on planning the interval of loads that can be controlled to operate during on-peak periods. One way to manage consumption and generate power is to use storage resources. These resources have many applications in various fields, such as integrating renewable resources, peak shaving, helping sustainability, and increasing produced power quality. Therefore, it is necessary to introduce a source (Arani, Gharehpetian, & Abedi, 2019). In this context, power storage resources are discussed in Chapter 11. Several energy storage systems (ESSs) have been trained and studied in this source, and their control methods for efficient system design are described. Another example (Nakabi & Toivanen, 2019) is a residential area under consideration. In this study, regardless of the type of loads used by each unit, a unique load curve is presented using an intelligent algorithm. This method performs planning based on the price of electricity in each interval and presents the schedule by separating the bases based on changing the hours of use.

11.1.1 Renewable and hybrid energy system

Consumption of fossil fuels such as oil and gas does irreparable damage through increasing greenhouse gases, acid rain, water vapor production, carbon dioxide, and toxic gases, thus increasing the temperature of the Earth, causing climate change. To avoid these adverse effects on the environment, renewable energy such as solar energy, wind energy, and geothermal energy is recommended. Optimal use of renewable energy in the sustainable development of countries has consistently been recognized as a fundamental goal (Kanoğlu, Çengel, & Cimbala, 2020).

A hybrid energy system (HES) aims to achieve synchronized power from various combinations of green, nuclear, and fossil fuels. Significant efforts to replace fossil fuels with green energy and the instability of environmentally unfriendly resources result in the tendency to use several different sources. Hence, as an intelligent solution, consideration must be given to designing an optimal system for each locality (Dong et al., 2021).

A significant reason to employ this system is the dependency of different sources on multiple factors, making them unpredictable and challenging to control. For example, solar energy is highly dependent on environmental conditions such as temperature and sun irradiation, which affect the amount of produced

294 Sustainable developments by artificial intelligence & machine learning

energy and make it an astatic source. In these conditions, a combination of sources must be coordinated to reach a sustainable output (Tress et al., 2019).

Worldwide, renewable sources are mainly used to develop hybrid renewable energy sources (HRES), which can be used at both large and small scales to meet the demands of a specific area. Fundamental factors such as sizing, controlling, and choosing appropriate sources to combine and manage the generated power are principal subjects in HRES, and many papers are published to provide suitable solutions (Roth, Boix, Gerbaud, Montastruc, & Etur, 2019).

11.1.2 Demand-side management

As mentioned earlier, one of the fundamental factors of a standard network is that at any given moment, the minimum consumption is less than the maximum production. Due to this issue, one of the critical tasks of DSM is to maintain the stability of the grid by checking the proportionality of electricity production with consumption. In violation of this principle, the network decides to stop supplying electricity to a part of the network to prevent problems for the entire network (Kumar & Saravanan, 2019). In the past, to maintain the stability of the network, much less capable power plants were used. The reason for this was the ability to increase production if demand increased, even when the network faced an increase in demand, even at one point (Groppi, Pfeifer, Garcia, Krajačić, & Duić, 2021). This increased costs and reduced system efficiency. However, today, with the help of DSM, consumption can be predicted, planned, and controlled so that there is no need to produce less than productive capacity. Another way to take advantage of DSM is to manage renewable resources. One of the disadvantages of renewable energies is their dependence on uncontrollable atmospheric conditions such as wind speed and solar intensity. Therefore, DSM should combine resources and use them so that the network does not have problems (Ahmadi-Nezamabad et al., 2019).

The primary purpose of DSM is to prevent fluctuations in the consumer load curve and to flatten it as much as possible (Zand et al., 2020c). For this purpose, various methods are used, such as energy efficiency, energy-saving, demand shifting, spinning reserve, virtual power plant, and demand response programs. For example, in the residential category, devices can be divided into several categories. Appliances such as refrigerators, known as permanent loads, cannot be cut off, although a unique program can be offered to help reduce peak time consumption. However, appliances such as washing machines or dishwashers can be used during off-peak hours. Therefore, with consumers' knowledge in each section and presenting a program to eliminate some consumer use at the peak period of electricity consumption, this can be done (Cai, Shen, Lin, Li, & Xiao, 2019). Furthermore, in some cases, power generators can be managed to respond to the needs of consumers at different times. In addition to maintaining the stability of the grid, financial issues are also considered. The following is a brief overview of some of these programs and techniques.

One of the most popular programs is load growth, and its function is to encourage consumers to increase consumption at times when supply is more than demand and storage is not possible. As its name implies, energy-saving refers to saving electricity consumption, and has been considered for many years by installing related devices. Energy efficiency refers to reducing losses by replacing loads with more efficient loads (Khan, 2019). In addition, some programs affect the payment costs, so controlling the consumption during the on-peak hours results in a reduction in bills. The opposite method is also possible, where if they increase consumption at the on-peak time, the consumer is subject to a fine (Hietaharju, Ruusunen, & Leiviskä, 2019).

11.2 Machine learning at a glance

In general, it can be assumed that machine learning enables the study of large amounts of data. Today, machine learning is adopted in most industries and businesses; renewable energies and related subjects are not an exception. Vital decisions in today's world are made based on data processing. Machine learning uses different methods, which fall into four major categories: supervised learning, unsupervised learning, semisupervised learning, and reinforcement learning (Zand et al., 2020a).

Supervised methods: In a supervised model, the algorithm's input and output are known from the beginning. The supervised machine learning algorithm begins to analyze the input data (Train Data | Train Set). After training on a set of data with specific outputs, the supervised learning algorithm invents a pattern, or rather, it is a model, based on which the input data is converted into output data (Boza & Evgeniou, 2021). The trained machine will use the inferred model to predict the output of new data samples with a specific error coefficient (Farhoumandi, Zhou, & Shahidehpour, 2021). This algorithm can compare its output with the correct and predetermined output and find the existing errors to correct the model accordingly (Zand et al., 2021a).

Unsupervised methods: Unsupervised algorithms are used when the input and output of the system are not clear from the beginning. The goal is not to find the input–output connection; instead, the machine learning algorithm looks for a function to describe the hidden and specific structure of the data (meaning input data whose output is unknown) (Zhou & Zheng, 2020a). In unsupervised machine learning methods, the algorithm obtains a pattern for clustering the data by examining the similarities and differences between the data and, based on this pattern, it can predict the output for the new data sample (Zaadnoordijk, Besold, & Cusack, 2020).

Semisupervised methods: Semisupervised algorithms can be placed between supervised learning and unsupervised learning. In semiobservational learning, both data types are used, namely data with specified output (Labeled | Labeled data) and data with no specific output (Labeled | Unlabeled Data), for the model training phase. Typically, while teaching a semisupervised learning model, a small amount of labeled data and a large amount of unlabeled data are used. Systems that use this method significantly improve learning accuracy (Van Engelen & Hoos, 2020).

Reinforcement learning methods: In reinforcement learning, the "intelligent agent" communicates with its environment by taking actions in the environment and observing the corresponding result. The outcome of the agent's involvement with the environment can be an error or a reward. The goal of reinforcement learning methods is to maximize rewards. The agent acts to maximize the reward (points) and otherwise receives an error (punishment). In reinforcement learning, the intelligent agent is trained to make successive decisions, or in other words, the agent learns to reach the goal of maximizing rewards in an unknown and complex environment (Jin, Yang, Wang, & Jordan, 2020; Wen, Zhou, Li, & Wang, 2020).

11.2.1 Machine learning meets model-based control

Model-based control methods such as model predictive control have seen progressing favor in developing complicated engineering utilization. Numerous model-based control systems meet hurdles associated with the rigor of modeling complex arrangements or the necessity for control approaches with provably reliable and robust performance that have moderate online computational and memory needs. Recent years have seen the maturation of machine learning methods in computer science and the growing availability of data and new computation, sensing, and communication aptitudes. The combination of machine learning with model-based control suggests notable advantages linked to controller characteristics, such as performance under uncertainty, constraint satisfaction, convergence, and stability (Zand et al., 2020b).

With the accelerated growth of AI, machine learning has been extensively adopted in building energy systems. Several examples of such applications are going to be explored in the following. According to research by Sanjeevikumar, Zand, Nasab, Hanif, and Bhaskar (2021), building energy use has been predicted in a large-size office building by comparative analysis between various data-driven models. Gaussian mixture model regression reached the most remarkable accuracy in this research. Peng, Rysanek, Nagy, and Schlüter (2018) focused on indoor occupancy in an office building employing unsupervised and supervised learning approaches for online room set-points. It has been demonstrated that an energy saving of up to 52% is likely using a demand-driven control approach. In the case of DSM, Pallonetto, De Rosa, Milano, and Finn (2019) considered the expense and emission of a heating system in a conventional residence. Based on a predictive-based (machine learning) approach, electricity end-use expense remained 41.8%, and carbon emission equaled 37.9% of the reference case. In other researches (Azimi Nasab, Zand, Padmanaban, Dragicevic, & Khan, 2021; Zhou & Zheng, 2020b), surrogate model development for system optimization has been taken into account;

as a result, cumulative heat gain (Zhou et al., 2019) and similar combined energy production Azimi Nasab, Zand, Padmanaban, et al., 2021 were the subject of prediction. Both implemented supervised machine learning with the cross-entropy method. It was concluded that cumulative heat gain could be diminished by 7.2% (Zhou et al., 2019); commensurate combined energy production can be improved by 2.6% Azimi Nasab, Zand, Padmanaban, et al., 2021.

11.2.2 The application of machine learning in hybrid demand-side controllers

A comprehensive study of this topic is covered by Zhou and Zheng (2020a). The majority of researches mainly concentrate on thermal energy of systems' energy flexibility. In contrast, limited attention is paid to the quantification method of energy flexibility offered by combined thermal and electrical storage systems in buildings. There is no universal flexibility quantification outline for modern building energy systems.

Furthermore, the nonlinearity and complexity of building energy forecasts, including modeling growing and computational load, have not been adequately resolved, particularly combining multilevel parameters' uncertainty. Precise forecasts of building energy with uncomplicated models adopting machine learning schemes are full of promising prospects in promoting energy flexible buildings. In addition, the extent of high-level controllers with short-term building energy forecasts under high-level scenario uncertainties can be investigated more deeply. The impact of advanced controllers on building energy flexibility is worthy of being well researched. A review of machine learning for the improvement of high-level controllers is presented in Fig. 11.1. The diagram represents an inevitable consequence of an antecedent sufficient case scenario.

Fig. 11.2 pictures a scenario including uncertainty: a supervised learning-based punishment for the controller. The process of training in such schemes is outlined in Fig. 11.3.

The column vector of weights following machine learning can be captured and then be executed in the situation, with high-level uncertainty for the short-term forecast. The short-term forecast outcomes by the data-driven model will be achieved in the improvement of the controller. The progressive training is presented in Fig. 11.3. As presented in Fig. 11.1, while training, the learning section measures the error among the exact data and the foretold amounts and modifies the weight to lessen the error.

Taking the example of a typical house, these models are illustrated in Figs. 11.4 and 11.5. Fig. 11.4 describes the complete arrangement of the supervised machine-learning model. There are three layers in the artificial neural network, comprising one input layer, one hidden layer, and one output layer, including one node. Within the input layer, there are 10 situation parameters: the ambient heat, the solar air heat, the supplied and return chilled water heats in the air-handling unit cooling system, the power consumption of the air-

298 Sustainable developments by artificial intelligence & machine learning

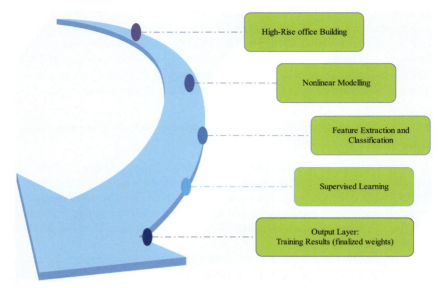

FIG. 11.1 A review of the machine learning for the improvement of high-level controllers.

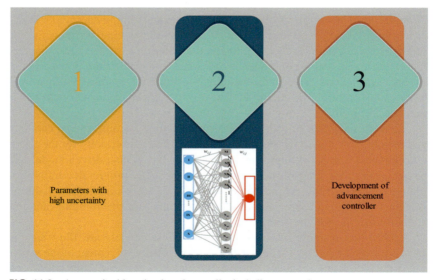

FIG. 11.2 A supervised learning-based controller including uncertainty.

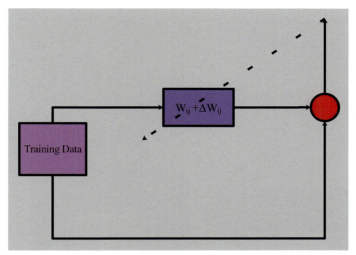

FIG. 11.3 The process of training of a supervised learning-based controller.

handling unit cooling chiller, the supply and return chilled water heats in the space cooling system, the power usage of the space cooling chiller, the in-house gain, and the indoor air set-point heat.

The final purpose of precisely foretelling the building energy demand is to efficiently employ the building energy flexibility with high-level controllers. Fig. 11.5 describes a holistic sketch of the flexibility quantification fashion concerning various high-level controllers, considering the uncertainty of the situation parameters. The initial action is the short-term forecast of building energy demand under high situation uncertainty. The following action is the evolution of high-level controllers concerning various signals. The next action is to execute the advanced controllers for DSM. The final action is the building energy flexibility evaluation. The rule-based controller tracks the signal within the current building energy demand and the current renewable production.

In the example given, the application of machine learning in hybrid demand-side controllers was given in an example of a residential building. Many controllers can take advantage of machine learning techniques in the field. This cooperation will be made possible by combining machine learning methods and different parts of a controller. Later in this chapter, various up-to-date machine learning methods are introduced that have a high ability to analyze data types for controllers.

11.2.3 Support vector machine

The support vector machine (SMV) is one of the most popular classification methods (Fig. 11.6). In this way, with the help of vectors, it sorts the data. What

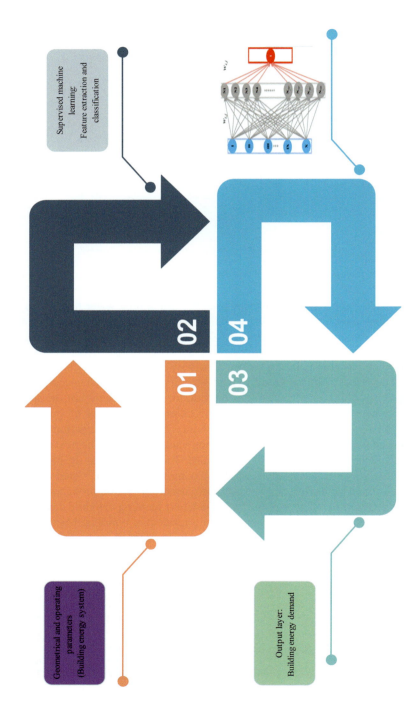

FIG. 11.4 A supervised machine-learning model for a typical residential house.

Machine learning-based hybrid demand-side controller **Chapter | 11** 301

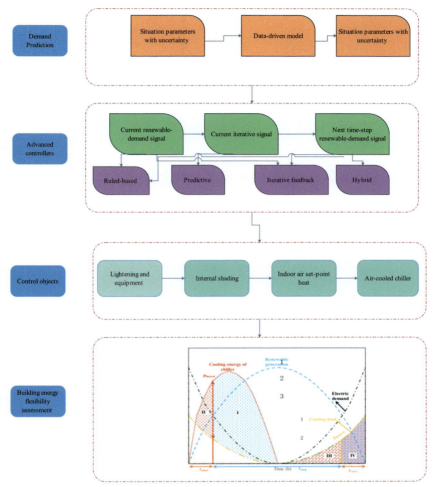

FIG. 11.5 A holistic sketch of the flexibility quantification fashion concerning various high-level controllers, considering the uncertainty of the situation parameters.

distinguishes this method from other methods is that it considers a margin around the dividing line of the categories, so that the data which is closest to the separating line should have the most distance from the margin.

If we want the algorithm to categorize data so that none crosses the margin, it is called hard margin classification. Doing this mainly results in a specific decrease in the width of the margins. Only a tiny number of data are out of the range, placed in the margins, which can be neglected. However, this method has poor performance in the case of high-dispersion data. Thus, another method, called soft-margin classification, is proposed, aiming to create a relative balance

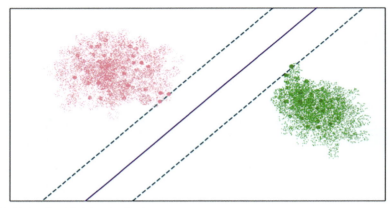

FIG. 11.6 Classified data by SVM.

between maintaining boundaries and placing some data in that range (Dun, Xu, Chen, & Wu, 2020; Nie, Roccotelli, Fanti, Ming, & Li, 2021).

11.2.4 K-means clustering

The purpose of clustering data is to place more similar data in a cluster; the k-means method is an example of such an application and purpose, and is a form of unsupervised learning. In this method, we need a score function to evaluate the quality of clustering (Chellamani & Chandramani, 2020).

First, the point, k, is randomly selected and forms the center of the cluster. Then, to form clusters, the distance of each datum from all centers is calculated, and finally, the data join the cluster with the shortest distance to its center:

$$x_i \in c_j \Big| j = argmin_{j=1}^{k} \{\|x_j - \mu_i\|\}$$

In the function of calculating the minimum distance of point x from the centers of cluster C, data ith belongs to cluster j. After assigning the data to the cluster, the new average of each cluster is calculated, and that average is placed as the center of the cluster. This algorithm is repeated until the mean distance between the two consecutive steps is less than the desired value, as shown below:

$$\sum_{i=1}^{k} \left\|\mu_j^t - \mu_{j-1}^{t-1}\right\|^2 \leq \varepsilon \quad \varepsilon > 0$$

The advantages of this method are high speed and ease of use, and the possibility of implementation for large data, but the need to have the exact number of clusters (k) is a disadvantage (Chellamani & Chandramani, 2020; Ghasemi & Akbari, 2019).

11.2.5 Extreme learning machine

The extreme learning machine (EML) is based on a single feedforward neural network (SLFN) with tree fundamental layer (one input layer, one output, and one hidden layer), but is distinct, due to the way its weights are tuning. Basically, as it has only one hidden layer. Two weight classes must be determined, which are input-hidden layer and hide layer-output. The first group is constant and assign randomly, while the second group trains. This algorithm shows great speed and accuracy during training and has better generalization (Zand et al., 2021b).

11.2.6 Linear regression

One of the learning machine techniques based on a simple mathematical equation is linear regression. Whenever we can find a linear relationship between two variables, we can use this type of regression to predict the values of these variables based on the value of the other variable. The linear relation is to see that, with the increase of one variable, the other variable increases (or decreases), and with its decrease, the second variable decreases (or increases), and this increase or decrease is a direct relation (simple coefficient) with the value of the first variable, which we call the independent variable. Therefore, this method first receives one variable and predicts another variable with the help of the existing linear relation. In Fig. 11.7, the blue dots are the main data and the red dots are the predicted data (Briggs, Fan, & Andras, 2020; Dun et al., 2020).

It should be noted that in order to use this method, the data must be in a normal distribution, there must not be much data outside the range, and for each direct variable, there must be one variable with a linear relationship.

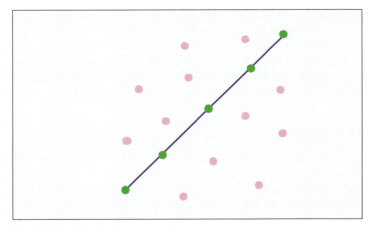

FIG. 11.7 Classified data by linear regression.

11.2.7 Partial least squares

In cases where the data, unlike the previous method, does not have a normal distribution, another method called partial least squares is used. This method finds linear regression instead of finding the relationship between variables. Also, linear regression cannot determine the appropriate answer if the number of variables to be predicted is greater than the number of leading variables. In such cases, this method is used (da Silva, Ribeiro, Moreno, Mariani, & dos Santos Coelho, 2021; Mehmood, Sæbø, & Liland, 2020).

11.2.8 Challenges and future research direction

Numerous areas of science are employing machine learning to boost their capabilities; However, although they take advantage of its benefits, some restrictions must be taken into account, and hybrid demand-side controllers are no exception. Generally speaking, statistics is the foundation of conventional machine learning. Without being programmed in a clear and detailed manner, the algorithm works entirely on reasoning and patterns. The challenge associated with conventional machine learning schemes is that they resemble or are characteristic of a machine. They necessitate plenty of human interference and solely suit what they are intended for. Deep learning brings a bit more hope for a hybrid demand-side controller.

In conventional machine learning procedures, the largest of the utilized characteristics need to be distinguished by a field specialist to overcome the data complication and compose enhanced apparent patterns to feed the learning phase of algorithms. On the other hand, deep learning is perceived essentially as a mature and mathematically complicated evolution of machine learning methods. The topic has been gaining much notice recently because novel improvements have attended outcomes that were not imagined to be feasible earlier.

Furthermore, machine learning and deep learning procedures expose a notable contrast, which is how they tackle problem-solving. Machine learning necessitates splitting the problem statements into various parts to be determined and the following results to be merged at the ultimate step. In contrast, deep learning manages to resolve the issue in an end-to-end manner, making it an excellent alternative to assist a hybrid demand-side controller.

11.3 Conclusion

It is crucial to analyze possible energy sources, and a variety of consumers and their consumption patterns, to come up with a program for each zone that balances the combination of sources at different times based on complex parameters such as season or the type of consumers. This analysis must also be considered in DSM. The primary purpose of DSM is to prevent fluctuations in the consumer load curve and to flatten it as much as possible. Researchers have made great efforts to design the best controller for networks and

Machine learning-based hybrid demand-side controller **Chapter | 11** **305**

consumers by using new and creative methods to make the best decisions and provide intelligent solutions, in addition to guaranteeing system stability and increasing plant productivity. As a result of trying to make controllers smarter and increase the accuracy of their behavior in the face of problems. In general, it can be assumed that machine learning enables the study of large amounts of data. Today, machine learning is adopted in most industries and businesses; renewable energies and related subjects are no exception. When machine learning schemes are utilized as precise energy estimators of a structure based on models with a lower level of complexity, they assure the likelihood of improving energy usage. A notable challenge associated with conventional machine learning is that they are characteristic of a machine and promising features of deep learning schemes make them excellent alternatives to assist a hybrid demand-side controller in future studies.

References

Ahmadi-Nezamabad, H., et al. (2019). Multi-objective optimization based robust scheduling of electric vehicles aggregator. *Sustainable Cities and Society*, *47*, 101494.

Antonopoulos, I., Robu, V., Couraud, B., Kirli, D., Norbu, S., Kiprakis, A., et al. (2020). Artificial intelligence and machine learning approaches to energy demand-side response: A systematic review. *Renewable and Sustainable Energy Reviews*, *130*, 109899.

Arani, A. K., Gharehpetian, G. B., & Abedi, M. (2019). Review on energy storage systems control methods in microgrids. *International Journal of Electrical Power & Energy Systems*, *107*, 745–757.

Azimi Nasab, M., Zand, M., Eskandari, M., Sanjeevikumar, P., & Siano, P. (2021). Optimal planning of electrical appliance of residential units in a smart home network using cloud services. *Smart Cities*, *4*, 1173–1195. https://doi.org/10.3390/smartcities4030063.

Azimi Nasab, M., Zand, M., Padmanaban, S., Dragicevic, T., & Khan, B. (2021). Simultaneous long-term planning of flexible electric vehicle photovoltaic charging stations in terms of load response and technical and economic indicators. *World Electric Vehicle Journal*, *12*, 190. https://doi.org/10.3390/wevj12040190.

Boza, P., & Evgeniou, T. (2021). Artificial intelligence to support the integration of variable renewable energy sources to the power system. *Applied Energy*, *290*, 116754.

Briggs, C., Fan, Z., & Andras, P. (2020). Privacy preserving demand forecasting to encourage consumer acceptance of smart energy meters. *arXiv Preprint*. arXiv:2012.07449.

Cai, H., Shen, S., Lin, Q., Li, X., & Xiao, H. (2019). Predicting the energy consumption of residential buildings for regional electricity supply-side and demand-side management. *IEEE Access*, *7*, 30386–30397.

Canale, L., Di Fazio, A. R., Russo, M., Frattolillo, A., & Dell'Isola, M. (2021). An overview on functional integration of hybrid renewable energy systems in multi-energy buildings. *Energies*, *14*(4), 1078.

Chellamani, G. K., & Chandramani, P. V. (2020). An optimized methodical energy management system for residential consumers considering price-driven demand response using satin bowerbird optimization. *Journal of Electrical Engineering and Technology*, *15*(2), 955–967.

da Silva, R. G., Ribeiro, M. H. D. M., Moreno, S. R., Mariani, V. C., & dos Santos Coelho, L. (2021). A novel decomposition-ensemble learning framework for multi-step ahead wind energy forecasting. *Energy*, *216*, 119174.

306 Sustainable developments by artificial intelligence & machine learning

Dong, Z., Li, B., Li, J., Guo, Z., Huang, X., Zhang, Y., et al. (2021). Flexible control of nuclear cogeneration plants for balancing intermittent renewables. *Energy*, *221*, 119906.

Dun, M., Xu, Z., Chen, Y., & Wu, L. (2020). Short-term air quality prediction based on fractional grey linear regression and support vector machine. *Mathematical Problems in Engineering*, *2020*.

Farhoumandi, M., Zhou, Q., & Shahidehpour, M. (2021). A review of machine learning applications in IoT-integrated modern power systems. *The Electricity Journal*, *34*(1), 106879.

Ghasemi, M., & Akbari, E. (2019). An efficient modified HPSO-TVAC-based dynamic economic dispatch of generating units. *Electric Power Components and Systems*, *47*(19–20), 1826.

Groppi, D., Pfeifer, A., Garcia, D. A., Krajačić, G., & Duić, N. (2021). A review on energy storage and demand side management solutions in smart energy islands. *Renewable and Sustainable Energy Reviews*, *135*, 110183.

Hietaharju, P., Ruusunen, M., & Leiviskä, K. (2019). Enabling demand side management: Heat demand forecasting at city level. *Materials*, *12*(2), 202.

Hosseini-Fashami, F., Motevali, A., Nabavi-Pelesaraei, A., Hashemi, S. J., & Chau, K. W. (2019). Energy-life cycle assessment on applying solar technologies for greenhouse strawberry production. *Renewable and Sustainable Energy Reviews*, *116*, 109411.

Iqbal, R., Maniak, T., Doctor, F., & Karyotis, C. (2019). Fault detection and isolation in industrial processes using deep learning approaches. *IEEE Transactions on Industrial Informatics*, *15*(5), 3077–3084.

Jin, C., Yang, Z., Wang, Z., & Jordan, M. I. (2020). Provably efficient reinforcement learning with linear function approximation. In *Conference on learning theory* (pp. 2137–2143). PMLR.

Jurasz, J., Ceran, B., & Orłowska, A. (2020). Component degradation in small-scale off-grid PV-battery systems operation in terms of reliability, environmental impact and economic performance. *Sustainable Energy Technologies and Assessments*, *38*, 100647.

Jyoti Saharia, B., Brahma, H., & Sarmah, N. (2018). A review of algorithms for control and optimization for energy management of hybrid renewable energy systems. *Journal of Renewable and Sustainable Energy*, *10*(5). https://doi.org/10.1063/1.5032146, 053502.

Kanoğlu, M., Çengel, Y. A., & Cimbala, J. M. (2020). *Fundamentals and applications of renewable energy*. McGraw-Hill Education.

Khan, I. (2019). Energy-saving behaviour as a demand-side management strategy in the developing world: The case of Bangladesh. *International Journal of Energy and Environmental Engineering*, *10*(4), 493–510.

Kiptoo, M. K., Adewuyi, O. B., Lotfy, M. E., Ibrahimi, A. M., & Senjyu, T. (2020). Harnessing demand-side management benefit towards achieving a 100% renewable energy microgrid. *Energy Reports*, *6*, 680–685.

Kumar, K. P., & Saravanan, B. (2019). Day ahead scheduling of generation and storage in a microgrid considering demand Side management. *Journal of Energy Storage*, *21*, 78–86.

McIlwaine, N., Foley, A. M., Morrow, D. J., Al Kez, D., Zhang, C., Lu, X., et al. (2021). A state-of-the-art techno-economic review of distributed and embedded energy storage for energy systems. *Energy*, 120461.

Mehmood, T., Sæbø, S., & Liland, K. H. (2020). Comparison of variable selection methods in partial least squares regression. *Journal of Chemometrics*, *34*(6), e3226.

Nakabi, T. A., & Toivanen, P. (2019). An ANN-based model for learning individual customer behavior in response to electricity prices. *Sustainable Energy, Grids and Networks*, *18*, 100212.

Nie, P., Roccotelli, M., Fanti, M. P., Ming, Z., & Li, Z. (2021). Prediction of home energy consumption based on gradient boosting regression tree. *Energy Reports*, *7*, 1246–1255.

Pallonetto, F., De Rosa, M., Milano, F., & Finn, D. P. (2019). Demand response algorithms for smart-grid ready residential buildings using machine learning models. *Applied Energy*, *239*, 1265–1282.

Peng, Y., Rysanek, A., Nagy, Z., & Schlüter, A. (2018). Using machine learning techniques for occupancy-prediction-based cooling control in office buildings. *Applied Energy*, *211*, 1343–1358.

Rohani, A., Joorabian, M., et al. (2019). Three-phase amplitude adaptive notch filter control design of DSTATCOM under unbalanced/distorted utility voltage conditions. *Journal of Intelligent & Fuzzy Systems*, *37*(1), 847.

Roth, A., Boix, M., Gerbaud, V., Montastruc, L., & Etur, P. (2019). A flexible metamodel architecture for optimal design of Hybrid Renewable Energy Systems (HRES)—Case study of a stand-alone HRES for a factory in tropical island. *Journal of Cleaner Production*, *223*, 214–225.

Sanjeevikumar, P., Zand, M., Nasab, M. A., Hanif, M. A., & Bhaskar, M. S. (2021). Spider community optimization algorithm to determine UPFC optimal size and location for improve dynamic stability. In *2021 IEEE 12th energy conversion congress & exposition—Asia (ECCE-Asia)*, doi:10.1109/ECCE-Asia49820.2021.9479149.

Sarker, E., Halder, P., Seyedmahmoudian, M., Jamei, E., Horan, B., Mekhilef, S., et al. (2021). Progress on the demand side management in smart grid and optimization approaches. *International Journal of Energy Research*, *45*(1), 36–64.

Tian, Y., Zhang, F., Yuan, Z., Che, Z., & Zafetti, N. (2020). Assessment power generation potential of small hydropower plants using GIS software. *Energy Reports*, *6*, 1393–1404.

Tightiz, L., et al. (2020). An intelligent system based on optimized ANFIS and association rules for power transformer fault diagnosis. *ISA Transactions*, *103*, 63–74.

Tress, W., Domanski, K., Carlsen, B., Agarwalla, A., Alharbi, E. A., Graetzel, M., et al. (2019). Performance of perovskite solar cells under simulated temperature-illumination real-world operating conditions. *Nature Energy*, *4*(7), 568–574.

Van Engelen, J. E., & Hoos, H. H. (2020). A survey on semi-supervised learning. *Machine Learning*, *109*(2), 373–440.

Wen, L., Zhou, K., Li, J., & Wang, S. (2020). Modified deep learning and reinforcement learning for an incentive-based demand response model. *Energy*, *205*, 118019.

Zaadnoordijk, L., Besold, T. R., & Cusack, R. (2020). The next big thing (s) in unsupervised machine learning: Five lessons from infant learning. *arXiv Preprint*. arXiv:2009.08497.

Zand, M., et al. (2019). Fault locating transmission lines with thyristor-controlled series capacitors by fuzzy logic method. In *2020 14th International Conference on Protection and Automation of Power Systems (IPAPS)* (p. 62).

Zand, M., et al. (2020a). Energy management strategy for solid-state transformer-based solar charging station for electric vehicles in smart grids. *IET Renewable Power Generation*, *14*(18), 3843–3852.

Zand, M., et al. (2020b). Using adaptive fuzzy logic for intelligent energy management in hybrid vehicles. In *2020 28th Iranian conference on electrical engineering (ICEE)*IEEE.

Zand, M., et al. (2020c). A hybrid scheme for fault locating in transmission lines compensated by the TCSC. In *2020 15th International Conference on Protection and Automation of Power Systems (IPAPS)*.

Zand, M., et al. (2021a). In *Vol. 1. Maximum power point tracking of photovoltaic renewable energy system using a new method based on turbulent flow of water-based optimization (TFWO) under partial shading conditions* (p. 285). Springer.

Zand, M., et al. (2021b). Robust speed control for induction motor drives using STSM control. In *2021 12th power electronics, drive systems, and technologies conference (PEDSTC)*.

Zhou, Y., & Zheng, S. (2020a). Machine-learning based hybrid demand-side controller for high-rise office buildings with high energy flexibilities. *Applied Energy*, *262*, 114416.

Zhou, Y., & Zheng, S. (2020b). Machine learning-based multi-objective optimisation of an aerogel glazing system using NSGA-II—Study of modelling and application in the subtropical climate Hong Kong. *Journal of Cleaner Production*, *253*, 119964.

Chapter 12

Prediction of energy generation target of hydropower plants using artificial neural networks

Krishna Kumar[a], Gaurav Saini[b], Narendra Kumar[c], M. Shamim Kaiser[d], Ramani Kannan[e], and Rachna Shah[f]

[a]*Research & Development Unit, Uttarakhand Jal Vidyut Nigam (UJVN) Ltd., Dehradun, Uttarakhand, India,* [b]*School of Advanced Materials, Green Energy and Sensor Systems, Indian Institute of Engineering Science and Technology Shibpur, Howrah, West Bengal, India,* [c]*School of Computing, DIT University, Dehradun, Uttarakhand, India,* [d]*Institute of Information Technology, Jahangirnagar University, Dhaka, Bangladesh,* [e]*Electrical and Electronics Engineering Department, University of Technology PETRONAS (UTP), Seri Iskandar, Malaysia,* [f]*Indian Institute of Information Technology, Guwahati, Assam, India*

12.1 Introduction

Due to the ever-increasing demand for energy, various countries are inclined toward energy generation with the use of renewable sources of energy, keeping in view the climate change and pollution scenarios. India has vast potential for various renewable energy-based sources, i.e., solar, wind, hydro, geothermal, and bioenergy. Among the various renewable energy sources, hydro has been used for centuries to generate power (Saini & Saini, 2020). In hydropower plants, the hydro turbine is the heart of the system. A hydro turbine is a rotary hydraulic machine that converts the energy of water (kinetic and potential energy) into rotational (mechanical) energy. Furthermore, the mechanical energy can be directly supplied to a generator through coupling to generate electricity (Saini & Saini, 2018). The typical structure of a hydropower plant is shown in Fig. 12.1. The generated power from the hydro turbine depends on the net head, discharge, and conversion efficiency of the generator and turbine. The generated power of the system can be calculated by following Eq. (12.1):

$$P = \rho \times g \times Q \times H \times \eta \tag{12.1}$$

where P is the generated power, ρ is the density of water, g is gravitation acceleration, Q is the discharge, H is the net head, and η is the turbine generator's overall efficiency.

Sustainable Developments by Artificial Intelligence and Machine Learning for Renewable Energies.
https://doi.org/10.1016/B978-0-323-91228-0.00005-7
Copyright © 2022 Elsevier Inc. All rights reserved.

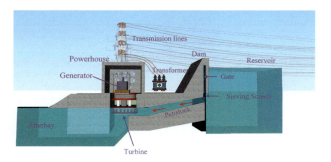

FIG. 12.1 Typical Structure of a hydropower plant.

For the optimization of power generation systems, different technologies are being used (Kumar & Singh, 2020; Kumar, Singh, & Singh, 2020), of which artificial intelligence is one. Machine learning techniques are popularly used to better understand the insights available in the data (Kumar & Saini, 2020).

12.2 Artificial neural network (ANN)

Artificial intelligence allows computers and machines to mimic the human mind's perception, learning, problem-solving, and decision-making abilities. A typical neuron scheme and artificial neural network are depicted in Fig. 12.2. The inputs X_n are connected to neurons that multiply the weightage to form the product $W_n X_n$. All the weighted inputs are added and the result is the argument of the transfer function (f). Most common ANN architectures consist of one input and output layer, with more than one hidden layer.

The training of the network is carried out to do the task by adjusting the weights to achieve the desired output. The most commonly used transfer functions are step, linear, and sigmoid, as shown in Fig. 12.3.

In the hydropower energy sector, various investigations are being carried out to optimize the performance of hydropower plants using ANN. The feedforward neural network architecture with two algorithms, i.e., Levenberg-Marquardt and resilient back-propagation, has been used for the prediction of the monthly river discharge data over 58 years. Mean square error (MSE) and correlation coefficients were used to measure the performance of the system, and it was observed that more accurate prediction is available for the Levenberg-Marquardt algorithm with sigmoid activation function at the hidden layer, and for the linear activation function at the output layer (Maxwell, 2014).

Using a DWT (discrete wavelet transform) on neural network-based monthly streamflow prediction shows that WANN (wavelet transform and artificial neural network) and WHAFIS (wave height analysis for flood insurance studies) models have better results than ANN and ANFIS (adaptive neuro-fuzzy inference system) (Meque Uamusse, 2015). However, ANN can also forecast

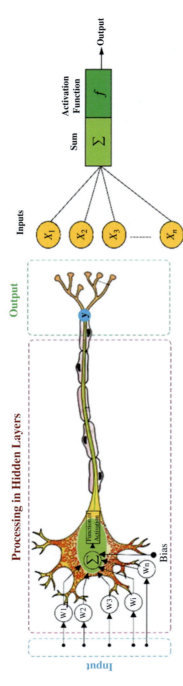

FIG. 12.2 Biological neurons versus artificial neural network.

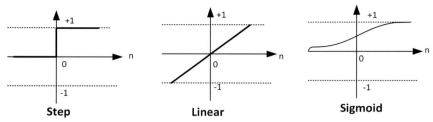

FIG. 12.3 Commonly used transfer functions.

the river flow rate, volume, and duration of rainfall accurately (Ichiyanagi, Kobayashi, & Matsumura, 1993). Prediction of river discharge data and validation using genetic algorithms gives a better prediction model (Panagoulia, Tsekouras, & Kousiouris, 2017).

Experimental data of the Kaplan turbine was used to train the models using ANN, and the results show that it has a good prediction of the turbine parameters under unknown operating conditions (Božić & Jovanović, 2016). The optimum operation policy of reservoirs using an SDP (service delivery platform) model for each month shows that ANNs are a good substitution for simulation models (Haddad, Alimohammadi, & Narmak, 2005). Forecasting of future water storage using a neural network (NN) model with the values of the correlation coefficient indicated that the model fits the variables fairly well (Abdulkadir, Salami, Fatai, & Josiah, 2015). The impact of climate change prediction on the runoff-based hydrometeorological data was used with an ANN model (Nn, Salami, Mohammed, & Okeola, 2014). Based on past project cost data of existing hydropower plants, three different types of ANN—FFBPN (feed-forward back propagation network), GRNN (general regression neural network), and RBFNN (radial basis function neural network)—were compared, and it was found that among the different models, FFBPNN showed the best applicability (Tayyab, Zhou, Zeng, & Adnan, 2016). ANN models can also be used to estimate the reservoir level fluctuations, providing a better way to understand the autoregressive and autoregressive moving average models to provide more scattered results compared to ANNs for all input combinations (Unes, Demirci, & Kişi, 2015).

Wavelet transforms combined with ANN and compared with an adaptive neuro-fuzzy system can be used to predict precipitation. The original time series using wavelet theory was decomposed to multiple subtime series. These subseries were applied as input for ANN and the results revealed that wavelet models and neural networks together with the adaptive neuro-fuzzy system have better performance (Solgi, Nourani, & Pourhaghi, 2014). Abhishek, Singh, Ghosh, and Anand (2012) forecasted the weather using ANN, and concluded that increasing the number of neurons gave a sharp decrease in MSE.

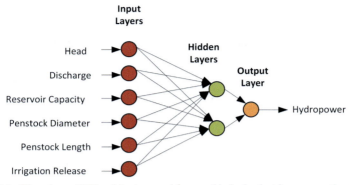

FIG. 12.4 Three-layer ANN architecture used for monthly hydroelectric energy estimation.

Prediction of load using wavelet and particle swarm optimization to optimize the initial weights and thresholds of neural networks gives a better approximation (Li, Deng, Tan, Yang, & Zheng, 2016). ANN architecture for predicting hydropower potential is based on gross head, reservoir capacity, irrigation, monthly inflow, penstock diameter, and length, which are considered as input, whereas energy generation is considered output, as shown in Fig. 12.4.

The improvement in performance characteristics of a hydropower plant is possible with proper scheduling of operation and planning (Gaffar & Aisjah, 2019). For estimation of the potential for hydropower generation, using the GMDH (group method of data handling) and ANN techniques provides an improvement in the results. The self-organizing feature of the GMDH network was found to be appropriate for selecting the best input arguments of the network to optimize the model result (Lopes et al., 2019). ANN approach with feed-forward back propagation was utilized for better performance predictions of a hydropower plant (Hammid, Sulaiman, & Abdalla, 2018). Lee, Jung, and Lee (2019) utilized the energy consumption prediction in South Korea with a correlation coefficient value of 0.6, and a model with an MSE of 1.0203×10^4 was achieved.

Based on the aforementioned literature review, a summary has been prepared and given in Table 12.1.

Based on the various research outcomes, it has been found that much work is still required to achieve the effective utilization of hydropower resources.

12.3 Performance measurement parameters

The mean square error (MSE) measurement is commonly used for the prediction of errors in overall training vectors. The MSE is expressed in Eq. (12.2):

$$MSE = \frac{1}{n}\sum_{i}^{n}(h_i - h_m)^2 \qquad (12.2)$$

314 Sustainable developments by artificial intelligence & machine learning

TABLE 12.1 Summary of research work carried out by various researchers.

Findings of the study	References
Prediction of daily silt load	Kumar et al. (2020)
Prediction of river discharge	Maxwell (2014), Meque Uamusse (2015), Ichiyanagi et al. (1993), Panagoulia et al. (2017)
Prediction of operating condition	Božić and Jovanović (2016)
Operation policy of reservoir	Haddad et al. (2005)
Forecasting of future storage of water	Abdulkadir et al. (2015)
Forecasting of the climate change impact	Nn et al. (2014)
Forecasting of project cost	Tayyab et al. (2016)
Forecasting of reservoir level	Unes et al. (2015)
Precipitation forecasting	Solgi et al. (2014), Abhishek et al. (2012)
Plant load prediction	Li et al. (2016)
Potential estimation	Lopes et al. (2019)

where h_i and h_m are the real and forecasted output values, respectively, for all the ith training vectors and "n" is the entire value of training.

The correlation coefficient (R^2) is used to check the accuracy level of output and is expressed in Eq. (12.3):

$$R^2 = 1 - \sqrt{\frac{\sum_1^n |(h_n - h_m)|^2}{h_n}} \tag{12.3}$$

12.4 Modeling and analysis

Various neural networks were used in this study to train and forecast energy generation. A learning algorithm such as Levenberg-Marquardt was trained in the research to test the selection of predictive models. Feed-forward neural networks (FFNNs) are used in the present analysis. The statistics of the input and output data were grouped for the analysis using ANNs based on 9 inputs and 1 output. Financial year past energy generation data of the plants and energy generation target for the financial year 2020–21 (as a target) are shown in Figs. 12.5 and 12.6, respectively.

There are no standard guidelines for the selection of the numbers of neurons or the numbers of hidden layers. Trial and error may be the correct way. Some numbers of neurons are present and internal layers were checked and evaluated. Each model is tested for execution and the best one is selected, as shown in

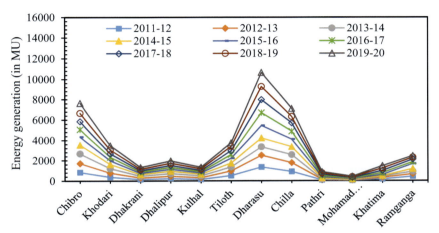

FIG. 12.5 Financial year cumulative energy generation (in MU).

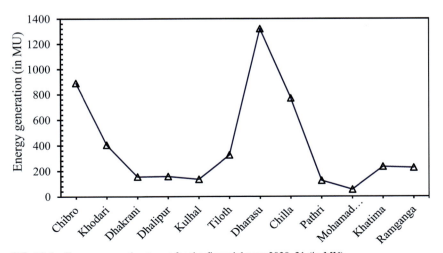

FIG. 12.6 Energy generation target for the financial year 2020–21 (in MU).

316 Sustainable developments by artificial intelligence & machine learning

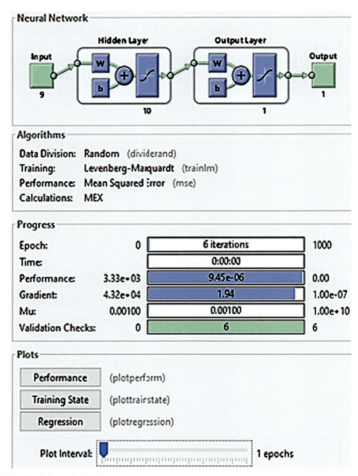

FIG. 12.7 Training of ANN.

Fig. 12.7. In order to train the network, both input and output data were used. Fig. 12.8 shows the different ANN combination characters that were checked to train the network until the full regression coefficient was achieved. Thereafter, several tests were conducted with different hidden layer numbers and neuron numbers in each layer of the ANN and these were chosen to maintain the ANN's maximum regression coefficient (R^2) in each layer.

The algorithm is used to regulate everyone's weighted average. Conversely, Figs. 12.9 and 12.10 show the mean square error (MSE), and the best validation

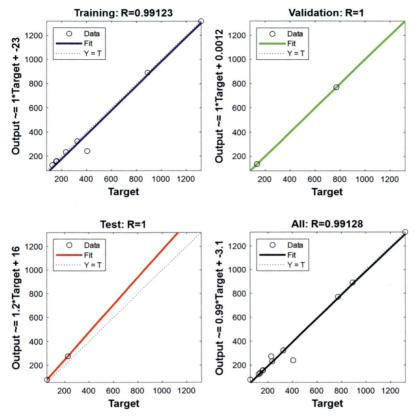

FIG. 12.8 Regression plot.

output value is found to be 0.0000060716, corresponding to the epoch gradient as 1.9375.

The explanations for the disparity over the years in the comparative studies represent various types of significant changes. Firstly, the selection of dataset numbers is used to train the ANN; secondly, there are clear differences in opinion about analytical methods and strategies; and thirdly, it is necessary to understand that how to reach the identified destination for an R^2 value of 1. Based on the above modeling, the generation target for each hydropower plant has been predicted. Fig. 12.11 shows the energy generation target and predicted energy generation of individual hydropower plants.

318 Sustainable developments by artificial intelligence & machine learning

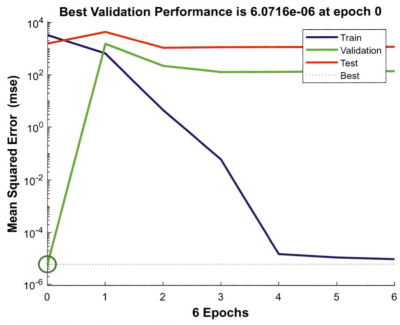

FIG. 12.9 Training, testing, and validation of MSE.

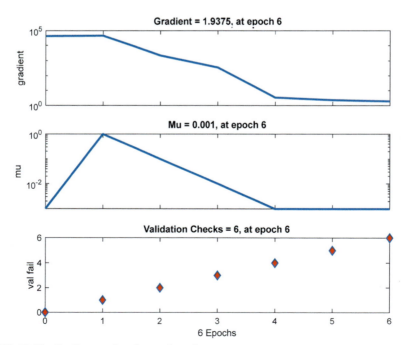

FIG. 12.10 Gradient epochs of network performance.

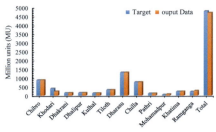

FIG. 12.11 Comparison of target and predicted energy generation.

12.5 Conclusion

In this study, the successful execution of energy generation prediction with the use of ANNs including single-layer perception (SLP) with feed-forward back propagation algorithm was carried out. It has been particularly modeled using the actual yearly energy generation data observed from 2011 to 2020.

The model shows that it has a better correlation coefficient and minimum MSE compared to similar work done by different researchers. The correlation coefficient value for this model is 0.99128 with mean square value (MSE) of 0.0000060716.

References

Abdulkadir, T. S., Salami, A. W., Fatai, B., & Josiah, S. (2015). Neural network based model for forecasting reservoir storage for hydropower dam operation. *International Journal of Engineering Research and General Science, 3*(5), 639–647.

Abhishek, K., Singh, M. P., Ghosh, S., & Anand, A. (2012). Weather forecasting model using artificial neural network. *Procedia Technology, 4,* 311–318. https://doi.org/10.1016/j.protcy.2012.05.047.

Božić, I., & Jovanović, R. (2016). Prediction of double-regulated hydraulic turbine on-cam energy characteristics by artificial neural networks approach. *FME Transactions, 44*(2), 125–132. https://doi.org/10.5937/fmet1602125B.

Gaffar, A. G., & Aisjah, A. S. (2019). Load forecasting for power system planning and operation using artificial neural network (a case study on Larona Hydro Power in the nickel smelting plant)., doi:10.1063/1.5095321.

Haddad, O. B., Alimohammadi, S., & Narmak, T. (2005). *Combining stochastic dynamic programming (SDP) and artificial neural networks (ANN) in optimal reservoir operation* (pp. 985–998). Retrieved from: http://www.academia.edu/download/31023546/496-258.pdf%5Cnhttp://iwtc.info/2005_pdf/14-2.pdf.

Hammid, A. T., Sulaiman, M. H. B., & Abdalla, A. N. (2018). Prediction of small hydropower plant power production in Himreen Lake dam (HLD) using artificial neural network. *Alexandria Engineering Journal, 57*(1), 211–221. https://doi.org/10.1016/j.aej.2016.12.011.

Ichiyanagi, K., Kobayashi, H., & Matsumura, T. (1993). *Application of artificial neural network to forecasting methods of time variation of the flow rate into a dam for a hydro-power plant* (pp. 349–354).

Kumar, K., & Saini, R. P. (2020). Materials today: Proceedings application of machine learning for hydropower plant silt data analysis. *Materials Today: Proceedings*. https://doi.org/10.1016/j.matpr.2020.09.375.

Kumar, K., & Singh, A. K. (2020). *Synchronous condensing mode operation of hydropower plants* (pp. 486–488).

Kumar, K., Singh, A. K., & Singh, R. P. (2020). Power system stabilization tuning and step response test of AVR: A case study. In *2020 6th International Conference on Advanced Computing and Communication Systems ICACCS 2020* (pp. 482–485). https://doi.org/10.1109/ICACCS48705.2020.9074156.

Lee, S., Jung, S., & Lee, J. (2019). Prediction model based on an artificial neural network for user-based building energy consumption in South Korea. *Energies*, *12*. https://doi.org/10.3390/en12040608.

Li, M., Deng, C. H., Tan, J., Yang, W., & Zheng, L. (2016). Research on small hydropower generation forecasting method based on improved BP neural network. In *ICMEMTC* (pp. 1085–1090). https://doi.org/10.2991/icmemtc-16.2016.214.

Lopes, M. N. G., da Rocha, B. R. P., Vieira, A. C., de Sá, J. A. S., Rolim, P. A. M., & da Silva, A. G. (2019). Artificial neural networks approaches for predicting the potential for hydropower generation: A case study for Amazon region. *Journal of Intelligent & Fuzzy Systems*, *36*(6), 5757–5772. https://doi.org/10.3233/jifs-181604.

Maxwell, C. O. (2014). *Prediction of river discharge using neural networks*. Master thesis project University of Nairobi.

Meque Uamusse, M. (2015). Monthly stream flow prediction in Pungwe River for small hydropower plant using wavelet method. *International Journal of Energy and Power Engineering*, *4*(5), 280. https://doi.org/10.11648/j.ijepe.20150405.17.

Nn, A., Salami, A. W., Mohammed, A. A., & Okeola, O. G. (2014). Evaluation of climate change impact on runoff in the kainji lake basin using an artificial neural network model (ANN). *Malaysian Journal of Civil Engineering*, *26*(1), 35–50.

Panagoulia, D., Tsekouras, G. J., & Kousiouris, G. (2017). A multi-stage methodology for selecting input variable in ANN forecasting of river flows. *Global NEST Journal*, *19*(1), 49–57.

Saini, G., & Saini, R. P. (2018). Numerical investigations on hybrid hydrokinetic turbine for electrification in remote area. In *All India seminar on renewable energy for sustainable development*. India: Institution of Engineers.

Saini, G., & Saini, R. P. (2020). Study of installations of hydrokinetic turbines and their environmental effects. *AIP Conference Proceedings*, *2273*. https://doi.org/10.1063/5.0024338, 050022.

Solgi, A., Nourani, V., & Pourhaghi, A. (2014). Forecasting daily precipitation using hybrid model of wavelet-artificial neural network and comparison with adaptive neurofuzzy inference system (case study: Verayneh station, Nahavand). *Advances in Civil Engineering*, *2014*. https://doi.org/10.1155/2014/279368.

Tayyab, M., Zhou, J., Zeng, X., & Adnan, R. (2016). Discharge forecasting by applying artificial neural networks at the Jinsha River Basin, China. *European Scientific Journal*, *12*(9), 108. https://doi.org/10.19044/esj.2016.v12n9p108.

Unes, F., Demirci, M., & Kişi, Ö. (2015). Prediction of Millers Ferry dam reservoir level in USA using artificial neural network. *Periodica Polytechnica Civil Engineering*, *59*(3), 309–318. https://doi.org/10.3311/PPci.7379.

Chapter 13

Response surface methodology-based optimization of parameters for biodiesel production

Pijush Dutta[a], Bittab Biswas[a], Biplab Pal[a], Madhurima Majumder[b], and Amit Kumar Das[c]

[a]*Department of Electronics and Communication Engineering, Global Institute of Management and Technology, Krishnagar, West Bengal, India,* [b]*Department of Electrical & Electronics Engineering, Mirmadan Mohanlal Government Polytechnic, Gobindapur, West Bengal, India,* [c]*Department of Physics, Global Institute of Management and Technology, Krishnagar, West Bengal, India*

13.1 Introduction

Due to the rapid growth of populations and industries, the requirement for energy increases day by day, as well as the reserves of traditional energy becoming depleted (Saluja, Kumar, & Sham, 2016). The major problem of fossil fuel affects the environment substantially, and emission of greenhouse gasses may cause major human health issues; as a result, we have to choose alternative, renewable energy sources (Verma & Sharma, 2016). Today, we are using solar energy, biodiesel, geothermal energy, etc. as nonconventional energy resources to solve the energy crisis (Dalla Longa, Nogueira, Limberger, Wees, & van der Zwaan, 2020; Milosavljević, Pavlović, Mirjanić, & Divnić, 2016; Mohd Noor, Noor, & Mamat, 2018).

Biodiesel is one of the alternative pure renewable fuels that reduce the emission of carbon monoxide and are nondestructive and biodegradable (Joshi, Pandey, Rana, & Rawat, 2017; Mohd Noor et al., 2018; Oumer, Hasan, Baheta, Mamat, & Abdullah, 2018; Saluja et al., 2016; Sitepu, Heimann, Raston, & Zhang, 2020; Yilmaz & Atmanli, 2017). Biodiesel is a product of the transesterification process, where glycerol and biodiesel are produced at specific temperatures after the reaction between lipid and alcohol in the presence of a catalyst (Kirubakaran & Arul Mozhi Selvan, 2018; Narula et al., 2017; Yuvaraj et al., 2019). The quality and amount of biodiesel nonrenewable

---

Sustainable Developments by Artificial Intelligence and Machine Learning for Renewable Energies.
https://doi.org/10.1016/B978-0-323-91228-0.00002-1
Copyright © 2022 Elsevier Inc. All rights reserved.

321

energy depends upon a number of potential inputs, including crude materials, kinds of compound, working conditions, and methanol to oil ratio (Ayoola et al., 2019; Mohamadzadeh Shirazi, Karimi-Sabet, & Ghotbi, 2017).

There are several ideal transesterification processes proposed such as palm oil sludge (POS) (Gu et al., 2018; Luangpaiboon & Aungkulanon, 2020) different types of feedstocks (Banerjee, Barman, Saha, & Jash, 2018), Amari tree seed (Kakati, Gogoi, & Pakshirajan, 2017), eucalyptus oil (Khan, Khan, Yadav, & Sharma, 2017), rubber seed oil (RSO) (Onoji, 2018), refined cottonseed oil, soybean oil (Dai, Kao, & Chen, 2017), waste cooking oil (Degfie, Mamo, & Mekonnen, 2019), algal oil (Garg & Jain, 2020), *Parinari polyandra* seed oil (Ogunkunle & Ahmed, 2019), papaya seed oil (Anwar, Rasul, & Ashwath, 2017), and microalga (Nassef et al., 2019) to biodiesel were investigated. However, biodiesel is also be produced in a constant microchannel through the arrangement of a heterogeneous impetus CaO and CaO/Al_2O_3 from demineralized water plant sediments (Narula et al., 2017) and RSO by lipase-catalyzed transesterification (Aarathi, Harshita, Nalinashan, Ashok, & Prasad, 2019). To obtain the optimum value of input parameter response surface methodology (RSM) cum Box-Behnken Design (BBD) used (Wong, Isa, & Bashir, 2018). Biodiesel is also be used by transesterification process from waste cooking oil and catalyzed $(Ca(OCH_3)_2)$ with the assistance of RSM (Mehta, Divya, & Jha, 2019).

An experiment on a mixture of diesel, mahua methyl ester (MME), and methanol utilized on the IDI motor (Prasada Rao & Appa Rao, 2017). A method for methyl ester creation from melon oil. RSM, given a five-level, four variable central composite design (CCD) to identify the impact of the process parameter (Onoji, 2018). RSM model also used for finding the optimum process parameters of Pongamia biodiesel (Mehta et al., 2019) and finally compared with experimental output.

There are several nonlinear models that have been utilized for parametric optimization to produce biodiesel, such as response methodology (Ayoola et al., 2020; Bello, Ogedengbe, Lajide, & Daniyan, 2016; Kumar, Jain, & Kumar, 2017; Mehta et al., 2019; Ogunkunle & Ahmed, 2019; Venkatesan, Nalarajan, Sivamani, & Thomai, 2020; Yatish, Lalithamba, Suresh, Arun, & Kumar, 2016), Taguchi (Dhawane, Bora, Kumar, & Halder, 2017), multivariable analysis (Zahed, Zakeralhosseini, Mohajeri, Bidhendi, & Mesgari, 2018), Box-Behnken RSM (Wong et al., 2018), and an artificial neural network (ANN) linear model (Kumar et al., 2017).

Different advanced techniques like antlion algorithm (ALO) (Dinkar & Deep, 2019), the fuzzy logic model were proposed for optimized parameters of waste cooking oil to produce biodiesel (Dutta & Kumar, 2017, 2018; Inayat et al., 2019). To upgrade the transesterification process parameters, both RSM and ANN models were applied (Garg & Jain, 2020; Selvaraj, Moorthy, Kumar, & Sivasubramanian, 2019). Biodiesel also be produced from the transesterification process of castor oil (Obayomi, Bello, Ogundipe, & Olawale, 2020). To upgrade the transesterification process parameters, ANN (Nayak,

Dhanarajan, Dineshkumar, & Sen, 2018) and multiobjective NSGA-II (Jaliliantabar, Ghobadian, Najafi, Mamat, & Carlucci, 2019) were proposed. The results show good accuracy. There are several hybrid optimization techniques were used for parametric optimization of palm oil to produce the biodiesel products (Luangpaiboon & Aungkulanon, 2020), parametric optimization of fuzzy logic controller modeled particle swarm optimization (PSO) (Nassef et al., 2019), genetic algorithm optimized neural network model (Sivamani, Selvakumar, Rajendran, & Muthusamy, 2019), adaptive fuzzy inference system (Najafi, Faizollahzadeh Ardabili, Shamshirband, Chau, & Rabczuk, 2018), genetic algorithm coupled adaptive fuzzy inference (Ogaga Ighose et al., 2017).

Biodiesel is produced with the addition of three mixes of fish oil: B15 (15% fish oil + 85% diesel), B30 (30% fish oil + 70% diesel), and B45 (45% fish oil + 55% diesel) (Yuvaraj et al., 2019). Results have been confirmed by a 95% confidence level. From the experimental results on the transesterification process of biodiesel from waste frying vegetable oil (WFVO) and waste frying palm oil (WFPO) (Aworanti, Ajani, & Agarry, 2019), it has been seen that the optimum biodiesel is produced when the ratio of WFVO and WFPO is 97:90. RSM with CCD has been carried out to find out the ideal working conditions to upgrade biodiesel (Wong et al., 2018). The anticipated output has been found in acceptable agreement with the experimental value, with R2 = 0.9902.

Elephant swarm water search algorithm (ESWSA) is one of the modern metaheuristic optimization techniques, and was motivated by the water asset search techniques of herds of intelligent and social elephants during the dry season (Mandal, Dutta, & Kumar, 2019). This optimization technique can be effectively applied in different process systems for parameters tuning, due to the presence of the least number of parameters setting being required. Until now, ESWSA has been applied to numerous domains, for example, displaying the fluid stream process, parametric enhancement of the photovoltaic cell, and demonstrating the welding process (Mandal et al., 2019).

The fundamental goal of this work is to propose a model so that we can demonstrate biodiesel generation more precisely and dependably. In this work, we used a novel metaheuristic optimization technique, ESWSA, by presenting the changing switching probability and Levy-based global search. Here, RSM is utilized for the numerical models (which are upgraded utilizing improved adaptations of ESWSA) for the biodiesel creation process. The remainder of this chapter is composed as follows. Section 13.2 demonstrates the biodiesel process. In this segment, the test arrangement of biodiesel is portrayed to acquire the trial information. In Section 13.3, RSM-based nonlinear numerical models for the process are explained. The philosophy of the ESWSA metaheuristic is also explained. In Section 13.4, reenacted results, conversations, and examinations are given. The points of interest and disadvantages of these improved forms of ESWSA are additionally examined. Finally, Section 13.5 concludes the chapter and is followed by the references.

13.2 Problem formulation

Waste cooking oil was acquired from Afe Babalola University, Nigeria, and was separated and prewarmed at a temperature of 100°C to expel polluting influences. Methanol has been pick as an asset of liquor because of inexpensive and responds quickly. The grouping of HCl and sodium hydroxide was set up in the range of 1.0–2.0 wt%. Waste cooking oil was separated and prewarmed at a temperature to evacuate water and other unpredictable contaminants. HCl at a concentration of 1.4 wt% was added to methanol, with methanol the larger part in a molar proportion of 6:1 to oil. The blend was delicately mixed in the blending tank for 10 minutes to frame methanolic HCl. The blend was then mixed at 305 rpm and a temperature of 60°C for 3 hours. The product was released into the isolating channel and stored for 24 hours, at the end of which the product contained two particular layers. The top layer was a blend of water and methanol, and the base layer, the transesterified oil. Next, this transesterified oil was returned to the reactor and blended sodium methoxide was added. Once again, blend mixing was carried out in the reactor at 305 rpm and 60°C for 3 hours. After that the final product contained two different particular layers: the lighter top layer was biodiesel and the thick base, glycerol. Table 13.1 shows the name and ranges of the input parameters.

In this research, the effect of output (percentage of biodiesel production) of the process system was designed using four-level and five-factor potential independent parameters: reaction time (hours), temperature (°C), stir speed (rpm), catalyst concentration (wt%), and methanol-oil ratio. The following input process parameters were varied. Sixteen sets of input and output parameters combinations are shown in Table 13.2.

13.3 Mathematical model of biodiesel production

Due to the nonlinear characteristics of biodiesel production, a variation of methanol to oil ratio and catalyst concentration has been observed with the change in reaction time and temperature. Moreover, biodiesel depends upon the methanol to oil ratio, catalyst concentration, stir speed, and reaction temperature. So it is

TABLE 13.1 Input parameters ranges.

Si no.	Name of the potential input parameters	Ranges
1	Reaction time	1–5 h
2	Reaction temperature	40–100°C
3	Stir speed	200–400 rev/min
4	Catalyst concentration	1–2 wt%
5	Methanol to oil ratio	4:1–9:1

TABLE 13.2 Input and output of process parameters sets.

Si no.	Reaction time (h)	Reaction temperature (°C)	Stir speed (rpm)	Catalyst concentration (wt%)	Methanol to oil	Yield (%)
1	A	B	C	D	E	Y
2	1	40	400	2	9	88.8
3	1	40	200	2	4	89.9
4	5	40	400	1	9	90
5	5	90	400	1	4	81.5
6	3	102.8	300	1.5	6.5	80
7	1	90	400	2	9	84.2
8	1	40	200	2	9	91
9	5	40	200	2	4	70
10	5	90	200	2	9	78
11	5	40	400	2	4	86.7
12	1	40	200	2	4	90.2
13	3	65	300	1.5	6.5	78
14	5	90	400	2	4	90
15	5	90	200	1	4	82.8
16	5	90	400	2	9	81

326 Sustainable developments by artificial intelligence & machine learning

very tedious and hard to recalibrate the analysis each time if any change is made to the predominant input parameter.

To improve the calibration, a mathematical model is implemented by using an empirical tool—RSM or ANOVA (analysis of variance)—and finding the optimal values of prevailing input parameters by using computational intelligence. These numerical models are based on nonlinear connections between input process parameters: reaction time, temperature, catalyst concentration, methanol to oil ratio, and stir speed, with the output process parameter being production of biodiesel. These models will assist with discovering the ideal working condition and to foresee the production of biodiesel under a specific condition (for example for various estimations of reaction time, temperature, catalyst concentration, methanol to oil ratio, and stir speed) without recalibration. There are various nonstraight scientific models to depict a procedure, however, the most well-known nonlinear models are RSM (Dutta & Kumar, 2017, 2018) and ANOVA (Dutta, Mandal, & Kumar, 2018, 2019).

In the present research, we utilized RSM (Dutta et al., 2018; Dutta & Kumar, 2020; Mandal et al., 2019) by and large for experimental model structure and breaking down an issue. The broadest utilization of RSM is in the specific circumstances where a few information factors conceivably impact the execution or reaction of the procedure. In RSM, a higher request polynomial condition is utilized to depict the connections among factors. The ensuing solicitation models are versatile and can take on the wide grouping of utilitarian structures. Therefore, in this present research utilizing RSM, production of biodiesel can be expressed in terms of reaction time, temperature, catalyst concentration, methanol to oil ratio, and stir speed using an RSM-based model as follows:

$$Y = \beta_0 + \sum_{j=1}^{k} \beta_j X_j + \sum_{j=1}^{k} \beta_j X_j^2 + \sum_{j=1}^{k} \sum_{i=1}^{k} \beta_{ij} X_i X_i \qquad (13.1)$$

where $Y =$ Response (solar parabolic collector yield), β_0, β_j, and β_{ij} are the regression coefficients, with $i, j = 1, 2, \ldots, k$ and X_i are the k input variables.

13.3.1 Optimization of the mathematical model

Discovering the exact estimations of the coefficients of a solar parabolic collector needs an efficient optimization technique. Ordinarily, metaheuristics can be utilized for this sort of improvement to such an extent that the determined attribute of the sun-oriented parabolic collector fits with the trial one. Hence the temperature and liquid discharged should satisfy the three variable RSM mathematical equations:

$$f(A, B, C) = \beta_0 + \beta_1 \cdot A + \beta_2 \cdot B + \beta_3 \cdot C + \beta_{11} \cdot A^2 + \beta_{22} \cdot B^2 + \beta_{33} \cdot C^2 + \beta_{12} \cdot A \cdot B$$
$$+ \beta_{23} \cdot B \cdot C + \beta_{13} \cdot A \cdot C \qquad (13.2)$$

$$X = \{\beta_0, \beta_1, \beta_2, \beta_2, \beta_{11}, \beta_{22}, \beta_{33}, \beta_{12}, \beta_{23}, \beta_{13}\} \qquad (13.3)$$

13.3.2 Proposed methodology

From Eqs. (13.2) and (13.3), it is seen that the RSM mathematical equation of production of biodiesel contains multiple numbers of coefficients that can be obtained from it. In the next stage, our objective is to apply a proficient meta-heuristic technique to obtain the optimum value of the input process parameter corresponding to a given target output: percentage of biodiesel (by evaluating the ideal or best arrangement of qualities for the model parameters). In this exploration work, we have utilized a fundamental ESWSA advancement strategy (Mandal et al., 2019) for streamlining or demonstrating a sun-oriented parabolic collector control issue.

13.3.3 Basic elephant swarm water search algorithm (ESWSA)

The elephant swarm water search algorithm was proposed by Mandal (2018) and Mandal et al. (2019). This optimization was, for the most part, based on the water search methodology of elephant herds (swarms) during a dry spell, with the assistance of various correspondence procedures.

Let the optimization problem be d-dimensional and the elephant groups randomly placed. Now, the position of the i-th elephant group of a swarm after t-th iteration is given as:

$$X_{i,d}^t = (x_{i1}, x_{i2}, ..., x_{id})$$ (13.4)

and velocity is given by:

$$V_{i,d}^t = (v_{i1}, v_{i2}, ..., v_{id})$$ (13.5)

Locally, the best solution by i-th elephant group at current iteration is:

$$P_{best,i,d}^t = (P_{i1}, P_{i2}, ..., P_{id})$$ (13.6)

Globally, the best solution is denoted by:

$$G_{best,d}^t = (G_1, G_2, ..., G_d)$$ (13.7)

These values are updated according to the following equations depending on the switching probability (p):

$$V_{i,d}^{t+1} = V_{i,d}^t * \omega^t + \epsilon \odot (G_{best,d}^t - X_{i,d}^t)$$ (13.8)

$$V_{i,d}^{t+1} = V_{i,d}^t * \omega^t + \epsilon \odot (P_{best,i,d}^t - X_{i,d}^t)$$ (13.9)

where the range of ϵ is:

$$\omega^t = \omega_{max} - \left\{ \frac{\omega_{max} - \omega_{min}}{t_{max}} \right\} \times t$$ (13.10)

where t_{max}, ω_{max}, and ω_{min} are the values of maximum iteration number, upper boundary (0.6) and lower boundary (0.4) of the inertia weight, respectively

$$X_{i,d}^{t+1} = V_{i,d}^{t+1} + X_{i,d}^{t} \tag{13.11}$$

The optimum output is obtained from the best position value of the algorithm. It has been found from the literature (Mandal et al., 2019) that $p = 0.6$ gives unrivaled execution for ESWSA. In this way, we have additionally utilized this incentive for our current research.

13.4 Methodology

To confirm its results, the proposed elephant swarm water search algorithm (ESWSA) is tested against parameters or coefficients estimation issues for determining the process of biodiesel production. Here, ESWSA is optimized and compared to the RSM-based model as depicted in previous sections. The trial dataset has been acquired from the examination as referenced in the prior segment. The dataset comprises 16 input process parameters of reaction temperature, reaction time, stir speed, catalyst concentration, and methanol to oil ratio. This exploratory dataset has been utilized to design the RSM-based nonlinear model of biodiesel production. Fig. 13.1 shows the objective function obtained by the difference between target response percentage of the biodiesel and RSM model. Finally, this objective function is applied to the metaheuristic optimization to find the optimal input parameters when the root mean square of the objective function is minimum. For the optimization of the RSM-based ESWSA algorithm, we used MATLAB 2013b version, and the specification of the PC is 4 GB RAM, Intel(R) Core (TM) i3processor with Windows7 operating System.

In this study, the optimization problem model of biodiesel production shown in Fig. 13.1 (Bello et al., 2016) represents the regression coefficient of potential independent components of biodiesel production by using the RSM model with

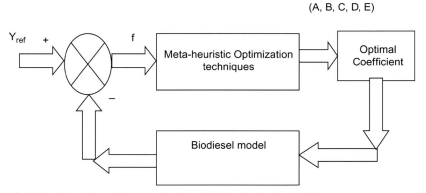

FIG. 13.1 Generation of objective function. *(No Permission Required.)*

Response surface methodology (RSM)-based optimization technique **Chapter | 13 329**

the help of Minitab 17. Eqs. (13.13) and (13.14) represent nonlinear regression equations for biodiesel where the target biodiesel output is set at $Y_{ref} = 90.6\%$. Finally, Eq. (13.14) is optimized by the novel elephant swarm water search algorithm.

$$f = Y_{ref} - f_1 \tag{13.12}$$

$$\begin{aligned} f = \big(Y_{ref} - (178.82 - 19.076 * A - 0.8864 * B - 0.02283 * C \\ - 37.54 * D + 0.6875 * E + 1.6247 * A^{\wedge}2 + 0.003622 * B^{\wedge}2 \\ + 0.03950 * A * B + 0.023625 * A * C - 0.4975 * A * E \\ - 0.001370 * B * C + 0.4160 * B * D + 0.02150 * C * D)\big) \end{aligned} \tag{13.13}$$

where $Y_{ref} = 90.6$ and

$$\begin{aligned} f_1 = 90.6 - (178.82 - 19.076 * A - 0.8864 * B - 0.02283 * C \\ - 37.54 * D + 0.6875 * E + 1.6247 * A^{\wedge}2 + 0.003622 * B^{\wedge}2 \\ + 0.03950 * A * B + 0.023625 * A * C - 0.4975 * A * E \\ - 0.001370 * B * C + 0.4160 * B * D + 0.02150 * C * D) \end{aligned} \tag{13.14}$$

13.5 Reaction conditions by RSM

In this research, RSM was employed to make a relationship between the response and potential independent input variables of the biodiesel process shown in Eqs. (13.13) and (13.14). The coded and uncoded potential input parameters for response are shown in Table 13.3. Here, our main objective is to obtain the optimum input parameters corresponding to a target response percentage in biodiesel production. The target output response variable, biodiesel, is set to 90.6%. Table 13.4 shows the optimum input parameters of biodiesel by using RSM and RSM-based ESWSA. From the calculation, it seems that the error obtained from the RSM model is 34.866%, 29.48%, 30.93%, 16.42%, and 33.33% for reaction time, reaction temperature, stir speed, catalyst concentration, and methanol to oil ratio, respectively. But if this RSM-based nonlinear model is optimized by the elephant swarm water search algorithm (ESWSA), then the error of these potential input parameters is reduced to 3.66%, 3.23%, 1.52%, 5.71%, and 1.66%, respectively. The response and observed values showed a better correlation coefficient, which implies that this quadratic model was highly significant.

13.6 Surface plot by different combinations in RSM model

Figs. 13.2– 13.11 represent the 3D surface plot of biodiesel production with respect to different combinations of input variables (reaction time, reaction temperature, methanol to oil ratio, catalyst concentration) in Minitab. Each of the figures shows how the output—biodiesel production—was affected corresponding to the combination of two input variables.

330 Sustainable developments by artificial intelligence & machine learning

TABLE 13.3 Different coefficient of an RSM model from Minitab.

Si no.	Terms	Coefficient
1	Constant	178.82
2	Reaction time	−19.076
3	Reaction temperature	−0.8864
4	Stir speed	−0.02283
5	Catalyst concentration	−37.54
6	Methanol to oil ratio	0.6875
7	Reaction time * Reaction time	1.624
8	Reaction temperature * Reaction temperature	0.0036
9	Reaction time * Reaction temperature	0.0395
10	Reaction time * stir speed	0.0236
11	Reaction time * Methanol to oil	−0.4975
12	Reaction Temperature * Stir speed	−0.00137
13	Reaction temperature * Catalyst concentration	0.416
14	Stir speed * Catalyst concentration	0.0215

TABLE 13.4 Comparison study between RSM and ESWSA-based RSM models.

Variables	True value	RSM		RSM-based ESWSA	
		Measured value	% Error	Measured value	% Error
Reaction time (h) (A)	3	1.954	34.866	2.89	3.66
Reaction temperature (°C)	58	40.99	29.48	56.125	3.23
Stir speed (rpm)	305.5	400	30.93	310.17	1.52
Catalyst concentration (wt%)	1.4	1.63	16.42	1.48	5.71
Methanol to oil yield (%)	6:1	4:1	33.33	5:1	1.66

Response surface methodology (RSM)-based optimization technique Chapter | 13 **331**

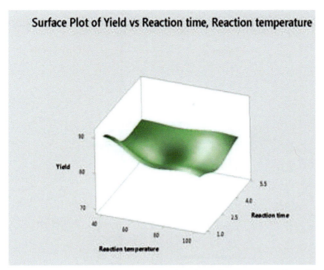

FIG. 13.2 Surface plot of biodiesel vs stir speed and catalyst concentration. *(No Permission Required.)*

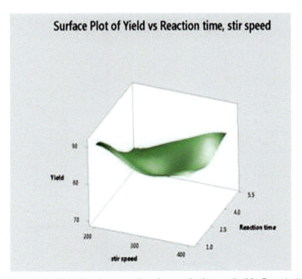

FIG. 13.3 Surface plot of biodiesel vs reaction time and stir speed. *(No Permission Required.)*

13.7 Conclusion

Production of biodiesel is dependent upon the properties of potential independent parameters: methanol to oil ratio, catalyst concentration, stir speed, reaction time, and reaction temperature. In this research, we used waste cooking oil to produce biodiesel by transesterification. To complete the transesterification,

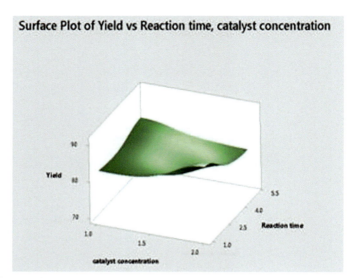

FIG. 13.4 Surface plot of biodiesel vs stir speed and methanol to oil ratio. *(No Permission Required.)*

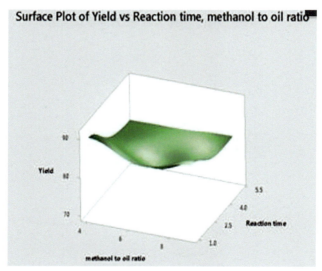

FIG. 13.5 Surface plot of biodiesel vs reaction temperature and catalyst concentration. *(No Permission Required.)*

other ingredients such as methanol to oil ratio, catalyst concentration, stir speed, reaction time, and reaction temperature are also required. So, to obtain the desired biodiesel it is necessary to achieve the optimum contribution of potential independent input parameters of the process.

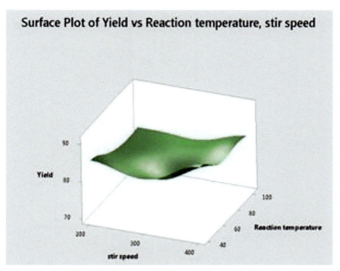

FIG. 13.6 Surface plot of biodiesel vs reaction temperature and methanol to oil ratio. *(No Permission Required.)*

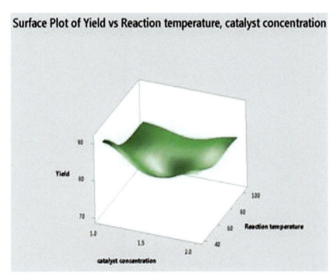

FIG. 13.7 Surface plot of biodiesel vs reaction temperature and stir speed. *(No Permission Required.)*

In the first phase of the research, we used RSM to determine the nonlinear model of the production of biodiesel. For a given target value of biodiesel (as a percentage) we calculated the potential input parameters. In the second phase, a nonlinear model of RSM of biodiesel parametric optimization was done and the potential input parameters calculated. A comparative study was done between

334 Sustainable developments by artificial intelligence & machine learning

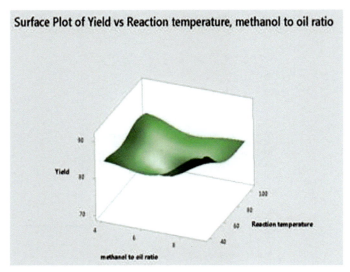

FIG. 13.8 Surface plot of biodiesel vs reaction time and methanol to oil ratio. *(No Permission Required.)*

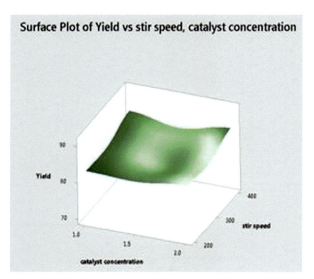

FIG. 13.9 Surface plot of biodiesel vs reaction time and catalyst concentration. *(No Permission Required.)*

the actual potential input parameters and the input parameters by calculated the RSM model and optimized RSM-based ESWSA model. From the resulting analysis, it is seen that the RSM-based ESWSA optimized model has maximum accuracy (96.863%) compared to the RSM based model (71.24%). It is also found that the identified optimum parameter levels were experimentally confirmed.

Response surface methodology (RSM)-based optimization technique Chapter | 13 **335**

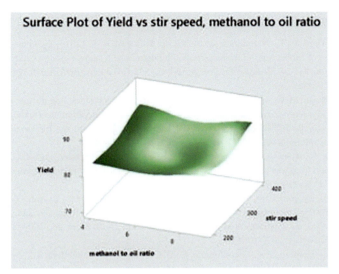

FIG. 13.10 Surface plot of biodiesel vs catalyst concentration and methanol to oil ratio. *(No Permission Required.)*

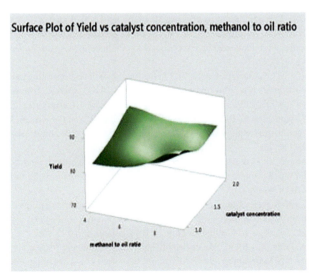

FIG. 13.11 Surface plot of biodiesel vs reaction time and reaction temperature. *(No Permission Required.)*

In the future, a better predictive model can be implemented by using nonlinear models like analysis of variance and ANN optimized by other metaheuristic optimization techniques.

336 Sustainable developments by artificial intelligence & machine learning

References

Aarathi, V., Harshita, E., Nalinashan, A., Ashok, S., & Prasad, R. K. (2019). Synthesis and characterisation of rubber seed oil trans-esterified biodiesel using cement clinker catalysts. *International Journal of Sustainable Energy, 38*(4), 333–347. https://doi.org/10.1080/14786451.2017.1414052.

Anwar, M., Rasul, M. G., & Ashwath, N. (2017). Optimization of biodiesel production process from papaya (*Carica papaya*) seed oil. In *2017 IEEE 7th international conference on power and energy systems, ICPES 2017* (pp. 131–134). Institute of Electrical and Electronics Engineers Inc. https://doi.org/10.1109/ICPESYS.2017.8215935. Vols. 2017-.

Aworanti, O. A., Ajani, A. O., & Agarry, S. E. (2019). Process parameter estimation of biodiesel production from waste frying oil (vegetable and palm oil) using homogeneous catalyst. *Journal of Food Processing & Technology, 10*, 1–10.

Ayoola, A. A., Hymore, F. K., Omonhinmin, C. A., Babalola, P. O., Fayomi, O. S. I., Olawole, O. C., et al. (2020). Response surface methodology and artificial neural network analysis of crude palm kernel oil biodiesel production. *Chemical Data Collections, 28*, 100478. https://doi.org/10.1016/j.cdc.2020.100478.

Ayoola, A. A., Hymore, F. K., Omonhinmin, C. A., Olawole, O. C., Fayomi, O. S. I., Babatunde, D., et al. (2019). Analysis of waste groundnut oil biodiesel production using response surface methodology and artificial neural network. *Chemical Data Collections, 22*, 100238. https://doi.org/10.1016/j.cdc.2019.100238.

Banerjee, N., Barman, S., Saha, G., & Jash, T. (2018). Optimization of process parameters of biodiesel production from different kinds of feedstock. *Materials Today: Proceedings, 5*(11), 23043–23050. Elsevier Ltd. https://doi.org/10.1016/j.matpr.2018.11.033.

Bello, E. I., Ogedengbe, T. I., Lajide, L., & Daniyan, I. A. (2016). Optimization of process parameters for biodiesel production using response surface methodology. *American Journal of Energy Engineering, 4*(2), 8–16.

Dai, Y. M., Kao, I. H., & Chen, C. C. (2017). Evaluating the optimum operating parameters of biodiesel production process from soybean oil using the Li2TiO3 catalyst. *Journal of the Taiwan Institute of Chemical Engineers, 70*, 260–266. https://doi.org/10.1016/j.jtice.2016.11.001.

Dalla Longa, F., Nogueira, L. P., Limberger, J., Wees, J. D.v., & van der Zwaan, B. (2020). Scenarios for geothermal energy deployment in Europe. *Energy, 206*. https://doi.org/10.1016/j.energy.2020.118060.

Degfie, T. A., Mamo, T. T., & Mekonnen, Y. S. (2019). Optimized biodiesel production from waste cooking oil (WCO) using calcium oxide (CaO) Nano-catalyst. *Scientific Reports, 9*(1). https://doi.org/10.1038/s41598-019-55403-4.

Dhawane, S. H., Bora, A. P., Kumar, T., & Halder, G. (2017). Parametric optimization of biodiesel synthesis from rubber seed oil using iron doped carbon catalyst by Taguchi approach. *Renewable Energy, 105*, 616–624. https://doi.org/10.1016/j.renene.2016.12.096.

Dinkar, S. K., & Deep, K. (2019). Process optimization of biodiesel production using antlion optimizer. *Journal of Information and Optimization Sciences*, 1281–1294. https://doi.org/10.1080/02522667.2018.1491821.

Dutta, P., & Kumar, A. (2017). Intelligent calibration technique using optimized fuzzy logic controller for ultrasonic flow sensor. *Mathematical Modelling of Engineering Problems*, 91–94. https://doi.org/10.18280/mmep.040205.

Dutta, P., & Kumar, A. (2018). Design an intelligent flow measurement technique by optimized fuzzy logic controller. *Journal Européen Des Systèmes Automatisés, 51*, 89–107. https://doi.org/10.3166/jesa.51.89-107.

Dutta, P., & Kumar, A. (2020). Modelling of liquid flow control system using optimized genetic algorithm. *Statistics, Optimization and Information Computing, 8*(2), 565–582. https://doi.org/10.19139/soic-2310-5070-618.

Dutta, P., Mandal, S., & Kumar, A. (2018). Comparative study: FPA based response surface methodology & ANOVA for the parameter optimization in process control. *Advances in Modelling and Analysis C, 73*(1), 23–27. https://doi.org/10.18280/ama_c.730104.

Dutta, P., Mandal, S., & Kumar, A. (2019). Application of FPA and ANOVA in the optimization of liquid flow control process. *Review of Computer Engineering Studies*, 7–11. https://doi.org/10.18280/rces.050102.

Garg, A., & Jain, S. (2020). Process parameter optimization of biodiesel production from algal oil by response surface methodology and artificial neural networks. *Fuel, 277*, 118254. https://doi.org/10.1016/j.fuel.2020.118254.

Gu, J., Gao, Y., Xu, X., Wu, J., Yu, L., Xin, Z., et al. (2018). Biodiesel production from palm oil and mixed dimethyl/diethyl carbonate with controllable cold flow properties. *Fuel, 216*, 781–786. https://doi.org/10.1016/j.fuel.2017.09.081.

Inayat, A., Nassef, A. M., Rezk, H., Sayed, E. T., Abdelkareem, M. A., & Olabi, A. G. (2019). Fuzzy modeling and parameters optimization for the enhancement of biodiesel production from waste frying oil over montmorillonite clay K-30. *Science of the Total Environment, 666*, 821–827. https://doi.org/10.1016/j.scitotenv.2019.02.321.

Jaliliantabar, F., Ghobadian, B., Najafi, G., Mamat, R., & Carlucci, A. P. (2019). Multi-objective NSGA-II optimization of a compression ignition engine parameters using biodiesel fuel and exhaust gas recirculation. *Energy, 187*. https://doi.org/10.1016/j.energy.2019.115970.

Joshi, G., Pandey, J. K., Rana, S., & Rawat, D. S. (2017). Challenges and opportunities for the application of biofuel. *Renewable and Sustainable Energy Reviews, 79*, 850–866. https://doi.org/10.1016/j.rser.2017.05.185.

Kakati, J., Gogoi, T. K., & Pakshirajan, K. (2017). Production of biodiesel from Amari (Amoora Wallichii King) tree seeds using optimum process parameters and its characterization. *Energy Conversion and Management, 135*, 281–290. https://doi.org/10.1016/j.enconman.2016.12.087.

Khan, O., Khan, M. E., Yadav, A. K., & Sharma, D. (2017). The ultrasonic-assisted optimization of biodiesel production from eucalyptus oil. *Energy Sources, Part A: Recovery, Utilization and Environmental Effects, 39*(13), 1323–1331. https://doi.org/10.1080/15567036.2017.1328001.

Kirubakaran, M., & Arul Mozhi Selvan, V. (2018). A comprehensive review of low cost biodiesel production from waste chicken fat. *Renewable and Sustainable Energy Reviews, 82*, 390–401. https://doi.org/10.1016/j.rser.2017.09.039.

Kumar, S., Jain, S., & Kumar, H. (2017). Process parameter assessment of biodiesel production from a Jatropha–algae oil blend by response surface methodology and artificial neural network. *Energy Sources, Part A: Recovery, Utilization and Environmental Effects, 39*(22), 2119–2125. https://doi.org/10.1080/15567036.2017.1403514.

Luangpaiboon, P., & Aungkulanon, P. (2020). Hybrid approach for optimizing process parameters in biodiesel production from palm oil. In *Vol. 834. Key Engineering Materials* (pp. 16–23). Trans Tech Publications Ltd. https://doi.org/10.4028/www.scientific.net/KEM.834.16.

Mandal, S. (2018). Elephant swarm water search algorithm for global optimization. *Sadhana—Academy Proceedings in Engineering Sciences, 43*(1). https://doi.org/10.1007/s12046-017-0780-z.

Mandal, S., Dutta, P., & Kumar, A. (2019). Modeling of liquid flow control process using improved versions of elephant swarm water search algorithm. *SN Applied Sciences, 1*(8). https://doi.org/10.1007/s42452-019-0914-5.

Mehta, K., Divya, N., & Jha, M. K. (2019). Application of RSM for optimizing the biodiesel production catalyzed by calcium methoxide. In *Vol. 30. Lecture Notes in Civil Engineering* (pp. 75–83). Springer. https://doi.org/10.1007/978-981-13-6717-5_8.

338 Sustainable developments by artificial intelligence & machine learning

Milosavljević, D. D., Pavlović, T. M., Mirjanić, D. L. J., & Divnić, D. (2016). Photovoltaic solar plants in the Republic of Srpska—Current state and perspectives. *Renewable and Sustainable Energy Reviews*, *62*, 546–560. https://doi.org/10.1016/j.rser.2016.04.077.

Mohamadzadeh Shirazi, H., Karimi-Sabet, J., & Ghotbi, C. (2017). Biodiesel production from Spirulina microalgae feedstock using direct transesterification near supercritical methanol condition. *Bioresource Technology*, *239*, 378–386. https://doi.org/10.1016/j.biortech.2017.04.073.

Mohd Noor, C. W., Noor, M. M., & Mamat, R. (2018). Biodiesel as alternative fuel for marine diesel engine applications: A review. *Renewable and Sustainable Energy Reviews*, *94*, 127–142. https://doi.org/10.1016/j.rser.2018.05.031.

Najafi, B., Faizollahzadeh Ardabili, S., Shamshirband, S., Chau, K. W., & Rabczuk, T. (2018). Application of anns, anfis and rsm to estimating and optimizing the parameters that affect the yield and cost of biodiesel production. *Engineering Applications of Computational Fluid Mechanics*, *12*(1), 611–624. https://doi.org/10.1080/19942060.2018.1502688.

Narula, V., Khan, M. F., Negi, A., Kalra, S., Thakur, A., & Jain, S. (2017). Low temperature optimization of biodiesel production from algal oil using CaO and CaO/Al2O3 as catalyst by the application of response surface methodology. *Energy*, *140*, 879–884. https://doi.org/10.1016/j.energy.2017.09.028.

Nassef, A. M., Sayed, E. T., Rezk, H., Abdelkareem, M. A., Rodriguez, C., & Olabi, A. G. (2019). Fuzzy-modeling with particle swarm optimization for enhancing the production of biodiesel from microalga. *Energy Sources, Part A: Recovery, Utilization and Environmental Effects*, *41*(17), 2094–2103. https://doi.org/10.1080/15567036.2018.1549171.

Nayak, M., Dhanarajan, G., Dineshkumar, R., & Sen, R. (2018). Artificial intelligence driven process optimization for cleaner production of biomass with co-valorization of wastewater and flue gas in an algal biorefinery. *Journal of Cleaner Production*, *201*, 1092–1100. https://doi.org/10.1016/j.jclepro.2018.08.048.

Obayomi, K. S., Bello, J. O., Ogundipe, T. A., & Olawale, O. (2020). Extraction of castor oil from castor seed for optimization of biodiesel production. *IOP Conference Series: Earth and Environmental Science*, *445*(1), 012055. https://doi.org/10.1088/1755-1315/445/1/012055.

Ogaga Ighose, B., Adeleke, I. A., Damos, M., Adeola Junaid, H., Ernest Okpalaeke, K., & Betiku, E. (2017). Optimization of biodiesel production from *Thevetia peruviana* seed oil by adaptive neuro-fuzzy inference system coupled with genetic algorithm and response surface methodology. *Energy Conversion and Management*, *132*, 231–240. https://doi.org/10.1016/j.enconman.2016.11.030.

Ogunkunle, O., & Ahmed, N. A. (2019). Response surface analysis for optimisation of reaction parameters of biodiesel production from alcoholysis of *Parinari polyandra* seed oil. *International Journal of Sustainable Energy*, *38*(7), 630–648. https://doi.org/10.1080/14786451.2018.1554661.

Onoji, S. E. (2018). *Synthesis of biodiesel from rubber seed oil for internal compression ignition engine.*

Oumer, A. N., Hasan, M. M., Baheta, A. T., Mamat, R., & Abdullah, A. A. (2018). Bio-based liquid fuels as a source of renewable energy: A review. *Renewable and Sustainable Energy Reviews*, *88*, 82–98. https://doi.org/10.1016/j.rser.2018.02.022.

Prasada Rao, K., & Appa Rao, B. V. (2017). Parametric optimization for performance and emissions of an IDI engine with Mahua biodiesel. *Egyptian Journal of Petroleum*, *26*(3), 733–743. https://doi.org/10.1016/j.ejpe.2016.10.003.

Saluja, R. K., Kumar, V., & Sham, R. (2016). Stability of biodiesel—A review. *Renewable and Sustainable Energy Reviews*, *62*, 866–881. https://doi.org/10.1016/j.rser.2016.05.001.

Selvaraj, R., Moorthy, I. G., Kumar, R. V., & Sivasubramanian, V. (2019). Microwave mediated production of FAME from waste cooking oil: Modelling and optimization of process parameters by RSM and ANN approach. *Fuel*, *237*, 40–49. https://doi.org/10.1016/j.fuel.2018.09.147.

Sitepu, E. K., Heimann, K., Raston, C. L., & Zhang, W. (2020). Critical evaluation of process parameters for direct biodiesel production from diverse feedstock. *Renewable and Sustainable Energy Reviews, 123*. https://doi.org/10.1016/j.rser.2020.109762.

Sivamani, S., Selvakumar, S., Rajendran, K., & Muthusamy, S. (2019). Artificial neural network–genetic algorithm-based optimization of biodiesel production from *Simarouba glauca*. *Biofuels, 10*(3), 393–401. https://doi.org/10.1080/17597269.2018.1432267.

Venkatesan, H., Nalarajan, B., Sivamani, S., & Thomai, M. P. (2020). Optimizing the process parameters to maximize palm stearin wax biodiesel yield using response surface methodology. *Energy Sources, Part A: Recovery, Utilization and Environmental Effects*, 1–19. https://doi.org/10.1080/15567036.2020.1790695.

Verma, P., & Sharma, M. P. (2016). Review of process parameters for biodiesel production from different feedstocks. *Renewable and Sustainable Energy Reviews, 62*, 1063–1071. https://doi.org/10.1016/j.rser.2016.04.054.

Wong, L. P., Isa, M. H., & Bashir, M. J. K. (2018). Disintegration of palm oil mill effluent organic solids by ultrasonication: Optimization by response surface methodology. *Process Safety and Environmental Protection, 114*, 123–132. https://doi.org/10.1016/j.psep.2017.12.012.

Yatish, K. V., Lalithamba, H. S., Suresh, R., Arun, S. B., & Kumar, P. V. (2016). Optimization of scum oil biodiesel production by using response surface methodology. *Process Safety and Environmental Protection, 102*, 667–672. https://doi.org/10.1016/j.psep.2016.05.026.

Yilmaz, N., & Atmanli, A. (2017). Sustainable alternative fuels in aviation. *Energy, 140*, 1378–1386. https://doi.org/10.1016/j.energy.2017.07.077.

Yuvaraj, D., Bharathiraja, B., Rithika, J., Dhanasree, S., Ezhilarasi, V., Lavanya, A., et al. (2019). Production of biofuels from fish wastes: An overview. *Biofuels, 10*(3), 301–307. https://doi.org/10.1080/17597269.2016.1231951.

Zahed, M. A., Zakeralhosseini, Z., Mohajeri, L., Bidhendi, G. N., & Mesgari, S. (2018). Multivariable analysis and optimization of biodiesel production from waste cooking oil. *Environmental Processes, 5*(2), 303–312. https://doi.org/10.1007/s40710-018-0299-2.

Chapter 14

Reservoir simulation model for the design of irrigation projects

Siva Ramakrishna Madeti[a], Gaurav Saini[b], and Krishna Kumar[c]

[a]University of Santiago Chile, Santiago, Chile, [b]School of Advanced Materials, Green Energy and Sensor Systems, Indian Institute of Engineering Science and Technology Shibpur, Howrah, West Bengal, India, [c]Department of Hydro and Renewable Energy, Indian Institute of Technology Roorkee, Roorkee, Uttarakhand, India

14.1 Introduction

Water is a basic human need and a prime natural resource. In the past few decades, India has witnessed immense urbanization and industrialization. These economic developments, compounded with an increase in the population, have resulted in a continuous demand for water for various purposes. Although the expansion in consumption of available water has included increased agricultural production, the returns from the investments made have been disappointing. The overall efficiency of major irrigation projects remains as low as 30%–35% (Singla, Patel, Pal, & Hussain, 2018). This compels one to believe that there is a lack of proper planning and management of the available water resources in the country and, secondly, that the available irrigation water is not being efficiently or optimally utilized for increasing agricultural production. For the success of any irrigation project, fulfillment of both these objectives is absolutely essential.

The inputs for this simulation model are reservoir inflow, potential evapotranspiration, rainfall in the irrigated area, and cropping pattern. The authors (Guariso & Sangiorgio, 2020; Trivedi & Shrivastava, 2020) addressed a necessity of real-time multiple reservoir operation. They have used linear programming (LP) for optimization from period to period, and dynamic programming (DP) for the selection of an optimal reservoir storage policy to analyze the Shasta and Trinity subsystems of the California Central Valley (CCV) project.

Sustainable Developments by Artificial Intelligence and Machine Learning for Renewable Energies.
https://doi.org/10.1016/B978-0-323-91228-0.00009-4
Copyright © 2022 Elsevier Inc. All rights reserved.

342 Sustainable developments by artificial intelligence & machine learning

Li et al. (2020) developed a mathematical planning model to find optimum cropping patterns in irrigation in the context of river basin development. A multiobjective deterministic LP model was applied to a system of four reservoirs in India to maximize the net economic benefit and the irrigated cropped area. The problem is solved by using the constraint method and the tradeoffs were discussed. Awchi and Srivastava (2009) studied an existing reservoir with water supply and multiirrigation demands and showed the feasibility of simulation for reservoir planning. The LP and simulation technique (Chakraborti, 2003) for irrigation reservoir and for multipurpose reservoir were used (Sadeghian, Hosseini, Zare Javid, Ahmadi Angali, & Mashkournia, 2021) for planning purposes to test the project provisions.

Feng, Niu, Cheng, and Wu (2017) and Srivastava and Patel (1992) used optimization (LP and DP) simulation models for the systems' analysis of the Karjan irrigation reservoir project in India. Gong, Zhang, Ren, Sun, and Yang (2020) derived a steady state reservoir operating policy using stochastic dynamic programming (SDP). The objective of the SDP was to maximize the expected sum of relative yields of all crops in a year. Zhu et al. (2020) developed a model for irrigation of multiple crops at Malaprabha irrigation reservoir in Karnataka, India. Rangarajan, Surminski, and Simonovic (1999) proposed a reliability programming model, which incorporates a four-step simulation algorithm to derive the loss function. The performance of the model was demonstrated through a case study. Ahmed and Sarma (2005) proposed that hydraulic numerical models of irrigation canals. The optimal operating policy obtained using GA is similar to that obtained by linear programming. This model can be used for optimal utilization of the available water resources of any reservoir system to obtain maximum benefits.

Wang et al. (2020) discussed the role of water resource system models in planning. Sinha, Rollason, Bracken, Wainwright, and Reaney (2020) and Myo Lin et al. (2020) carried out an integrated study for planning and operation for the proposed Pampa-Achankovil-Vaippar link diversion system (PAV system), which includes three reservoirs located in the southern part of India. The authors used system analysis techniques to do multilevel planning and operation for the system. The operating policy derived from a synthetic monthly stream flow series of 100 years is compared with that of the SDP model on the basis of its performance in reservoir simulation. The primary objective of this paper is to demonstrate the applicability of the simulation approach for optimization on the Harabhangi irrigation reservoir system located at Orissa, with a complex objective function using MATLAB software.

This chapter is organized as follows. A detailed description of the study area is presented in Section 14.2. In Section 14.3, cost-benefit functions are presented. The methodology adopted is discussed in Section 14.4. Simulation computations and results are discussed in Sections 14.5 and 14.6, respectively. The response to the case study is given in Section 14.7. Finally, Section 14.8 summarizes the conclusion of this chapter.

14.2 System description

The "Harabhangi irrigation project" is an interbasin irrigation project in Orissa state, India. It has been used as a case study to demonstrate the simulation study to obtain the near-optimal value and application of it in reservoir planning by sampling data pertaining to design variables of single-purpose reservoir systems. Fig. 14.1 is a map of the Harabhangi irrigation project. The study area comprises parts of two adjoining river basins in the southern part of Orissa state, namely, the Rushikulya basin and the Vansadhara basin. Harabhangi reservoir, which serves as the source of water for the irrigation project, was formed by the construction of a dam on the river Harabhangi, a tributary of the river Vansadhara in the Vansadhara basin. The dam at Adava is located at about longitude 84.08° east and latitude 19.30° north with the MWL (mean water level) of the reservoir at 387.5 m above sea level. The catchment area at the dam site is 503.8 km^2 and extends from longitude 84.03° to 84.22° east and latitude 19.17° to 19.34° north. The elevation of the command area under the Soroda and Badagada blocks varies from 130 m to 90 m above sea level.

14.3 Cost-benefit functions

For the simulation model of the reservoir, along with the hydrological data, the cost-benefit functions are required as inputs into the model. These functions are the capital cost of a reservoir, the capital cost of irrigation works, irrigation benefit, loss in irrigation benefit due to irrigation deficit, and operation and maintenance costs. The cost and benefit values of the design were available for only one capacity. Therefore, the cost and benefit value for different possible ranges of the project on the basis of appropriate engineering approaches and suitable functions were estimated (Fig. 14.2).

Figs. 14.3 and 14.4 represent the variation of annual irrigation with respect to capital cost, operation, and maintenance cost benefits, respectively, whereas Fig. 14.5 represents percentage irrigation shortage with respect to percentage benefit loss. Fig. 14.6 shows reservoir capacity versus capital cost.

Fig. 14.7 represents the variation of the reservoir capacity with respect to capital cost, operation, and maintenance cost, respectively. Cost and benefit functions expressed as continuous functions of input and output, such as volume, target, capacities, flows, and allocations, are given as in Srivastava and Patel (1992).

Reservoir capacity, Irrigation demand (Y, I_r) in terms of benefits B_1, based on the data estimated, has been correlated as:

$$B_1 = e^{4.5512} \times Y^{-0.6829} \times I_r^{0.7322} \qquad (14.1)$$

where B_1 = present value of net benefits in Rs., Y = reservoir capacity, and I_r = irrigation demand.

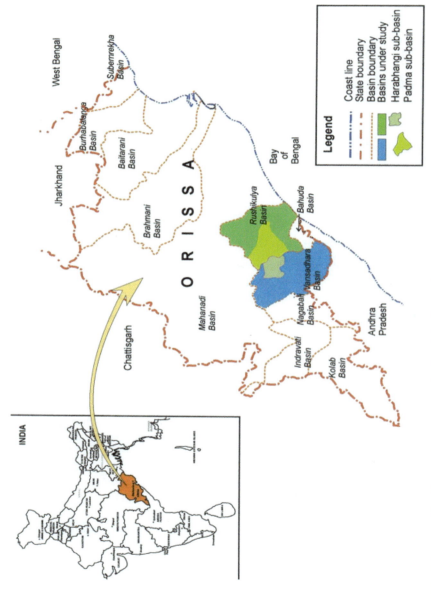

FIG. 14.1 Schematic of Harabhangi irrigation project. *(No Permission Required.)*

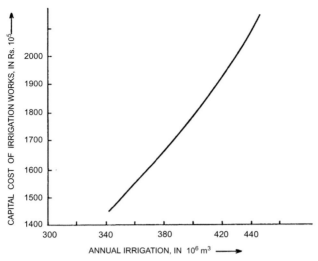

FIG. 14.2 Annual irrigation vs capital. *(No Permission Required.)*

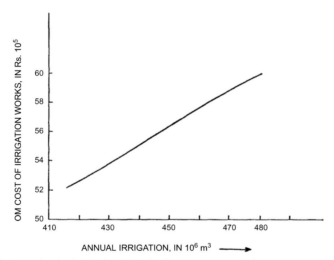

FIG. 14.3 Annual irrigation vs OM cost. *(No Permission Required.)*

14.4 Methodology

14.4.1 Linear programming model (LP model)

The objective function is to maximize the total annual net benefits from irrigation water. The benefits can be calculated as:

$$Maximize = (B_1 - (C_1 - C_2) - (Om_1 + Om_2)) \qquad (14.2)$$

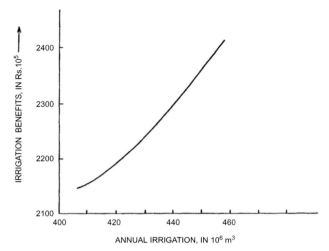

FIG. 14.4 Annual irrigation vs benefits. *(No Permission Required.)*

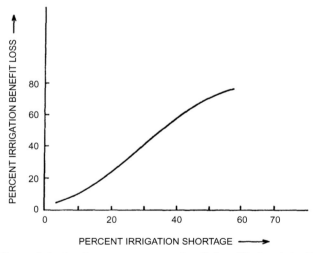

FIG. 14.5 Percent irrigation shortage vs percentage benefit loss. *(No Permission Required.)*

The first subscripts, 1 and 2, represent irrigation and reservoir, respectively, where:

$$B_1 = a_1 I_r \qquad (14.3)$$

$$C_1 = C'_1 I_r \qquad (14.4)$$

$$C_2 = C'_2 Y \qquad (14.5)$$

$$Om_1 = Om'_1 I_r \qquad (14.6)$$

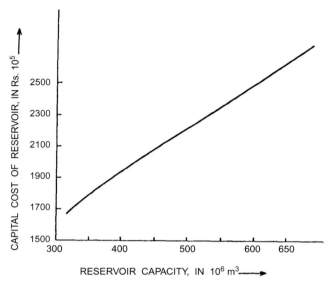

FIG. 14.6 Reservoir capacity vs capital cost. *(No Permission Required.)*

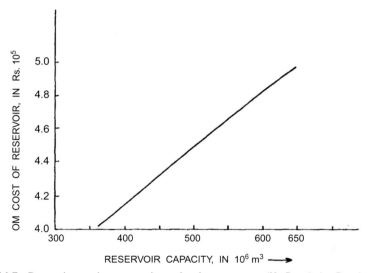

FIG. 14.7 Reservoir capacity vs operation and maintenance cost. *(No Permission Required.)*

where B_1 = gross annual irrigation benefits, C_1 = annual capital cost of irrigation, C_2 = annual capital cost of reservoir, Om_1 = annual operation and maintenance (OM) cost of irrigation, Om_2 = annual OM cost of the reservoir, a_1 = long-term benefit function for irrigation, C_1' = annual capital cost function for irrigation, C_2' = annual capital cost function for the reservoir, Om_1' = annual OM cost function for irrigation.

348   Sustainable developments by artificial intelligence & machine learning

The maximization of the objective function is subject to the following constraints:

(a) The value of the water released from the reservoir must be sufficient to meet the irrigation demand during that period, i.e.:

$$O_t + I_t'' = K_t I_r + (O_t)^-, \text{ for all } t, \text{ where } t = 1, 2, \ldots, N \qquad (14.7)$$

where O_t = total water release from the reservoir in time t, I_t'' = water that joins the main river just above the irrigation diversion canal in time t, I_r = annual irrigation water target, K_t = proportion of annual irrigation target I_r to be diverted for irrigation in time t, O_t = secondary water release from the reservoir in time t, and N = number of time periods in the planning horizon.

(b) The continuity equation for the reservoir is defined as:

$$S_t = S_{t-1} + I_t + I_t' + P_t - E_{lt} - O_t - O_t', \text{ for all } t \qquad (14.8)$$

For evapotranspiration consideration, the above equation can be modified as:

$$K_t' S_t = S_{t-1} + I_t + P_t + I_t' - O_t - O_t', \text{ for all } t \qquad (14.9)$$

where S_{t-1} = reservoir storage at the beginning of time t, I_t = inflow into reservoir during time t, I_t' = local inflow to the reservoir from the surrounding area in time t; P_t = Precipitation in the reservoir in time t, E_{lt} = evaporation losses from the reservoir in time t, O_t = total outflow (release) from the reservoir in time t, O_t' = release to natural channel from reservoir in time t, S_t = reservoir storage at the end of time t, and K_t' = amount by which K_t' exceeds unity as the fraction of the end storage that is assigned to reservoir evaporation loss computed from two trial working tables prepared with and without evaporation losses, respectively.

The above equation is subjected to the following constraints:

$$O_t \leq S_{t-1} + I_t + I_t' + P_t - E_{lt} - O_t' - Y_{\min_t}, \text{ for all } t \qquad (14.10)$$

$$O_t = Or_t' + Oa_t' + Oa_t'' + Sp_t, \text{ for all } t \qquad (14.11)$$

where Or_t' = actual release of water supply from reservoir in time t, Oa_t' = actual irrigation release from reservoir in time t, Oa_t'' = additional release from reservoir to fulfill energy demand in time t, and Sp_t = reservoir spill in time t.

(c) The contents of the reservoir at any period cannot exceed the capacity of the reservoir. Furthermore, the dead storage of the reservoir puts a lower limit on the reservoir storage. The storage capacity is also incorporated for conservation purposes, i.e.:

$$Y_d \leq Y_{\min_t} \leq S_{t-1} \leq Y_{\max_t} \leq Y, \text{ for all } t \qquad (14.12)$$

Reservoir simulation model for the design of irrigation projects **Chapter | 14** **349**

where S_{t-1} = gross reservoir storage, Y_{min_t} = minimum storage of reservoir in time t, and Y_{max_t} = maximum storage of reservoir in time t.

$$0 \leq Y_{min_t}{}' \leq S_{t-1} \leq Y_{max_t}{}' \leq Y_a, \text{ for all } t \qquad (14.13)$$

where $Y_{min_t}{}'$ = live capacity up to the minimum pool level of the reservoir in time t, $Y_{max_t}{}'$ = live capacity up to the normal pool level of the reservoir in time t, Y_a = total live capacity of the reservoir at the maximum pool level, and S_{t-1} is the gross reservoir storage.

(d) Further, if necessary, bounds may be put on individual design/operating variables:

$$L_1 \leq I_r \leq U_1; L_2 \leq Y \leq U_2 \qquad (14.14)$$

$$Omin_t \leq O_t \leq Omax_t \qquad (14.15)$$

14.4.2   Reservoir simulation

(a) The simulation starts in the month of June in the first year of the study, and the initial reservoir content is taken as dead storage, i.e., zero live storage.
(b) The release from the reservoir in any month is made from the total available water, i.e., the sum of the initial reservoir content in that month plus the inflow minus the evaporation from the reservoir during the month.
(c) The continuity equation holds good for each month.
(d) The reservoir content in any month cannot be more than the reservoir capacity.

14.4.2.1   The simulation model

The simulation problem for the reservoir system may be defined as follows. Sampling the annual target level of irrigation output and the reservoir capacity by using search techniques and determining where monthly runoff values and a suitable operating procedure for storing and releasing water from the reservoir are given.

14.4.2.2   System design variables, parameters, and constants

(i) Major design variables are gross and live capacity of the reservoir and target outputs:
   (a) target outputs for irrigation for command area (yearly),
   (b) a 12-element vector of annual target output for irrigation (monthly percentage of annual values).
(ii) Cost and benefit functions for irrigation area are:
   (a) annual target output for irrigation vs unit gross irrigation benefits relation,

(b) annual irrigation shortage vs irrigation loss relation,
(c) annual target output for irrigation vs capital costs of irrigation diversion, distribution, and pumping work relation, and
(d) annual target output for irrigation vs annual OM cost of irrigation diversion, distribution, and pumping work relation.
(iii) Reservoir costs and characteristics are:
(a) capacity of reservoir vs capital costs of reservoir relation, and
(b) capacity of reservoir vs annual OM cost of reservoir relation.
(iv) Other functions such as interest rates and formula used for discounting.

14.5 Simulation computations

Fig. 14.8 shows the flow chart of steps involved in the developed method. It consists of the main program, two subroutines, and one function subprogram. Two design variables—reservoir capacity and annual irrigation requirement—were sampled using systematic and random techniques. A number of sampled combinations were simulated and tested.

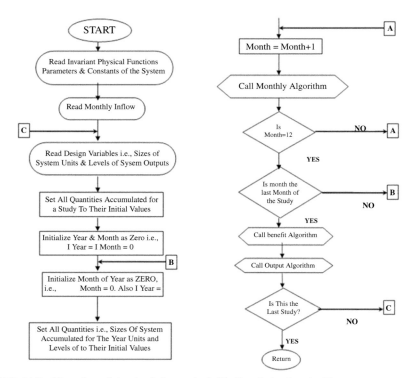

FIG. 14.8 Flow chart of the simulation method. *(No Permission Required.)*

Reservoir simulation model for the design of irrigation projects **Chapter | 14** **351**

The analysis period was 42 years, for which the river flows were taken from the article (Awchi & Srivastava, 2009). The simulation model was run for monthly periods. Loss in irrigation benefits due to deficit in irrigation, and evaporation losses from the reservoir were considered. All nonlinearities in the model were maintained as far as possible. All costs and benefits were linearized as given earlier. The various terms in $g_t\,(S_t, O_t)$:

$$B_{1,t} = a_1 \times Od_t' - L'(Od_t' - O_t),\ \text{if}\ O_t < Od_t' \tag{14.16}$$

$$B_{1,t} = a_1 \times Od_t',\ \text{if}\ O_t \geq Od_t' \tag{14.17}$$

$$C_{1,t} = (C_1'/12) \times O_t,\ \text{if}\ O_t < Od_t' \tag{14.18}$$

$$C_{1,t} = (C_1'/12) \times Od_t',\ \text{if}\ O_t \geq Od_t' \tag{14.19}$$

$$C_{2,t} = (C_2'/12) \times Y \tag{14.20}$$

$$Om_{1,t} = (Om_1'/12) \times O_t,\ \text{if}\ O_t < Od_t' \tag{14.21}$$

$$Om_{1,t} = (Om_1'/12) \times Od_t',\ \text{if}\ O_t \geq Od_t' \tag{14.22}$$

$$Om_{2,t} = (Om_2'/12) \times Y \tag{14.23}$$

where Od_t' = irrigation demand in period t and L' = loss in irrigation benefits due to a deficit in supply.

In the first systematic sampling, the ranges for the reservoir capacity and annual irrigation requirement selected are given in Table 14.1, keeping in view the present existing reservoir capacity and the future proposed irrigation

TABLE 14.1 Results of a simulation model.

Trail no. (alternative)	Y	I_r	Number of failure years (annual irrg. deficit allowed)			Present value of net benefit in Rs. 10^5
			0%	5%	10%	
First search with variables Y and I_r						
1	100	200	18	11	9	197.51
2	120	200	10	5	4	174.39
3	140	200	6	5	2	156.96
4	160	200	4	3	1	143.29
5	180	200	2	1	1	132.21
6	200	200	1	1	1	123.03
7	100	220	24	12	10	211.79

Continued

352 Sustainable developments by artificial intelligence & machine learning

TABLE 14.1 Results of a simulation model.—cont'd

Trail no. (alternative)	Y	I_r	Number of failure years (annual irrg. deficit allowed)			Present value of net benefit in Rs. 10^5
First search with variables Y *and* I_r			*0%*	*5%*	*10%*	
8	120	220	16	10	9	186.99
9	140	220	14	10	9	168.61
10	160	220	14	8	6	153.64
11	180	220	12	6	5	141.76
12	200	220	10	6	4	131.93

$100 \leq Y \leq 200$ and $200 \leq I_r \leq 220$.

requirement of 141.25 MCM (million cubic meters) and 220 MCM, respectively. In the first search, a coarse grid of 17 alternatives was chosen. The gross reservoir capacity and the annual irrigation target were varied with increments of 20 and $5 \times 10^6 \mathrm{m}^3$, respectively, as per the ranges.

14.6 Results and discussion

Out of these 17 alternatives, the feasible ones, with project dependability of 80% (35 years out of 42 years) or more, are given in Table 14.1. A successful year was defined in terms of different allowances in deficits in annual irrigation targets. These allowances were 0%, 5%, and 10%. Looking at the results, it may not be advisable to select a reservoir capacity and annual irrigation target as 120 and $220 \times 10^6 \mathrm{m}^3$ respectively to trail No. 8 in Table 14.1, keeping in view the present existing reservoir capacity and the future proposed irrigation requirement of 141.25 MCM and 220 MCM, respectively, all 17 combinations were simulated and it was observed from the results that the net benefits (present worth) were obtained in all the combinations.

In the second search, the gross reservoir capacity was kept as equal to the project provision of $141.25 \times 10^6 \mathrm{m}^3$ and the annual irrigation target was varied from 200 to $220 \times 10^6 \mathrm{m}^3$ with the increment of $5 \times 10^6 \mathrm{m}^3$, as given in Table 14.2.

From the results, it is found that with a projected provision of Y equal to $141.25 \times 10^6 \mathrm{m}^3$ for 0% allowance in the annual irrigation deficit, about $205 \times 10^6 \mathrm{m}^3$ annual irrigation, I_r, can be satisfied with about 80% dependability. The value of I_r for a 5% allowance may be about $210 \times 10^6 \mathrm{m}^3$ and, for 10%, it may be even higher than a value of about $220 \times 10^6 \mathrm{m}^3$.

Reservoir simulation model for the design of irrigation projects **Chapter | 14 353**

TABLE 14.2 Second search with variables Y and I_r.

1	141.25	200	6	5	2	156.01
2	141.25	205	9	3	2	158.86
3	141.25	210	13	5	4	161.69
4	141.25	215	14	9	5	164.50
5	141.25	220	14	9	8	167.29

14.7 Response of Harabhangi irrigation project

It can be easily observed from the results obtained from the simulation run by the two search techniques that the net benefit (present worth) increases with lower reservoir capacity and higher annual irrigation requirement. The average annual irrigation deficit is also on the increase, whenever there is a combination of lower reservoir capacity with higher annual irrigation requirements. The spillage is also considerable for lower capacities. Figs. 14.9–14.11 show how the net benefit, the percentage average annual irrigation deficit, and the percentage average annual spill from the reservoir change with the change in reservoir capacity for the present annual irrigation requirement of 212.780 MCM, respectively. There is a rapid decrease in the benefit as compared to the other two items.

14.7.1 Support for the use of simulation

From the results, as shown in Fig. 14.9, it is seen that if the annual irrigation target is kept constant and reservoir capacity is increased, the net benefit

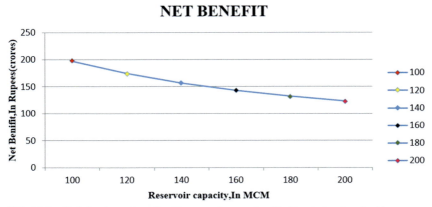

FIG. 14.9 Variation in net benefit with reservoir capacity. *(No Permission Required.)*

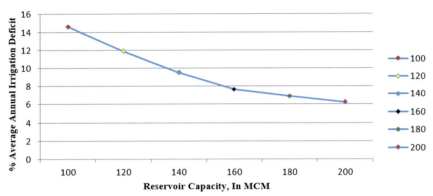

FIG. 14.10 Variation in irrigation deficit with reservoir capacity. *(No Permission Required.)*

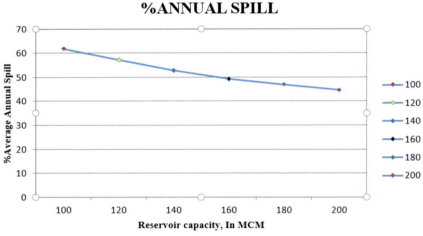

FIG. 14.11 Variation in annual spill with reservoir capacity. *(No Permission Required.)*

decreases and the variation is rapid. The same is the case with the average annual irrigation deficit and the average annual spill from the reservoir, but the variations are slow, as shown in Figs. 14.10 and 14.11, respectively. These behaviors seem to be due to the fact the cost of the reservoir is quite large as compared to the loss in irrigation benefit due to irrigation deficit. Now, compare the combination of the new proposal at series no. 12 (systematic sampling) in Table 14.2, having a reservoir capacity of 200 MCM and an annual irrigation requirement of 220 MCM, which gives a net benefit of Rs. 131.92 crores with the combination at series no. 1, having 100 MCM reservoir capacity and 220 MCM annual irrigation requirement, which gives a net benefit of

Rs. 211.79 crores. This comparison means that the new proposed annual irrigation requirement of 220 MCM may be satisfied just as well, even giving more net benefit, with a lower reservoir capacity than the existing capacity of 141.25 MCM. This could only be answered efficiently by using a simulation technique, as done here. Therefore, it is evident that if the planning of the reservoir by simulation studies had been carried out before the construction of this reservoir, it would have been more appropriate.

A simulation that predicts the behavior of the system in more detail, or rather is more descriptive than any of the mathematical techniques, may provide answers to many such problems as discussed above, which other methods of planning may fail to do so. In light of the above findings, the simulation technique may be a powerful tool for planning.

14.8 Conclusion

Simulation models were used here to investigate their applicability in reservoir planning and further analysis of a water resources system. The Harabhangi irrigation project was taken as the system. This project already existed and was analyzed for its further development. The simulation technique was used to simulate the various components of the reservoir to obtain the near-optimum values of the design variables by refining them by eliminating infeasible alternatives due to excess irrigation shortages based on a number of failure years in the project target. Two sampling techniques, namely, systematic and random, were used to explore the response (net benefit) surface. It was found that a lower capacity of 120 MCM may be sufficient to satisfy the future annual irrigation requirement of 220 MCM, whereas the existing reservoir capacity is already in excess of this (141.25 MCM, see Table 14.2). The near-perfect behavior of reservoirs for water conservation can only be predicted by simulation, making it the most feasible method for reservoir planning. Table 14.3 shows the applicability of previous techniques for reservoir simulation.

TABLE 14.3 Applicability of previous techniques for reservoir simulation.

	Techniques for reservoir simulation		
S. no.	Technique	Application	References
1	Linear programming and dynamic programming	LP—Used for period by period optimization DP—Selection of an optimal reservoir storage policy path through a specified number of policy periods	Feng et al. (2017)

Continued

TABLE 14.3 Applicability of previous techniques for reservoir simulation.—cont'd

		Techniques for reservoir simulation	
S. no.	*Technique*	*Application*	*References*
2	Mathematical planning model	To find optimal crop pattern in irrigation in context of river basin development	Myo Lin et al. (2020)
3	Linear programming and dynamic programming	LP—Used to find reservoir capacity DP—Further refining the output targets	Awchi and Srivastava (2009) and Srivastava and Patel (1992)
4	Water resource system model	Discussed about major challenges facing water resource system planners and managers	Chakraborti (2003)
5	Stochastic dynamic programming	Used to maximize the expected sum of relative yield of all crops in a year	Trivedi and Shrivastava (2020)
6	Operating policy model and allocation model	Operating policy model—Used to optimize reservoir release Allocation model—Used to optimize irrigation allocations	Zhu et al. (2020)
7	Reliability programming model	Used to find out the relationship between reliability and its economic losses	Rangarajan et al. (1999)
8	System analysis techniques	Used to develop multilevel planning and operation for the system	Guariso and Sangiorgio (2020)
9	GA and SDP	Two methods used to find out the optimal operating policy and compare the GA and SDP	Singla, Patel, Pal, & Hussain, 2018
10	Hydraulic numerical model tool (SIC and canal Man)	Used to simulate to predict the actual behavior and check its design	Sadeghian et al. (2021)
11	GA model	Used to optimize operating policy and optimal crop water allocations to maximize the benefits	Srivastava and Patel (1992)
12	MATLAB	Simulate the reservoir to obtain the near optimal value by sampling the data pertaining to design variables (reservoir capacity and irrigation demand)	Present study

References

Ahmed, J. A., & Sarma, A. K. (2005). Genetic algorithm for optimal operating policy of a multi-purpose reservoir. *Water Resources Management, 19*(2), 145–161. https://doi.org/10.1007/s11269-005-2704-7.

Awchi, T. A., & Srivastava, D. K. (2009). Analysis of drought and storage for Mula project using ANN and stochastic generation models. *Hydrology Research, 40*(1), 79–91. https://doi.org/10.2166/nh.2009.012.

Chakraborti, A. K. (2003). *Watershed prioritization—A case study in Salauli watershed of Zuari river basin, Goa* (pp. 42–44). ISRO. National Natural Resources Management Systems Bull.

Feng, Z.k., Niu, W.j., Cheng, C.t., & Wu, X.y. (2017). Optimization of hydropower system operation by uniform dynamic programming for dimensionality reduction. *Energy, 134*, 718–730. https://doi.org/10.1016/j.energy.2017.06.062.

Gong, X., Zhang, H., Ren, C., Sun, D., & Yang, J. (2020). Optimization allocation of irrigation water resources based on crop water requirement under considering effective precipitation and uncertainty. *Agricultural Water Management, 239*. https://doi.org/10.1016/j.agwat.2020.106264.

Guariso, G., & Sangiorgio, M. (2020). Performance of implicit stochastic approaches to the synthesis of multireservoir operating rules. *Journal of Water Resources Planning and Management, 146*(6), 04020034. https://doi.org/10.1061/(asce)wr.1943-5452.0001200.

Li, M., Fu, Q., Singh, V. P., Liu, D., Li, T., & Zhou, Y. (2020). Managing agricultural water and land resources with tradeoff between economic, environmental, and social considerations: A multi-objective non-linear optimization model under uncertainty. *Agricultural Systems, 178*. https://doi.org/10.1016/j.agsy.2019.102685.

Myo Lin, N., Tian, X., Rutten, M., Abraham, E., Maestre, J. M., & van de Giesen, N. (2020). Multi-objective model predictive control for real-time operation of a multi-reservoir system. *Water, 12* (7), 1898. https://doi.org/10.3390/w12071898.

Rangarajan, S., Surminski, H., & Simonovic, S. P. (1999). Reliability-based loss functions applied to Manitoba Hydro energy generation system. *Journal of Water Resources Planning and Management, 125*(1), 34–40. https://doi.org/10.1061/(ASCE)0733-9496(1999)125:1(34).

Sadeghian, M., Hosseini, S. A., Zare Javid, A., Ahmadi Angali, K., & Mashkournia, A. (2021). Effect of fasting-mimicking diet or continuous energy restriction on weight loss, body composition, and appetite-regulating hormones among metabolically healthy women with obesity: A randomized controlled, parallel trial. *Obesity Surgery, 31*(5), 2030–2039. https://doi.org/10.1007/s11695-020-05202-y.

Singla, S., Patel, N., Pal, S., & Hussain, M. (2018). Evaluation of the developed strategy at field level to enhance crop water productivity in Sirsa District. *Journal Homepage, 6*(6), 1–13.

Sinha, P., Rollason, E., Bracken, L. J., Wainwright, J., & Reaney, S. M. (2020). A new framework for integrated, holistic, and transparent evaluation of inter-basin water transfer schemes. *Science of the Total Environment, 721*. https://doi.org/10.1016/j.scitotenv.2020.137646.

Srivastava, D. K., & Patel, I. A. (1992). Optimization-simulation models for the design of an irrigation project. *Water Resources Management, 6*(4), 315–338. https://doi.org/10.1007/BF00872283.

Trivedi, M., & Shrivastava, R. (2020). Derivation and performance evaluation of optimal operating policies for a reservoir using a novel PSO with elitism and variational parameters. *Urban Water Journal, 17*(9), 774–784. https://doi.org/10.1080/1573062X.2020.1823431.

Wang, S. Y., Wang, S., Lyu, S., Zhou, J., Nakagami, M., Takara, K., et al. (2020). Historical assessment and future sustainability challenges of Egyptian water resources management. *Journal of Cleaner Production*, *263*, 121154.

Zhu, F., Zhong, P.a., Xu, B., Chen, J., Sun, Y., Liu, W., et al. (2020). Stochastic multi-criteria decision making based on stepwise weight information for real-time reservoir operation. *Journal of Cleaner Production*, *257*. https://doi.org/10.1016/j.jclepro.2020.120554.

Chapter 15

Effect of hydrofoils on the starting torque characteristics of the Darrieus hydrokinetic turbine

Gaurav Saini[a], Anuj Kumar[b], and R.P. Saini[c]

[a]School of Advanced Materials, Green Energy and Sensor Systems, Indian Institute of Engineering Science and Technology Shibpur, Howrah, West Bengal, India, [b]School of Mechanical Engineering, Vellore Institute of Technology, Vellore, India, [c]Department of Hydro and Renewable Energy, Indian Institute of Technology Roorkee, Roorkee, India

Nomenclature

2D	two dimensional
AoA	angle of attack
RNG	renormalization group
RPM	revolutions per minute
SIMPLE	semiimplicit pressure linked equations
TSR	tip speed ratio
URANS	unsteady Reynolds averaged Navier-Stoke
C_P	coefficient of power
C_T	coefficient of torque
U	fluid velocity
V	blade velocity/tangential velocity
W	resultant velocity
α	angle of attack
Ψ	velocity angle
Θ	azimuth angle

15.1 Introduction

Hydrokinetic technology has grown over last decade to transform the energy scenario provided by conventional hydropower (Güney & Kaygusuz, 2010).

Sustainable Developments by Artificial Intelligence and Machine Learning for Renewable Energies.
https://doi.org/10.1016/B978-0-323-91228-0.00008-2
Copyright © 2022 Elsevier Inc. All rights reserved.

Hydrokinetic technology is similar in working principle to wind energy technology. However, due to the density difference between air and water, hydrokinetic turbines face greater forces (lift and drag) as compared to wind turbines (Alam & Iqbal, 2010).

Hydrokinetic turbines are basically classified in two broad categories based on the orientation of the rotor shaft with respect to water flow direction. The first category of hydrokinetic turbine has the rotor axis always parallel to the water flow direction and are known as axial flow turbines. The second category of hydrokinetic turbines has its rotor shaft axis orthogonal to the water flow direction and are usually known as cross flow turbines (Saini & Saini, 2019). The axial flow turbine needs more depth for rotation due to its circular sweep area. Therefore, these turbines are mainly suitable for applications where depth is not a constraint, i.e., ocean or tidal applications (Lago, Ponta, & Chen, 2010). On the other hand, cross flow hydrokinetic turbines sweep a cylindrical area during rotation; therefore, these turbines can be installed in shallow depth applications like rivers, manmade channels, and canals (Kumar & Saini, 2016).

In terms of generating power from a cross-flow hydrokinetic turbine, the Darrieus rotor shows better performance over other existing cross flow hydrokinetic turbine rotors (e.g., Savonius rotor). However, the Darrieus rotor is associated with the problem of self-starting during the initial starting of the turbine (Asr, Nezhad, Mustapha, & Wiriadidjaja, 2016). Due to airfoil shaped blades, the lift/drag force ratio in a Darrieus rotor is greater (Zeiner-Gundersen, 2014). The variation of the forces over the blade profile with the variation in pitch angle can be seen in Fig. 15.1, where "α" is the angle of attack, "W" is the resultant velocity of the "U" (fluid velocity) and "V" (Blade velocity/tangential velocity), "Ψ" is the angle between "U" and "V," "N" is the result of lift and drag forces, and "$\ominus$" is the azimuth angle for rotation of turbine. To start the turbine, more drag force is required during initial conditions in order to

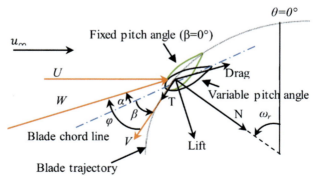

FIG. 15.1 Schematic of the forces over a blade profile. *(From Abdalrahman, G., Melek, W., & Lien, F. S. (2017). Pitch angle control for a small-scale Darrieus vertical axis wind turbine with straight blades (H-Type VAWT). Renewable Energy, 114, 1353–1362. https://doi.org/10.1016/j.renene.2017.07.068.)*

overcome the limiting friction and inertia of the rotor. The resistance created by friction and rotor inertia can be improved by increasing the drag force portion in the lift/drag ratio. Various investigations have been used in previous studies to improve the drag force.

Bianchini, Balduzzi, Ferrara, and Ferrari (2016) proposed a novel approach to study the virtual camber theory of airfoils. The results were predicted to study the flow past the airfoil rotating and having the flow orthogonal to its axis. The study was carried out to study the stall phenomena. Zhao, Yang, Liu, and Zhao (2013) carried out an experimental study to investigate the effects of preset angle of attack for a tidal current turbine and validate it with numerical simulations. The study revealed that on increasing the angle of attack from $-3°$ to $5°$, the hydrodynamics performance of the turbine is found to increase and the maximum power coefficient is found to be 0.348.

Abdalrahman, Melek, and Lien (2017) compared the performance of a variable pitch blade type turbine with the fixed pitch type blade. The results of the study show that the variable pitch type turbine has better performance than the fixed pitch type turbine and a maximum of 25% increment in performance was observed. Paillard, Astolfi, and Hauville (2015) investigated numerically the variable pitch Darrieus rotor and compared it with the fixed pitch rotor. The study concluded a 52% increment in the power harmonics in the case of the variable pitch rotor. Based on the flow behavior across the turbine, it was observed that the turbine performance can be improved by a slight recirculation across the turbine blades.

Kirke (2016) tested a variable pitch Darrieus rotor hydrokinetic turbine in a tow tank. The study concluded that for a small scale turbine, the Reynolds number is an important factor to decide the performance of the turbine. However, higher performance may be achieved for large scale turbines. Furthermore, it has been observed that parasitic forces are significant for turbine analysis. In order to improve the cyclical hydrodynamics forces and to improve the self-starting performance of the Darrieus rotor, a memetic algorithm was developed. The proposed algorithm results in better starting torque with maintained peak efficiency compared to fixed pitch blades (Lazauskas & Kirke, 2012). Along similar lines, Kirke (2016) studied the limitations of fixed pitch blades and challenges faced during implementation of variable pitch blades. The study revealed that the variable pitch blades result in better starting behavior with high efficiency and reduced vibration and shaking phenomena. However, the pitch control mechanism adds more complexity to the system (Kirke & Lazauskas, 2011).

The above mentioned literature review is to study the performance of the Darrieus rotor cross flow turbines under various operating conditions. For wind cross flow turbines and hydrokinetic turbines, it has been observed that the Darrieus rotor suffers from the problem of self-starting. Pitch control or variation in the angle of attack is suggested by various investigators in order to vary the lift drag ratio and to make the system self-start.

The aim of the present investigation is to study the behavior of Darrieus cross flow hydrokinetic turbine rotors under different angles of attack. The performance of the turbine was assessed in terms of power coefficient (CP), torque coefficient (CT), and flow behavior across the turbine vicinity. 2D computational simulations were carried out to obtain the abovementioned parameters and the following objectives were established for the study:

(a) To investigate the performance characteristics of the Darrieus rotor under different angles of attack under variable operating conditions.
(b) To study the self-starting behavior of the Darrieus rotor at different angles of attack.
(c) To study the flow behavior across the turbine vicinity and channel computational domain.

15.2 Investigated parameters for the Darrieus hydrokinetic turbine

In the present study, a Darrieus rotor with three airfoil shaped blades was selected to carry out the numerical simulation-based analysis. A straight, symmetrically shaped, S-1046 airfoil was selected for the turbine blades (Mohamed, 2012; Saini & Saini, 2019). Angle of attack is considered with steps of $-8°$, $-4°$, $0°$, $4°$, and $8°$ for complete simulation analysis. The complete design parameters for the Darrieus rotor are given in Table 15.1.

The velocity of water in the computational channel domain was considered constant at 0.5 m/s. In order to analyze the system under variable operating conditions, TSR was varied in a range varying from 0.5 to 1.7.

15.3 Numerical simulation analysis

For the numerical analysis in the current study, the commercially available software ANSYS (v.19.1) was used. ANSYS provides the tools for model development, grid generation, and solver (Fluent/CFX) to solve the unsteady Reynolds

TABLE 15.1 Parameters considered under the present investigation.

Parameter	Values/range
Airfoil profile	S-1046
Number of blades	3
Diameter of rotor (mm)	175
Angle of attack (°)	$-8°$, $-4°$, $0°$, $4°$, $8°$
Flow velocity (m/s)	0.5

Effect of hydrofoils on the Darrieus rotor starting torque characteristics **Chapter | 15** 363

averaged Navier-Stokes (URANS) equations. The results from the solver were processed through the CFD post module of the ANSYS workbench.

The numerical simulation study was carried out in two dimensions (2D) with Fluent as the solver. The Navier-Stokes equations were solved in conjunction with RNG k-ε turbulence model (Chen & Lian, 2015; Lee, Lee, & Lim, 2016; Sengupta, Biswas, & Gupta, 2016). An incompressible fluid (water) was considered for the fluid zones. Unsteady (transient) simulations were performed to investigate the turbine rotor under variable operating conditions.

15.3.1 Turbine model development

The objective of the present study was to investigate the performance of a Darrieus rotor turbine at different angles of attack (Fig. 15.2) and to study the variations in the water flow approaching the turbine. Therefore, five different models of Darrieus hydrokinetic turbine (2D) were created in the "DesignModeler" module of ANSYS as per the dimensions given in *Nomenclature*. In order to provide the rotation to the turbine blades, a circular domain (rotating) enclosing the Darrieus hydrokinetic turbine was created (Saini & Saini, 2020). The outer boundary of the rotating domain was described as "interface" to maintain the continuity of the fluid flow.

The Darrieus rotor turbine was modeled along with the 2D computational channel domain (stationary). The 2D domain of the channel is 2.5 m in length and 0.7 m in width. The rotor was placed at 1.0 m from the inlet section of the channel. The Darrieus turbine, along with the channel, is shown in Fig. 15.3.

The complete computational domain (rotating and stationary) was selected as the "fluid" in the fluid zone because the flowing water is modeled and the properties were taken for water as the working fluid. Further, the various sides of the channel were named as per the given boundary conditions.

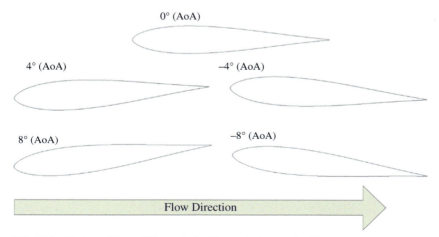

FIG. 15.2 Blade profiles at different AoA. *(No permission required.)*

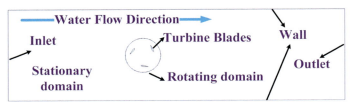

FIG. 15.3 Computational domain (2D) with different named selections. *(No permission required.)*

15.3.2 Grid generation

In order to discretize the computational domain, the ANSYS meshing module was used. The complete model was exported into the mesh interface and an unstructured mesh was generated, which provided better flexibility for automatic generation of mesh. For better flow analysis at the rotor boundaries (airfoils), dense mesh was generated to form the boundary layers. The height of the first prism layer was decided by a nondimensional parameter known as "$y+$." The value of $y+$ was considered as less than the one for the considered turbulence model (Balduzzi, Bianchini, Maleci, Ferrara, & Ferrari, 2016).

The mesh of the computational domain was refined several times to obtain the mesh independence results with the defined level of accuracy as shown in Table 15.2. The mesh of the domain was refined by varying the body size. The optimized mesh is known as the mesh independence limit (MIL). Four different levels of refinements having 201,532, 308,247, 373,619, and 489,636 elements, respectively, were considered and the third level of refinement selected for overall numerical simulations, having mesh properties as given in parameters (power coefficient and torque coefficient). These output parameters were assessed corresponding to considered velocity and TSR range. The complete mesh of the computational domain is shown in Fig. 15.4.

TABLE 15.2 Mesh properties of the selected level of refinement.

	Nodes	Elements	Skewness	Aspect ratio	Ortho-quality
Rotating domain	184,931	184,338	0.84	12.43	0.999
Stationary domain	190,446	189,281	0.54	2.30	1.0
Complete domain	375,377	373,619	0.84	12.43	1.0

Effect of hydrofoils on the Darrieus rotor starting torque characteristics Chapter | 15 365

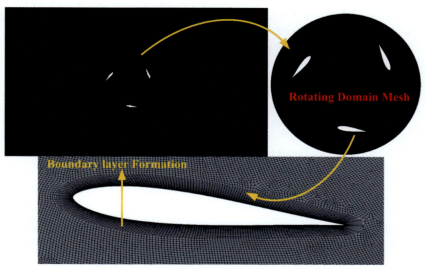

FIG. 15.4 Mesh details of the computational domain. *(No permission required.)*

15.3.3 Boundary conditions and turbulence modeling

In order to solve the URANS equations, constraints need to be applied at the various positions of computational domain though the boundary conditions. For the present numerical simulations, the boundary conditions were applied as given in Table 15.3. A "moving mesh" approach was used to obtain the actual rotating behavior under variable operating conditions (different RPM). An RNG k-ε turbulence model was used along with scalable wall function to capture the flow field at wall boundaries (Saini & Saini, 2018). For the present numerical simulations, 3% turbulence intensity and turbulence viscosity ratio of 10 were utilized.

TABLE 15.3 Considered boundary conditions of present numerical analysis.

Boundary position	Boundary type
Inlet	Velocity inlet (Drichlet condition)
Wall	Wall (stationary)
Interface	Interface
Turbine (rotor blades)	Wall (stationary)
Outlet	Pressure outlet

The SIMPLE (semiimplicit method for pressure linked equations) method was adopted along with second order upwind schemes to couple the pressure and velocity with higher accuracy. Convergence criteria for the residuals were defined as 1×10^{-5} for each time step. The time step was calculated for 10° rotation corresponding to each time step. In order to calculate stable torque, the turbine rotor was rotated for five complete rotations. Therefore, a total of 180 time steps were provided, along with 50 iterations per time step.

The coefficient of torque was monitored on the turbine blade surface and the torque of the last revolution was used to analyze the turbine performance. Turbine power was calculated by multiplying the average turbine torque with the corresponding angular velocity (rad/s).

15.4 Results and discussion

The characteristics of the Darrieus hydrokinetic turbine were monitored in terms of flow contours across the turbine vicinity and performance.

15.4.1 Performance characteristics

The effects of the variation in AoA were assessed in terms of power coefficient (CP) and torque coefficient (CT). Both of these parameters are dimensionless and depend on the third dimensionless parameter, known as TSR. Fig. 15.5 shows the variation of the power coefficient with respect to TSR corresponding

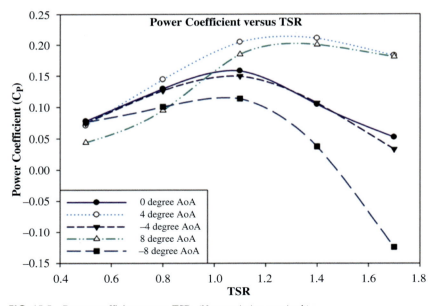

FIG. 15.5 Power coefficient versus TSR. *(No permission required.)*

to the considered range of AoA. The trends of the curves show that as the AoA increases from $-8°$ to $4°$, the power coefficient increases. Beyond the value of $4°$, the power coefficient tends to decrease. It has also been observed that negative values of AoA lead to a low value of power coefficient as compared to positive values of AoA. This is due to the fact that the increment in drag force created by negative AoA tends to reduce the resultant driving force (combination of lift and drag force). In the case of positive AoA, the increment in drag force by positive AoA assists the resultant driving force. However, beyond a certain limit of positive AoA, the drag force fraction increases the resultant force and the system tends to shift towards drag force devices, which have more torque and a low power coefficient. The maximum value of power coefficient was found to be 0.214, which corresponds to a 1.25 value of TSR and $4°$ of AoA.

In order to analyze the starting behavior of the Darrieus hydrokinetic turbine, the coefficient of torque (C_T) was plotted at each TSR and comparison made at every AoA. Figs. 15.6–15.10 show the variation of torque coefficient at different values of TSR. For the considered flow velocity of 0.5 m/s and TSR range, it has been observed that positive AoA tends to have a better coefficient of torque in comparison to negative AoA. For low TSR range, $4°$ AoA has the highest C_T among all the AoAs. However, for high TSR range, $4°$ and $8°$ AoA have almost the same value of C_T.

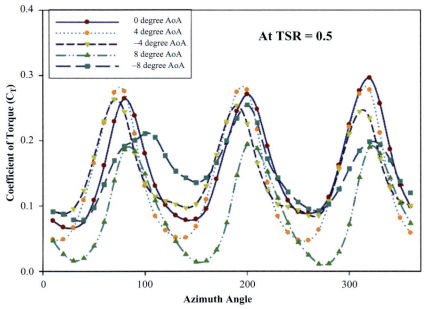

FIG. 15.6 Torque coefficient versus angular position at 0.5 value of TSR. *(No permission required.)*

368 Sustainable developments by artificial intelligence & machine learning

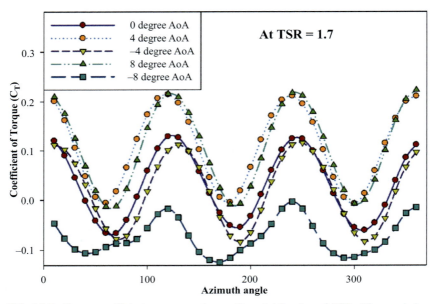

FIG. 15.7 Torque coefficient versus angular position at 1.7 value of TSR. *(No permission required.)*

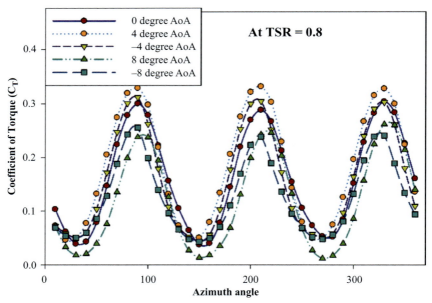

FIG. 15.8 Torque coefficient versus angular position at 0.8 value of TSR. *(No permission required.)*

Effect of hydrofoils on the Darrieus rotor starting torque characteristics Chapter | 15 **369**

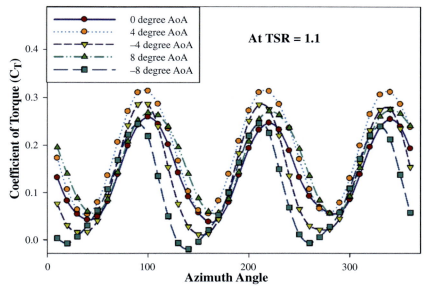

FIG. 15.9 Torque coefficient versus angular position at 1.1 value of TSR. *(No permission required.)*

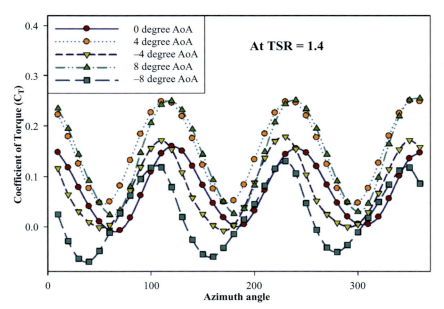

FIG. 15.10 Torque coefficient versus angular position at 1.4 value of TSR. *(No permission required.)*

370 Sustainable developments by artificial intelligence & machine learning

The Darrieus rotor shows the highest torque coefficient as 0.33 at the TSR value of 0.8, corresponding to the 4° AoA and 210° azimuth angle. It has also been observed that for all the considered cases, the coefficient of torque is better at 4° AoA as compared to 0° AoA. On the other hand, negative AoA reduces the coefficient of torque.

15.4.2 Flow contours

In order to analyze the flow behavior across the turbine blades and channel computational domain, pressure and velocity contours has been plotted and discussed corresponding to each angle of attack and turbine maximum performance. Pressure variations were captured by plotting the pressure contours across the turbine vicinity. For the considered value of velocity, pressure across the turbine vicinity varies due to variation in AoA. Fig. 15.11 shows the variation in pressure contour corresponding to each considered value of AoA. The red and blue color, respectively, show the maximum and minimum pressure areas in the contour, and corresponding values have been shown on the scale. It has been observed that a pressure difference is created across the turbine blade (airfoil) which results in lift generation and, hence, the resultant torque. A wake zone is also observed at the backside side of blade due to water circulation, except at −8° AoA. The water circulation (whirl phenomena) was reduced due to negative AoA, as can be seen in Fig. 15.12 (streamlines). In order to capture the velocity flow field across the turbine, velocity contours and streamlines are plotted. Similar to pressure contours, red and blue color, respectively, indicate the maximum and minimum value of velocity. Fig. 15.13 shows that the variation in the velocity corresponds to a considered range of AoA.

It can be observed from the velocity contours that the blade tip experiences the maximum velocity and this zone can be termed the high speed zone. However, the minimum value of velocity is observed at the trailing edge of the blade. Water approaches the turbine with the same velocity as that given in the boundary conditions (0.5 m/s). Therefore, turbine placement has almost no effect on velocity in the upstream channel. The placement of the turbine extracts a fraction of the flowing kinetic energy and the flow velocity reduces to some extent. The flow regains its original velocity due to gravity flow and the distance at which it regains its original velocity is known as the "wake recovery distance."

The velocity pattern in the rotating domain is similar for all AoAs. The difference in the velocity occurs across the blade profile, which further influences the lift and drag force generation. In order to visualize the flow direction of water, velocity streamlines have been drawn and it has been observed that for 4° of AoA, water interacts more smoothly with turbine blades and, hence, has better torque generation.

Furthermore, the Darrieus rotor with positive AoA generates more flow disturbances on the rotor blade as compared to the blades having negative AoA. The results of the present study have been found to correspond well with earlier

Effect of hydrofoils on the Darrieus rotor starting torque characteristics **Chapter | 15** **371**

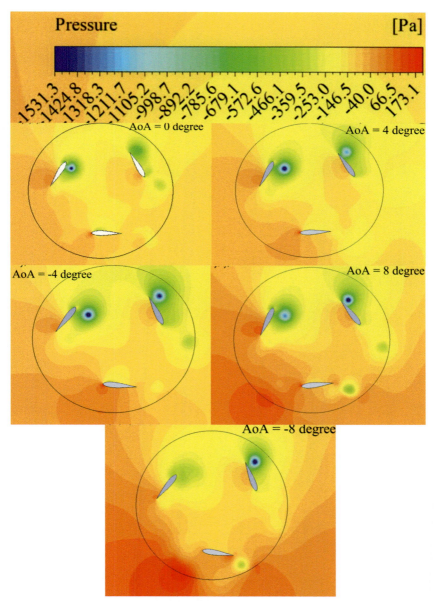

FIG. 15.11 Pressure contour corresponding to each AoA. *(No permission required.)*

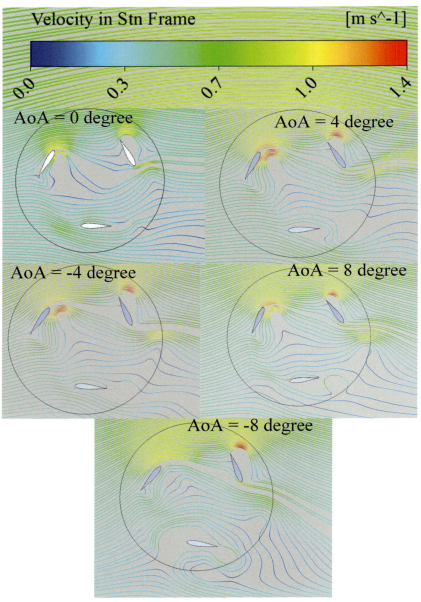

FIG. 15.12 Velocity streamlines across the turbine vicinity corresponding to each AoA. *(No permission required.)*

Effect of hydrofoils on the Darrieus rotor starting torque characteristics **Chapter | 15 373**

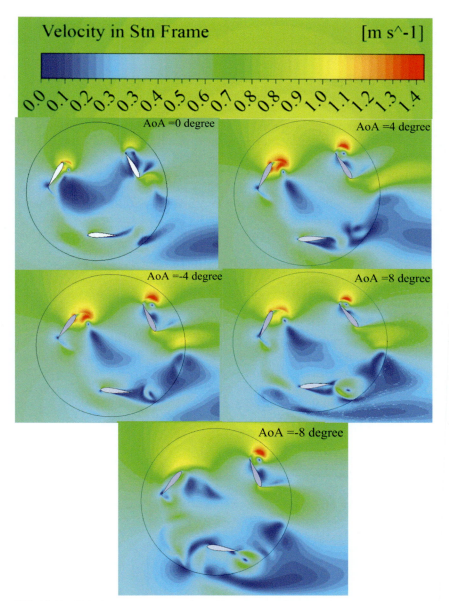

FIG. 15.13 Velocity contours across the turbine corresponding to each AoA. *(No permission required.)*

374 Sustainable developments by artificial intelligence & machine learning

studies (Mahdi, Mohammad Javad, & Seyed Rasoul, 2016). At negative AoA, a whirling phenomenon and backward flow were also observed during rotation (Figs. 15.12 and 15.13).

15.5 Conclusions

In order to improve the performance of the Darrieus hydrokinetic turbine, the turbine has been numerically investigated with different values of AoA under 0.5 m/s of flow velocity and by considering a TSR range varying from 0.5 to 1.7. The power coefficient was selected in order to assess the performance of turbine under different AoAs. The self-starting behavior of the turbine was investigated by plotting the coefficient of torque at every azimuth position corresponding to the considered range of AoA. Furthermore, the variations in the flow fields were analyzed by plotting the pressure, velocity, and streamlines across the turbine vicinity.

ANSYS (Fluent v.19.1) was used to solve the URANS along with RNG k-ε turbulence model. Based on the present numerical simulation-based study, it has been observed that positive AoA increases the performance and self-starting capability of the turbine. However, positive AoA results in detached flow, which may lead to stall phenomena. On the other hand, negative AoA reduces the turbine performance and self-starting capability. Based on the present study, it can be concluded that 4° AoA is the optimum AoA, which has the maximum value of power coefficient, as 0.214 corresponds to a TSR value of 1.25. The findings of the present study may be useful for further investigations on performance enhancement under different operating conditions.

References

Abdalrahman, G., Melek, W., & Lien, F. S. (2017). Pitch angle control for a small-scale Darrieus vertical axis wind turbine with straight blades (H-type VAWT). *Renewable Energy, 114*, 1353–1362. https://doi.org/10.1016/j.renene.2017.07.068.

Alam, M. J., & Iqbal, M. T. (2010). A low cut-in speed marine current turbine. *Journal of Ocean Technology, 5*(4), 49–62. http://www.journalofoceantechnology.com/?page_id=78&jot_download_article=213.

Asr, M. T., Nezhad, E. Z., Mustapha, F., & Wiriadidjaja, S. (2016). Study on start-up characteristics of H-Darrieus vertical axis wind turbines comprising NACA 4-digit series blade airfoils. *Energy, 112*, 528–537. https://doi.org/10.1016/j.energy.2016.06.059.

Balduzzi, F., Bianchini, A., Maleci, R., Ferrara, G., & Ferrari, L. (2016). Critical issues in the CFD simulation of Darrieus wind turbines. *Renewable Energy, 85*, 419–435. https://doi.org/10.1016/j.renene.2015.06.048.

Bianchini, A., Balduzzi, F., Ferrara, G., & Ferrari, L. (2016). A computational procedure to define the incidence angle on airfoils rotating around an axis orthogonal to flow direction. *Energy Conversion and Management, 126*, 790–798. https://doi.org/10.1016/j.enconman.2016.08.010.

Chen, Y., & Lian, Y. (2015). Numerical investigation of vortex dynamics in an H-rotor vertical axis wind turbine. *Engineering Applications of Computational Fluid Mechanics, 9*(1), 21–32. https://doi.org/10.1080/19942060.2015.1004790.

Güney, M. S., & Kaygusuz, K. (2010). Hydrokinetic energy conversion systems: A technology status review. *Renewable and Sustainable Energy Reviews*, *14*(9), X2996–X3004. https://doi.org/10.1016/j.rser.2010.06.016.

Kirke, B. (2016). Tests on two small variable pitch cross flow hydrokinetic turbines. *Energy for Sustainable Development*, *31*, 185–193. https://doi.org/10.1016/j.esd.2016.02.001.

Kirke, B. K., & Lazauskas, L. (2011). Limitations of fixed pitch Darrieus hydrokinetic turbines and the challenge of variable pitch. *Renewable Energy*, *36*(3), 893–897. https://doi.org/10.1016/j.renene.2010.08.027.

Kumar, A., & Saini, R. P. (2016). Performance parameters of Savonius type hydrokinetic turbine— A review. *Renewable and Sustainable Energy Reviews*, *64*, 289–310. https://doi.org/10.1016/j.rser.2016.06.005.

Lago, L. I., Ponta, F. L., & Chen, L. (2010). Advances and trends in hydrokinetic turbine systems. *Energy for Sustainable Development*, *14*(4), 287–296. https://doi.org/10.1016/j.esd.2010.09.004.

Lazauskas, L., & Kirke, B. K. (2012). Modeling passive variable pitch cross flow hydrokinetic turbines to maximize performance and smooth operation. *Renewable Energy*, *45*, 41–50. https://doi.org/10.1016/j.renene.2012.02.005.

Lee, J. H., Lee, Y. T., & Lim, H. C. (2016). Effect of twist angle on the performance of Savonius wind turbine. *Renewable Energy*, *89*, 231–244. https://doi.org/10.1016/j.renene.2015.12.012.

Mahdi, Z., Mohammad Javad, M., & Seyed Rasoul, V. (2016). Starting torque improvement using J-shaped straight-bladed Darrieus vertical axis wind turbine by means of numerical simulation. *Renewable Energy*, *95*(C), 109–126. https://doi.org/10.1016/j.renene.2016.03.069.

Mohamed, M. H. (2012). Performance investigation of H-rotor Darrieus turbine with new airfoil shapes. *Energy*, *47*(1), 522–530. https://doi.org/10.1016/j.energy.2012.08.044.

Paillard, B., Astolfi, J. A., & Hauville, F. (2015). URANSE simulation of an active variable-pitch cross-flow Darrieus tidal turbine: Sinusoidal pitch function investigation. *International Journal of Marine Energy*, *11*, 9–26. https://doi.org/10.1016/j.ijome.2015.03.001.

Saini, G., & Saini, R. P. (2018). A numerical analysis to study the effect of radius ratio and attachment angle on hybrid hydrokinetic turbine performance. *Energy for Sustainable Development*, *47*, 94–106. https://doi.org/10.1016/j.esd.2018.09.005.

Saini, G., & Saini, R. P. (2019). A review on technology, configurations, and performance of cross-flow hydrokinetic turbines. *International Journal of Energy Research*, *43*(13), 6639–6679. https://doi.org/10.1002/er.4625.

Saini, G., & Saini, R. P. (2020). A computational investigation to analyze the effects of different rotor parameters on hybrid hydrokinetic turbine performance. *Ocean Engineering*, *199*, 107019.

Sengupta, A. R., Biswas, A., & Gupta, R. (2016). Studies of some high solidity symmetrical and unsymmetrical blade H-Darrieus rotors with respect to starting characteristics, dynamic performances and flow physics in low wind streams. *Renewable Energy*, *93*, 536–547. https://doi.org/10.1016/j.renene.2016.03.029.

Zeiner-Gundersen, D. H. (2014). A vertical axis hydrodynamic turbine with flexible foils, passive pitching, and low tip speed ratio achieves near constant RPM. *Energy*, *77*, 297–304. https://doi.org/10.1016/j.energy.2014.08.008.

Zhao, G., Yang, R. S., Liu, Y., & Zhao, P. F. (2013). Hydrodynamic performance of a vertical-axis tidal-current turbine with different preset angles of attack. *Journal of Hydrodynamics*, *25*(2), 280–287. https://doi.org/10.1016/S1001-6058(13)60364-9.

Index

Note: Page numbers followed by *f* indicate figures, *t* indicate tables and *b* indicate boxes.

A

Absorption coefficient, 66–69, 67*f*, 92–93
Active cooling strategy, 257–259, 260*t*, 262–263, 263*f*
Active power loss (APL), 150–151, 165–166, 165–166*f*, 168
Active solar space-heating systems, 5
Agent-based P2P energy trading. *See* Peer-to-peer (P2P) energy trading
Agilent *I-V* measurement system, 114, 114*f*
Air pollution, 1
Air ventilation-based BIPV/PCM system, 260*t*, 261–262
Algal biofuel production, 193
Alternating current optimal power flow (ACOPF), 22
Alternating current power flow (ACPF), 22
Alternative, clean sources of energy (ACE) applications, 2
    biomass (*see* Biomass energy)
    geothermal energy (GE) (*see* Geothermal energy (GE))
    limitations, 17
    ocean and tidal energy (*see* Ocean energy (OE))
    small, mini, and micro-hydropower plants, 15–17, 16*f*
    solar energy (SE) (*see* solar energy (SE))
    wind energy (WE) (*see* Wind energy (WE))
Aluminum back surface field (Al BSF) solar cells, 63–65, 65*f*
Aluminum oxide (Al₂O₃), 63–65, 86–90
Anaerobic digestion, 11–12
Analysis of variance (ANOVA), 326
Angle of attack (AoA), effects on Darrieus turbine. *See* Darrieus hydrokinetic turbine, at different AoA
Antireflection coatings (ARC), 70–71, 70*f*, 84

Ant lion optimization algorithm (ALOA), 150, 168, 169–170*t*
Artificial intelligence (AI), 190, 221–229
    artificial neural network (ANN) model, 213, 214*f*
    deep learning (DL), 213
    foundation of, 206–213, 213*f*
    HVAC systems, application in, 248
    market prediction, technology for, 274, 283
    renewable energy sources (RESs), use in, 213, 215*f*
        bioenergy (BE), 213–214, 216–217*t*
        geothermal energy (GE), 214–215, 218–219*t*
        hydrogen energy (HE), 221, 222–223*t*
        hydropower energy (HPE), 197, 215–220, 220*t*
        ocean energy (OE), 221, 228*t*
        solar energy (SE), 221, 224*t*
        wind energy (WE), 221, 225–227*t*
Artificial neural network (ANN), 322–323
    architectural model of, 213, 214*f*
    drawbacks, 131
    heating load prediction, 239
    hydropower plant, load prediction (*see* Hydropower plants)
    load frequency control (LFC) problems, 131
Atomic layer deposition (ALD), 98–99, 106–107
Atomic layer deposition (ALD) Al₂O₃ process, 89–90, 90*f*
Auction-based pricing strategy, 276
Auger recombination, 75–76
Available transfer capability (ATC), 22, 44
Axial flow hydrokinetic turbines, 360

B

Back surface field (BSF) cells
    Al BSF solar cells, 63–65, 65*f*

377

378 Index

Back surface field (BSF) cells *(Continued)*
   market share of, 63, 64*f*
Backtracking search optimization (BTSO)
   algorithm, 151
Band-to-band recombination. *See* Radiative
   recombination
Battery energy storage system (BESS),
   136–137
Bifacial PERC technology, 97–98, 98*f*
Bill sharing (BS) mechanism, 273–274,
   276–277
Binary cycle power plants, 7
Biochar, 11
Biodiesel
   biofuel policy of India, aim of, 13
   definition, 13
   diesel engines, use in, 13
   production, oilseed crops for, 193
   transesterification
      process, 13
      reaction, 13, 14*f*
Biodiesel production, RSM
   carbon monoxide emission
      reduction, 321–322
   elephant swarm water search algorithm
      (ESWSA), 323
   formulation, 324
   hybrid optimization techniques, 322–323
   mathematical model of, 324–328
      basic elephant swarm water search
         algorithm, 327–328
      optimization of, 326
      proposed methodology, 327
   nonlinear models, parametric optimization,
      322
   reaction conditions, 329
   surface plot, by different combinations
      catalyst concentration and methanol to oil
         ratio, 329, 335*f*
      in Minitab, 329, 330*t*
      reaction temperature and catalyst
         concentration, 329, 332*f*
      reaction temperature and methanol to oil
         ratio, 329, 333*f*
      reaction temperature and stir speed, 329,
         333*f*
      reaction time and reaction temperature,
         329, 335*f*
      reaction time and stir speed, 329, 331*f*
      stir speed and catalyst concentration, 329,
         331*f*

      stir speed and methanol to oil ratio, 329,
         332*f*
Bioenergy (BE)
   artificial intelligence (AI), 213–214,
      216–217*t*
   benefits, 193
   categorization of, 193, 193*f*
   definition, 193
   electricity generation and installation
      capacity
      status of, 193, 195*f*
      world's top countries' contribution in, 193,
         196*f*
   production chain, 193, 194*f*
Bioethanol
   lignocellulose fermentation, 13
   sugar/starch fermentation, 12–13
Biofuels
   algal biofuel production, 193
   biofuel policy of India, 13
   classifications, 12, 12*t*
   definition, 12
   transportation sector, use in, 12
Biogeography-based optimization (BBO)
   algorithm, 151
Biomass energy
   anaerobic digestion, 11–12
   biodiesel, 13
   bioethanol production
      lignocellulose fermentation, 13
      sugar/starch fermentation, 12–13
   biofuels, 12
   definition, 10
   extraction methods
      direct combustion, 10
      pyrolysis, 10–11
      thermochemical/biochemical method, 10,
         10*f*
   gasification, 11
Blade velocity/tangential velocity, 360–361
Blockchain technology, in P2P energy trading
   blockchain-enhanced transaction model, 283
   challenges, 283
   cyber-enhanced transactive microgrid model,
      283
   decentralized platform, structure of, 282,
      282*f*
   functions of, 283
   layer functions, 274, 282
   microgrid energy market, 283
   vehicle-to-grid interaction, 283

Box-type solar cooker, 3–4, 4*f*
Bruker thickness profilometer, 113, 113*f*
Building energy consumption forecast model, 263–264, 264*f*
Building energy prediction, ML applications in
  cooling/heating/electrical load prediction, 237–239, 240–241*t*
  energy consumption prediction, 239, 242*f*
  feature extraction and classification, 239
  original data source and preparation, stage of, 239
  reinforcement learning, 242, 244*t*
  supervised learning
    building load predictions, 242, 244*t*
    energy performance prediction, 239–242, 243*f*
  on schedulable appliances, 247
  unsupervised learning
    building load predictions, 242, 244*t*
    energy performance prediction, 239–242, 243*f*
Building-integrated photovoltaics/phase change materials (BIPV/PCM), 259
Building load predictions, ML model for, 242, 244*t*
Buying price, P2P energy, 278–280, 282

C

California Central Valley (CCV) project, 341–342
Capacitance vs voltage *(C-V)* measurement system, 114, 114*f*
Carbonization, 10
Carrier transport equations
  assumptions, 79–80
  continuity equations, for electrons and holes, 79
  minority carrier diffusion equations, 80
  quasi-Fermi energy, 79
  solar cell parameters
    efficiency, 83
    fill factor, 82
    open circuit voltage, 81
    short-circuit current, 80
  total hole and electron currents, 79
Centroid defuzzification approach, 140
Chaotic differential evolution (CDE) technique, 151, 168–171, 169–170*t*
Climate change, 1
Coefficient of power (CP), 361–362, 364, 374
  low value of, 366–367

maximum value of, 366–367
  vs. tip speed ratio (TSR), 366–367, 366*f*
Coefficient of torque (CT), 362, 364, 366–367, 374
  vs. tip speed ratio (TSR), 370
  angular position at 0.8 value, 367, 368*f*
  angular position at 1.1 value, 367, 369*f*
  angular position at 1.4 value, 367, 369*f*
  angular position at 1.7 value, 367, 368*f*
Complex refractive index, 68–69
  base extinction coefficient, 68
  base refractive index, 68
  correction terms, 68
  free carrier absorption, 68–69
  temperature dependency of, 68–69
  wavelength dependency of, 68
Comprehensive teaching learning-based optimization (CTLBO) technique, 150
Continuous double auction (CDA) market, 283–284
Control accuracy and precision (CAP), 190
Convolutional neural network (CNN) model, 312
Cooling load prediction, ML applications in, 237–239, 240–241*t*
Copula models, 310
Cosine adapted whale optimization algorithm (CAWOA), 151–152
  cosine function, 155
  EDSs, PV-DG and WT-DG
    active and reactive branch currents, influence on, 171–172, 175–176*f*
    active power losses, 165–166, 165–166*f*
    bus voltages profile, 163–165, 163–164*f*
    bus voltage variation, 166, 167–168*f*
    loadability variation, impact of, 172–175, 177–181*f*
    optimization algorithms, comparison results of, 166–171, 169–170*t*
    optimization results, analysis of, 158–163, 162*t*
    parameters, 158, 161*t*
    test system, 155–157
  pseudo code of, 155, 157*b*
Cost-benefit allocation schemes, 274, 286
Cost-benefit analysis, 349
  annual irrigation
    vs. benefits, 343, 346*f*
    vs. capital, 343, 345*f*
    vs. OM cost, 343, 345*f*

380 Index

Cost-benefit analysis *(Continued)*
  of input and output, 343
  irrigation demand, 343
  percent irrigation shortage *vs.* percentage
    benefit loss, 343, 346*f*
  reservoir capacity, 343
  vs. capital cost, 343, 347*f*
  vs. operation and maintenance cost, 343,
    347*f*
Coyote optimization algorithm (COA), 151
Craziness based PSO (CRPSO) algorithm, 151
CRONE principle, 132
Cross-entropy method, 296–297
Cross flow hydrokinetic turbines
  Darrieus rotor *(see* Darrieus hydrokinetic
    turbine, at different AoA)
  Savonius rotor, 360–361
  shallow depth applications, 360
Cross-validation algorithm, 239
Current-voltage *(I-V)* measurement system,
  114, 114*f*, 118–120, 119*f*
Cyber-enhanced transactive microgrid
  model, 283

D

Dam suitability stream model (DSSM), 17
Darrieus hydrokinetic turbine, at different AoA,
  374
  airfoils, virtual camber theory of, 361
  blade profile, forces over, 360–361, 360*f*
  flow contours
    backward flow, 370–374
    pressure contours, 370, 371*f*
    velocity contours, 370, 373*f*
    velocity streamlines, 370, 372*f*
    wake recovery distance, 370
    whirling phenomenon, 370–374
  lift/drag force ratio, 360–361
  memetic algorithm, 361
  numerical simulation analysis, 362, 374
    ANSYS (v.19.1), 362–363
    blade profiles, 363, 363*f*
    boundary conditions, 365–366, 365*t*
    grid generation, 364
    2D computational channel domain, 363,
      364*f*
    unsteady Reynolds averaged Navier-
      Stokes (URANS) equations, 362–363
  objectives, 362
  parameters, 362, 362*t*
  parasitic forces, 361
  performance characteristics, 362

  power coefficient *vs.* tip speed ratio
    (TSR), 366–367, 366*f*
  torque coefficient *vs.* tip speed ratio (TSR)
    (see Coefficient of torque (CT))
  self-starting, problem of, 360–361
  variable pitch blade *vs.* fixed pitch type blade,
    361
  variable pitch rotor *vs.* fixed pitch rotor, 361
Data-driven smart charging strategy, 248
DC-P-OPF methodology, 32–34
Decentralized bilateral energy trading system,
  285
Decision tree model, 22
Decision trees, 312
Deep learning (DL), 213, 312
  electric load and renewable generation,
    predictions of, 238
  for load forecasting, 310
Deep Q-learning, 283–284
Deep reinforcement learning (DRL)
  technology, 283–284
Demand ratio (DR), 280
Demand shifting strategy, 243–247
Demand-side management (DSM), 273,
  291–294
  advanced controllers, development of
    for deterministic and stochastic scenarios,
      250–252, 250–251*f*
    rule-based and model-predictive control
      strategies, 250–252
    steps for, 252
    surrogate model, 250–252
  energy efficiency, 294–295
  energy-saving, 294–295
  flexible demand-side management strategies,
    238, 245–246*t*
    HVAC systems, 242–248
    plug-in loads and storages, 242–249
    smart appliances, 242–247
  load growth, 295
  power generators, 294
  purpose of, 294
Deterministic-based optimal PCMs-PV system,
  267
Developed double layer antireflection coating
  (DLARC), 71
Dielectric surface passivation, 86–87, 91–92
Diesel engine generator (DEG) model, 136
Differential learning with biogeography based
  optimization (DLBBO) algorithm,
  150
Direct steam power plants, 6

Index   381

Discrete-continuous hyperspherical search (DC-HSS) algorithm, 151
Distributed continuation power flow (DCPF), 22
Distributed generators (DG), in EDS. *See* Electrical distribution system (EDS), renewable DG in
Disturbance observer (DO), 132
Double auction mechanism, 279
Double flash steam power plants, 6–7
Dragonfly algorithm (DA) approach, 150
Dry oxidation, 101
Dynamic internal pricing model, 276
Dynamic programming (DP) simulation model, 341–342, 355–356*t*
Dynamic supply to demand ratio ($SDR_{i,t}$), 278–280

E

Electrical distribution system (EDS), renewable DG in, 151–152, 176–177
  mathematical problem formulation
    equality constraints, 153
    inequality constraints (*see* Inequality constraints)
    multiobjective function, 152–153
  motivation, 149–150
  parameters, before DG integration, 158, 161*t*
  PV-DG and WT-DG installation, CAWOA algorithm, 152
    active and reactive branch currents, influence on, 171–172, 175–176*f*
    active power losses, 165–166, 165–166*f*
    bus voltages profile, 163–165, 163–164*f*
    bus voltage variation, 166, 167–168*f*
    loadability variation, impact of, 172–175, 177–181*f*
    optimization algorithms, comparison results of, 166–171, 169–170*t*
    optimization results, analysis of, 158–163, 162*t*
    parameters, 158, 161*t*
    test system, 155–157
  techniques and algorithms, 150–151
Electrical energy, P2P energy trading with, 274–276, 275*t*
Electrical load prediction, ML applications in, 237–239, 240–241*t*
Electrochemical capacitance-voltage (ECV) measurements, 111
Electron cyclotron resonance (ECR) microwave discharge, 105–106

Elephant herding optimization (EHO) algorithm, 151
Elephant swarm water search algorithm (ESWSA), 323, 327–329
Energy consumption prediction, ML model for, 239, 242*f*
Energy sharing decision making, 274, 283
Enhanced geothermal systems (EGS), 197
Excess renewable-recharging strategy, 247–248
Extended state observer (ESO), 132
External quantum efficiency (EQE) measurement, 118
Extreme learning machine (EML), 303
Extreme learning machine (ELM)-based market prediction models, 284

F

Field-emission scanning electron microscopy (FE-SEM), 111, 112*f*
Fill factor *(FF)*, 74–75, 82
Finned double-pass double glass-covered concentrated photovoltaic thermal (PVTC), 259–261, 261*f*
Flower pollination algorithm (FPA) approach, 150, 168, 169–170*t*
Fluid velocity, 360–361
Flywheel energy storage system (FESS), 136–137
Fossil fuels, 1, 17
Four point probe measurement method, 111, 113*f*
Free carrier absorption (FCA), 69–70, 72–73
Fuel cell (FC), 136–137
Fuzzy logic controller (FLC), 131
  advantages, 137
  components
    defuzzification, 140
    fuzzification, 137
    fuzzy rule base inference system, 137, 139*t*
  disadvantages, 137
  gray wolf optimization (GWO) (*see* Gray wolf optimization (GWO) algorithm)
  IT2FLC-based feedback error learning (FEL) method, 132
  particle swarm optimization (PSO) (*see* Particle swarm optimization (PSO) algorithm)
Fuzzy membership function methods, 270
Fuzzy set theory, 131

G

Gasification, biomass, 11
Gaussian kernel regression model, 242

382 Index

Gaussian mixture model, 296–297
Geographical information system (GIS), 17
Geothermal energy (GE)
  artificial intelligence (AI), 214–215,
    218–219*t*
  binary cycle power plants, 7
  direct application and usage, 7, 7–8*f*
  direct steam power plants, 6
  double flash systems, 6–7
  electricity generation and installation
    capacity
      status of, 197, 198*f*
      world's top countries' contribution in, 197,
        199*f*
  forms, 6
  geothermal heating technologies, 197
  geothermal power generation, 6–7
  geothermal reservoirs, 193–197
  locations, in India, 197, 200*f*
  single flash system power plants, 6
Global warming, 1, 188–189
Glucose, 12–13
Grasshopper optimizer algorithm (GOA), 150
Gray wolf optimization (GWO) algorithm,
    132–133, 150, 168–171, 169–170*t*
  definition, 142–143
  features, 143
  frequency error, 143
  fuzzy rule base for, 137, 139*t*
  implementation, flow chart for, 144, 144*f*
  mathematical model of, 142–143
  microgrid (MG), frequency response of, 137,
    138*f*
  objective/fitness function, 143–144
  PI and FLC plus GWO, frequency change
    profile using, 145, 146*f*
Greenhouse gas (GHG) emission, 188–189

H

Harabhangi irrigation project, simulation
    model, 355–356
  computations, 350–352
  cost-benefit functions, 343–344
  flow chart of, 350, 350*f*
  linear programming (LP) model, 345–349
  map of, 343, 344*f*
  MATLAB, 342, 355–356*t*
  reservoir capacity
    annual spill, variation in, 353–355, 354*f*
    irrigation deficit, variation in, 353–355,
      354*f*
    net benefit, variation in, 353–355, 353*f*
  reservoir simulation, 349–350

results of, 351–353*t*, 352
study area description, 343
Harris Hawks optimization (HHO) algorithm,
    151
Heating load prediction, 237–239, 240–241*t*
Heat transfer mechanism, 261–262, 262*f*
Heuristic moment matching (HMM) technique,
    151
Hierarchical clustering methods, 239–242
High-grade temperature devices, 2
Human opinion dynamics (HOD) technique, 150
HVAC systems, 242–248, 245–246*t*
Hybrid demand-side controllers, ML
    applications in, 292, 304–305
  advanced controllers, development of
    for deterministic and stochastic scenarios,
      250–252, 250–251*f*
    rule-based and model-predictive control
      strategies, 250–252
    steps for, 252
    surrogate model, 250–252
  building energy prediction
    cooling/heating/electrical load prediction,
      237–239, 240–241*t*
    energy consumption prediction, 239, 242*f*
    feature extraction and classification, 239
    original data source and preparation, stage
      of, 239
    reinforcement learning, 242, 244*t*
    supervised learning, 239–242, 243*f*, 244*t*
    unsupervised learning, 239–242, 243*f*,
      244*t*
  challenges, 304
  demand-side management strategies
    (*see* Demand-side management (DSM))
  demand-side response and management, 238,
    242–243
  extreme learning machine (EML), 303
  future research, 304
  high-level controllers, improvement of, 297,
    298*f*
  k-means clustering, 302
  linear regression, 303, 303*f*
  partial least squares, 304
  residential building, 299
  supervised machine-learning model
    (*see* Supervised learning)
  support vector machine (SMV), 299–302
Hybrid energy system (HES), 293
Hybrid renewable energy sources (HRES), 294
Hydraulic numerical models, 342, 355–356*t*
Hydrogen based district energy sharing
    community, 248, 249*f*

Index **383**

Hydrogen energy (HE), 200
  artificial intelligence (AI), 221, 222–223*t*
  categories, 203
  hydrogen economy, key driver and
    challenges for, 203, 204*f*
  P2P energy trading with, 274–276, 275*t*, 277*f*
  production methods and process, 203, 203*f*
  sustainable development scenario (SDS),
    203, 205*f*
Hydrogen fuel, 188–189
Hydrogen vehicles (HVs), 248
Hydrokinetic technology, 359–360
Hydrokinetic turbines, 359–360
  axial flow turbines, 360
  cross flow turbines
    Darrieus rotor (*see* Darrieus hydrokinetic
      turbine, at different AoA)
    Savonius rotor, 360–361
    shallow depth applications, 360
Hydropower energy (HPE)
  artificial intelligence (AI), 197, 215–220,
    220*t*
  electricity generation and installation
    capacity
    status of, 200, 201*f*
    world's top countries' contribution in, 200,
      202*f*
  flowing water, energy extraction from, 197
  hydropower plant, 197
  machine learning, 197
Hydropower plants, 15–17, 16*f*, 313, 316–317
  daily plant load data, 310*f*, 313
    neural network training window data, 311*f*,
      313
    selforganizing map (SOM) clustering
      technique (*see* Selforganizing map
      (SOM), daily plant load data)
  decision trees, 312
  load prediction methods, 310–312
  monitoring and maintenance of, 309
  support vector machine (SMV), 310, 312

I

IEEE 30 bus system
  area, 27–28, 28*t*
  bus data of, 27–28, 29*t*
  conventional generator data of, 27–28, 27*t*
  line data of, 27–28, 29–30*t*
  one line diagram of, 27–28, 31*f*
  tie lines, 27–28, 28*t*
Iine flow capacity index, 151

Improved estimation of distribution algorithm
  (IEDA), 151
Indian Meteorological Department (IMD), 23
Indirect gap absorption process, 67
Individual peer energy trading price model, 276,
  279–280, 281*f*, 286
Indoor radiative cooling, 262–263
Inequality constraints
  of distributed generators (DG) units
    capacity limits, 154
    limits of constraint values, 154, 155*t*
    location, 154
    number, 154
    position, 154
    power factor, 154
  of distribution line
    bus voltage limits, 153
    line capacity constraint, 154
    voltage drop limit, 153
Integral of absolute error (IAE), 141–143
Integral of time multiplied by absolute error
  (ITAE), 143
Integral square error (ISE), 141–143
Integral square of absolute error (ISAE),
  141–143
Interband absorption, 72
Internal mode control (IMC), 132
Internet of energy things (IoET), 49–50
  energy management system, 51–54
  layers, 55*f*
    access layer, 54
    application layer, 55
    perception/physical layer, 54
    processing layer, 54
Internet of energy things-smart grid (IoET-SG),
  49–50, 60
  architecture, 55–56, 56*f*
  controlling, 56
  management, 57
  monitoring, 56
  regulation and market, 57
  research challenges and future guidelines, 59,
    59*f*
    big energy data, 58
    energy forecasting, 58
    IoT standard, 59
    load balancing/management, 57
    security and privacy, 58
    self-healing, 59
    smart meter (SM), 57
    user behavior prediction, 58
Internet of things (IoT), 49–54

384 Index

Interval type-2 fuzzy logic systems (IT2FLS), 132
Intraband absorption, 72
Irrigation projects
overall efficiency of, 341
simulation model
California Central Valley (CCV) project, 341–342
Harabhangi irrigation project (*see* Harabhangi irrigation project, simulation model)
Karjan irrigation reservoir project, 342
Pampa-Achankovil-Vaippar link diversion system (PAV system), 342

K

Karjan irrigation reservoir project, 342
Keithley C-V measurement system, 114, 114f
K-means clustering, 302
energy performance prediction, 239–242
heating load prediction, 239
Krill herd algorithm (KHA), 150, 168, 169–170t, 171

L

Laser ablation, 63–65, 84, 92–93, 107–108
Laser fire contact (LFC) process, 63–65, 83–84
Leader-followers Stackelberg game approach, 279, 283–284
Lignocellulose fermentation, 13
Linear matric inequality (LMI) method, 132
Linear programming (LP) simulation model, 341–342, 345–349, 355–356t
Linear regression, 303, 303f
Load flow balance equations, 153
Load frequency control (LFC)
artificial neural network (ANN), 131
fuzzy logic controller (FLC) (*see* Fuzzy logic controller (FLC))
internal mode control (IMC) framework, 132
sliding mode load frequency controller, 132
time delay issues, 132
Load scheduling, 242–243, 247
Locational marginal pricing (LMP), 22, 34
solar generation
performance index, 38, 38t
values for, 35–38, 37t
variation, 38, 39f
wind generation
performance index, 38, 38t
values for, 35–38, 36t
variation, 38, 39f

Long short-term memory (LSTM) model, 312
Loss sensitivity factors (LSF), 150
Low-grade temperature devices, 2
Low temperature oxidation (LTO), 101

M

Machine leading (ML), 292, 309
definition, 292
model-based control methods, 296–297
reinforcement learning (*see* Reinforcement learning)
semisupervised learning, 295–296
supervised learning algorithm (*see* Supervised learning)
unsupervised learning algorithm (*see* Unsupervised learning)
Machine learning (ML)
deep learning (DL), 206–213
hybrid demand-side controllers, applications in (*see* Hybrid demand-side controllers, ML applications in)
nonlinear systems, performance prediction of, 258
building energy consumption forecast model, 263–264, 264f
mathematical training, 264, 265f
supervised machine learning method, mechanism of, 263
PCMs-PV systems, application in
hybrid learning and advanced optimization algorithms, 258, 270
multicriteria decision-making, for trade-off solutions, 258, 270
multiobjective optimization, 258, 267
robust optimization, with multilevel uncertainty, 258, 267, 268–269f
single-objective optimization, 258, 266
stochastic sampling size and uncertainty-based optimization, 258, 268–270
thermal and electrical performance prediction, surrogate model for, 258, 264–265, 266f
uncertainty quantification and probability density function, 258, 268
in P2P energy trading
extreme learning machine (ELM)-based market prediction models, 284
near-optimal pricing strategy searching, 274, 283–284
optimal energy trading policy making, 283–284
trading price prediction, 283–284

Index **385**

Malaprabha irrigation reservoir, 342
Mamdani-type fuzzy inference, 140
Marine energy. *See* Ocean energy (OE)
Mathematical planning model, 341–342, 355–356*t*
MATLAB, 32, 145, 342, 355–356*t*
Mean absolute error, 239
Mechanical ventilation, 262–263
Medium grade temperature devices, 2
Mesh independence limit (MIL), 364
Metallization techniques, 109–110
Microgrid (MG) system
   conventional PI controller, frequency change profile using, 145, 145*f*
   fuzzy logic controller (FLC), 131
      advantages, 137
      defuzzification, 140
      disadvantages, 137
      fuzzification, 137
      fuzzy rule base inference system, 137, 139*t*
      gray wolf optimization (GWO) (*see* Gray wolf optimization (GWO) algorithm)
      IT2FLC-based feedback error learning (FEL) method, 132
      particle swarm optimization (PSO) (*see* Particle swarm optimization (PSO) algorithm)
   load change profile on, 145, 145*f*
   MATLAB, 145
   multiobjective fractional order fuzzy proportional integral derivative (MOFOFPID) controller, 132
   test system, 132–133
      battery energy storage system (BESS), 136–137
      block-diagram of, 133, 134*f*
      diesel engine generator (DEG) model, 136
      flywheel energy storage system (FESS), 136–137
      frequency response, 137, 138*f*
      fuel cell (FC), 136–137
      parameters values, 137, 139*t*
      photovoltaic (PV) model, 133–135, 134*f*
      wind energy, 135–136
   vehicle to grid (V2G) concept, 132
Midmarket rate (MMR) mechanism, 273–274, 276
Minority carrier diffusion equations, 80
Modified active disturbance rejection control (MADRC) scheme, 132
Modified black hole optimization algorithm (MBHOA), 132
Modified mothflame optimization (MMFO) algorithm, 151

Monocrystalline PERC cell, 93–95, 94–95*f*
Monte Carlo sampling-based simulation, 268
Moth-flame optimization (MFO) algorithm, 150–151
"Moving mesh" approach, 365
Multicriteria decision-making approach, 258, 270
Multicrystalline PERC cell, 93, 94*f*
Multilevel uncertainty-based optimization, 267, 268–269*f*
Multiobjective fractional order fuzzy proportional integral derivative (MOFOFPID) controller, 132
Multiobjective optimization approach, 258, 267, 270

N

Nanofluid-based PCMs-PVT systems, 259, 260*t*, 261–262, 262*f*
Nano-PCM PVT system, 264–265
Near-optimal pricing strategy searching, 274, 283–284
Net-zero energy community, P2P energy trading
   buying price, 280
   with hydrogen storages, 276, 277*f*
   selling price, 280
Nonparametric-based k-NN models, 242
Nonradiative Auger, 75*f*
Normalized mean bias error (NMBE), 242

O

Ocean energy (OE)
   artificial intelligence (AI), 221, 228*t*
   categorization of, 193*f*, 206
   conversion systems, types of
      ocean thermal energy converter (OTEC), 14–15
      tidal energy converter (TIC), 14–15
      wave energy converter (WEC), 14
   electricity generation and installation capacity
      status of, 206, 211*f*
      world's top countries' contribution in, 206, 212*f*
   ocean thermal energy (OTE), 206
   thermal and mechanical energy, 14
Ocean thermal energy (OTE), 206
Ocean thermal energy converter (OTEC), 14–15
Open circuit voltage, 81
Opposition-based tuned-chaotic differential evolution (OTCDE) technique, 151

386 Index

Optimal energy trading policy making, 283
Optimal power flow (OPF), 22

P

Pampa-Achankovil-Vaippar link diversion
  system (PAV system), 342
Parabolic collector, 4, 5*f*
Parabolic concentrating type solar cooker, 4, 5*f*
Pareto archive nondominated sorting genetic
  algorithm (NSGA-II algorithm), 267
Partial least squares, 304
Particle swarm optimization (PSO) algorithm,
  132–133, 150, 168, 169–170*t*
  advantages, 140
  computational flow chart of, 141–142, 142*f*
  fuzzy rule base for, 137, 139*t*
  microgrid (MG), frequency response of, 137,
    138*f*
  for m-variable optimization problem,
    140–141
  proportional integral (PI) controller
    and FLC plus PSO, frequency change
      profile using, 145, 146*f*
    optimum values, of parameters, 141–142
  velocity and position updates of particle, 141,
    141*f*
Passivated emitter and rear contact (PERC)
  solar cells, 69–70, 121
  Al_2O_3/Si_3N_4 dielectric layer stack, 63–65
  average production efficiency, 95
  bifacial PERC, 97–98, 98*f*
  carrier transport equations (*see* Carrier
    transport equations)
  characterization equipment
    capacitance vs voltage *(C-V)* measurement
      system, 114, 114*f*
    current-voltage *(I-V)* measurement system,
      114, 114*f*, 118–120, 119*f*
    external quantum efficiency (EQE)
      measurement, 118
    four point probe measurement, 111, 113*f*
    lifetime and Suns-V_{oc} measurement, 115,
      116*f*
    reflectance measurement, 116–118, 117*f*
    scanning electron microscopy (SEM), 111,
      112*f*
    thickness profilometer, 111–113, 113*f*
    X-ray photoelectron spectroscopy (XPS),
      114–115
  c-Si-based solar cell technologies, market
    share of, 63, 64*f*
  developement, 63–65, 83–84

electrical losses, 70–79
  recombination losses (*see* Recombination
    mechanisms, in solar cell)
  resistive losses, 74, 74*f*
fabrication
  atomic layer deposition (ALD), 106–107
  laser ablation, 63–65, 107–108
  metallization, 109–110
  phosphorus diffusion, 101
  plasma-enhanced chemical vapor
    deposition (PECVD), 104–106
  p-type PERC C-Si solar cells, process flow
    for, 98–99, 98*f*
  reactive ion etching (RIE), 103–104
  saw damage removal (SDR), 99–100, 100*f*
  texture etching, 99–100
  texturization, 99–100, 100*f*
  thermal oxidation, 101–103, 103*f*
  wafer cleaning, 99–100, 100*f*
  wet chemical bath, 99–100, 99*f*
improvements of, 95–97
industrial PERC cell, 83–85, 84*f*
lab PERC cell, 84–85, 84*f*
LBSF and rear local contact, 92–93
mono and multicrystalline cells, efficiency
  of, 93–95, 94–95*f*
optical losses, 70
  incomplete absorption, 71–73
  reflection loss, 70–71
  shadowing, 73
PERC/PERL/PERT/TOPCon solar cells,
  market share of, 63, 64*f*
photon absorption and optical generation, 66
  absorption coefficient, 66–69, 67*f*
  absorption length, 67
  complex refractive index, 68–69
  indirect gap absorption process, 67
  optical generation rate, 67
  phonon absorption/emission, 66–67, 66*f*
  photon absorption rate, 69
  photon flux, 67–68
  quantum yield model, 68
  ray tracing model, 69
pilot-line laser fired contact process, 63–65,
  83–84
process flow, 85*f*
p type Al-BSF solar cells, 63–65, 65*f*
p type PERC solar cells, 63–65, 65*f*
rear polishing, 93
surface passivation (*see* Surface passivation)
Passivated emitter rear totally-diffused (PERT)
  cells, 96–97, 97*f*
Passive cooling strategy, 257–261, 260*t*, 261*f*

Index **387**

Passive solar space-heating systems, 6
Peak power shaving, 242–243, 247–248
Peer-to-peer (P2P), 22
Peer-to-peer (P2P) energy trading, 273–274
  blockchain technology, 274, 282–283
  challenges, 286
  decentralized electricity market design, 274, 284–285
  in electrical energy forms, 274–276, 275*t*
  hydrogen energy (HE), 274–276, 275*t*, 277*f*
  internal pricing, 274, 286
    auction-based pricing strategy, 276
    bill sharing (BS) mechanism, 273–274, 276–277
    mathematical models, 278–282
    midmarket rate (MMR) mechanism, 273–274, 276
    supply and demand ratio (SDR), 273–274, 276–277
  machine learning technologies, 274, 283–284
  net-zero energy community, with hydrogen storages, 276, 277*f*
  techno-economic incentives, 274, 284–286
  in thermal energy forms, 274–276, 275*t*
Phase change materials integrated photovoltaic (PCMs-PV) systems, 258
  cooling systems
    active cooling systems, 259, 260*t*, 261–262, 262*f*
    integrated passive/active systems, 259, 260*t*, 262–263, 263*f*
    passive cooling systems, 259–261, 260*t*, 261*f*
  machine learning (ML)
    hybrid learning and advanced optimization algorithms, 258, 270
    multicriteria decision-making, for trade-off solutions, 258, 270
    multiobjective optimization, 258, 267
    robust optimization, with multilevel uncertainty, 258, 267, 268–269*f*
    single-objective optimization, 258, 266
    stochastic sampling size and uncertainty-based optimization, 258, 268–270
    thermal and electrical performance prediction, surrogate model for, 258, 264–265, 266*f*
    uncertainty quantification and probability density function, 258, 268
Phonons, 66–67, 66*f*
Phosphorus diffusion, 101
  drive-in step, 101
  phosphosilicate glass (PSG) removal, 101
  predeposition step, 101

  thermal diffusion, process flow for, 101, 102*f*
Phosphosilicate glass (PSG), 101
Photon absorption, 66–69
Photon flux, 67–70
Photovoltaic-based generators (PV-DG), in EDS. *See* Electrical distribution system (EDS), renewable DG in
Photovoltaic (PV) model, 133–135, 134*f*
Photovoltaic systems, 2
Plasma-enhanced chemical vapor deposition (PECVD)
  advantages of, 104–105
  aluminum oxide (Al_2O_3), 89–90
  electron cyclotron resonance (ECR) microwave discharge, 105–106
  glow discharge plasma, 105–106
  process steps of, 106
  silicon nitride, 71, 84, 87–88, 91
  for SiN deposition plasma oxidation, 105*f*, 106
Plug-in electrical vehicles (PEVs), 248–249, 249*f*
Poisson's equation, 80
Population-based incremental learning (PBIL) algorithm, 150
Power flow analysis, 27–28
Power generators, 294
Power system stabilizer (PSS), 312
Power-to-X conversions, 273
Precooling/preheating strategies, 247–248
Prediction-integration strategy optimization (PISO) model, 283–284
Predictive-based (machine learning) approach, 296–297
Probabilistic power flow (PPF), 32
Probability density function, 258, 268
Proportional integral (PI) controller, 132–133, 141–142
  frequency change profile using conventional controller, 145, 145*f*
  FLC plus GWO, 145, 146*f*
  FLC plus PSO, 145, 146*f*
  multiobjective fractional order fuzzy proportional integral derivative (MOFOFPID) controller, 132
Pyrolysis, 10–11

Q

Q-learning-based decision-making, 283–284
Quantum yield model, 68
Quasi-Fermi energy, 79
Quasisteady-state photoconductance (QSSPC) method, 115

388 Index

R

Radiative recombination, 75–76
Ray tracing model, 69
Reactive ion etching (RIE), 103–104, 105f
Rear polishing, PERC solar cells, 93
Recombination mechanisms, in solar cell,
    74–75
  Auger recombination, 75–76
  radiative recombination, 75–76
  Shockley-Read-Hall (SRH) recombination,
    75–77
  surface recombination, 75, 78–79, 78f
Reflectance measurement, 116–118, 117f
Reflection loss, of PERC solar cells, 70–71
Regression models, 310–312
Reinforcement learning, 295–296
  building load predictions, 242, 244t
  for load prediction, 310–312
  residential load scheduling, 247
Reliability programming model, 342, 355–356t
Remote plasma techniques, 105–106
Renewable energy sources (RESs), 41–43,
    221–229, 257, 292–294
  bioenergy (BE) (*see* Bioenergy (BE))
  congestion management, 22–23, 44
  data and uncertainty statistical analysis, 21
    Indian Meteorological Department
      (IMD), 23
    solar source analysis, 25–27
    wind source analysis, 24–25
  in electrical distribution system
    (*see* Electrical distribution system
    (EDS), renewable DG in)
  electricity generation, 189
    contribution, 190, 192f
  global total energy generation and
    installation capacity, 189–190, 191f
  energy consumption data, 189, 189f
  generation, 188–189
  geothermal energy (GE) (*see* Geothermal
    energy (GE))
  hybrid generation, optimization of, 22–23, 43
  hydrogen energy (HE) (*see* Hydrogen energy
    (HE))
  hydropower energy (HPE) (*see* Hydropower
    energy (HPE))
  ocean and tidal energy (*see* Ocean energy
    (OE))
  probabilistic nature and location, impact of
    on locational marginal pricing (LMP), 38
    on sensitivity factors, 35–38
    total transfer capacity (TTC), 39–40, 40f
    transmission reliability margin (TRM),
      39–40

solar energy (SE) (*see* solar energy (SE))
  sources of, 190, 193f
  test case modifications
    configuration of cases, 28–32, 31–32t
    linear sensitivity factors, 23, 33
    locational marginal pricing (LMP), 22, 34
    power flow analysis, 27–28
    reliability parameters, 34
    solution methodology, 32–33
    standard IEEE 30 bus system (*see* IEEE 30
      bus system)
  wind energy (WE) (*see* Wind energy (WE))
Reservoir capacity, 343, 350, 352
  annual spill, variation in, 353–355, 354f
  vs. capital cost, 343, 347f
  irrigation deficit, variation in, 353–355, 354f
  net benefit, variation in, 353–355, 353f
  vs. operation and maintenance cost, 343, 347f
  ranges for, 351–352, 351–352t
Reservoir simulation model. *See* Simulation
    model, for optimal reservoir system
Resistive loss, of PERC solar cells, 70, 74, 74f
Response surface methodology (RSM),
    biodiesel production
  carbon monoxide emission reduction,
    321–322
  elephant swarm water search algorithm
    (ESWSA), 323
  formulation, 324
  hybrid optimization techniques, 322–323
  mathematical model of, 324–328
  methodology, 328–329
  nonlinear models, parametric optimization,
    322
  reaction conditions, 329
  surface plot, by different combinations,
    329–330
Reynolds number, 361
RNG k-ε turbulence model, 363, 365, 374
Robust optimization, 258, 267, 268–269f

S

Scanning electron microscopy (SEM), 111,
    112f
Screen printing and firing process, 109–110,
    110f
Selforganizing map (SOM), daily plant load
    data, 310, 313
  input planes, 314, 315f
  neighbor connections, 313–314, 313f
  neighbor weight distances, 312f, 313–314
  sample hits plot, 314–315, 315f
  silt data, analysis of, 312
  supervised SOM framework, 312

Index 389

unsupervised SOM clustering technique, 313
weight positions, 315–316, 316f
Selling price, P2P energy, 278–280, 282
Semiimplicit method for pressure linked
    equations (SIMPLE), 366
Semisupervised learning, 295–296
Series resistance (R_s), 74
Shadowing, 70, 73
Shannon entropy, 270
Shockley-Read-Hall (SRH) recombination,
    75–77
Short-circuit current density, 74
Short-circuit current, of solar cell, 80
Shunt resistance (R_{sh}), 74
Silicon nitride (SiN_x), 63–65, 71, 86–88
Silicon oxide (SiO_2), 63–65, 86–88
Silicon (Si) solar cell, 70–71, 70f
Simulation model, for optimal reservoir system
    California Central Valley (CCV) project,
        341–342
    dynamic programming (DP) model,
        341–342, 355–356t
    Harabhangi irrigation project
        (see Harabhangi irrigation project,
        simulation model)
    hydraulic numerical models, 342, 355–356t
    inputs for, 341–342
    Karjan irrigation reservoir project, 342
    linear programming (LP) model, 341–342,
        355–356t
    mathematical planning model, 341–342,
        355–356t
    reliability programming model, 342,
        355–356t
    stochastic dynamic programming (SDP),
        342, 355–356t
    system analysis techniques, 342, 355–356t
    water resource system models, 342, 355–356t
Single feedforward neural network (SLFN), 303
Single flash system power plants, 6
Single junction solar cells, 73
Single-objective optimization approach, 258,
    266
Sliding mode pitch angle controller, 132
Small hydropower plant (SHP), 15–17, 16f
Smart charging strategy, 243–248
Smart grid (SG), 51
    advantages of, 51
    architecture, 51, 52f
    definition, 49
    digital sensing technology, 51
    features of, 51, 52–53t
    functionality of, 51, 54f

implementation, 49
IoET-SG systems (see Internet of energy
    things-smart grid (IoET-SG))
Smart meter (SM), 57
Solar cells
    cooling systems
        active cooling strategy, 257–259, 260t,
            262–263, 263f
        integrated passive/active systems,
            257–259, 260t, 262–263, 263f
        passive cooling strategy, 257–261, 260t,
            261f
    metallization, 109–110
    parameters
        efficiency, 83
        fill factor, 82
        open circuit voltage, 81
        short-circuit current, 80
    passivated emitter and rear contact (PERC)
        (see Passivated emitter and rear contact
        (PERC) solar cells)
Solar cookers
    box-type solar cooker, 3–4, 4f
    parabolic concentrating type solar
        cooker, 4, 5f
    use, 3
Solar energy (SE), 2, 257, 293–294
    artificial intelligence (AI), 221, 224t
    categorization of, 193f, 203–204
    electricity generation and installation
        capacity
        status of, 206, 207f
        world's top countries' contribution in, 206,
            208f
    electricity production, 189
    flow to Earth's surface, 204, 205f
    photovoltaic systems, 2
    solar cookers, 3–4
    solar space heating, 4–6
    solar thermal energy systems, 2
    solar water heating (SWH) systems, 2–3, 3f
    solar water pumps, 4
Solar source analysis
    OEM 10 kit, technical data, 26, 27t
    power output equations, 25–26
    random power output samples, 26–27
    solar insolation data, 25–26, 26t
Solar space heating, 4–5
    active space heating, 5
    passive space heating, 6
Solar thermal energy systems, 2
Solar water heating (SWH) systems, 2–3, 3f
Solar water pumps, 4

390 Index

Spider monkey optimization (SMO)
  algorithm, 151
Spring search algorithm (SSA), 151
Stochastic dynamic programming (SDP), 342,
  355–356t
Stochastic fractal search algorithm (SFSA),
  150, 168–171, 169–170t
Stochastic sampling size, 258, 268–270
Sugar/starch fermentation, 12–13
Suns-V_{oc} measurement technique, 115, 116f
Supervised learning, 295, 310–313
  building load predictions, 242, 244t
  cross-entropy method, 296–297
  energy performance prediction, 239–242,
    243f
  high-level controllers
    flexibility quantification fashion, 299, 301f
    including uncertainty, 297, 298f
    training process, 297, 299f
    typical residential house, 297–299, 300f
  nonlinear systems, performance prediction
    of, 263
  on schedulable appliances, 247
Supply and demand ratio (SDR), 273–274,
  276–280
Support vector machine (SMV), 299–302, 302f,
  310, 312
Surface passivation, 83–84
  aluminum oxide (Al_2O_3), 86–90
  dielectric materials, 86–87
  dielectric passivation, 86–87
  dielectric stack passivation, 91–92
  field-effect and chemical passivation, 85–87,
    86f
  final passivation, 87
  front surface passivation, 96
  interface defect and fix charge density,
    86–87, 86f
  intermediary passivation quality, 87
  silicon nitride (SiN_x), 86–88
  silicon oxide (SiO_2), 86–88
Surface recombination mechanism, 75, 78–79,
  78f
Surface recombination velocity (SRV), 87–88
Surplus ratio (SR), 280
Sustainable development scenario (SDS), 203,
  205f
Symbiotic organism search (SOS) algorithm,
  150
System analysis techniques, 342, 355–356t
System optimization, machine learning
  multiobjective optimization, 258, 267
  single-objective optimization, 258, 266

T

Tandem solar cell systems, 73
Teaching-learning-based optimization (TLBO)
  technique, 132
Texture etching, 99–100
Thermal ALD deposition technique, 107, 108f
Thermal diffusion, of phosphorus, 101, 102f
Thermal energy, P2P energy trading with,
  274–276, 275t
Thermal mass storage, 247–248
Thermal oxidation, 101–103, 103f
Thickness profilometer, 111–113, 113f
Tidal energy converter (TIC), 14–15
Time-of-use grid penalty cost model, 276
Time-of-use trading management
  strategy, 274–276, 286
Tip speed ratio (TSR), 362, 364, 374
  vs. power coefficient, 366–367, 366f
  vs. torque coefficient, 367, 368–369f, 370
Total active and reactive powers losses, 174,
  180–181f
Total operating cost (TOC), 153, 171
Total transfer capacity (TTC), 22, 34, 35f,
  39–40, 40f, 42–44
Traditional electricity grid
  advantages, 50
  definition, 49
  features of, 51, 52–53t
  limitation of, 50
Transaction price, 279
Transesterification
  process, 13
  reaction, 13, 14f
Transient photoconductance decay (transient
  PCD) method, 115
Transmission reliability margin (TRM), 22, 34,
  35f, 39–40, 42–44
Trap-assisted nonradiative recombination,
  76–77

U

Uncertainty-based optimization approach,
  267–270
Unified particle swarm optimization (UPSO),
  151
Uniform peer energy trading price model,
  279–280, 281f
Unsteady Reynolds averaged Navier-Stokes
  (URANS) equations, 362–363, 365, 374
Unsupervised learning, 295–297, 313–314
  building load predictions, 242, 244t
  energy performance prediction, 239–242, 243f

Index **391**

V

Vehicle to grid (V2G) concept, 132
Ventilative system
air ventilation-based BIPV/PCM system,
260t, 261–262
natural ventilated PCMs-PV system,
259–261, 261f
structural constitution of, 262–263, 263f
Voltage deviation (VD), 150–151, 166,
167–168f
Voltage deviation index, 151
Voltage stability index (VSI), 150–151

W

Wake recovery distance, 370
Waste frying palm oil (WFPO), 323
Waste frying vegetable oil (WFVO), 323
Water-based PCMs-PV system, 259, 260t,
261–262, 262f
Water cooling system, 312
Water resource system models, 342, 355–356t
Wave energy converter (WEC), 14
Wet oxidation, 101
Whale optimization algorithm (WOA)
approach
bubble-net hunting process, 154–155
cosine adapted WOA (CAWOA)
approach (*see* Cosine adapted
whale optimization algorithm
(CAWOA))
exploration mechanism implemented in, 155,
156f
humpback whales, natural body motions of,
154–155

Wind-based generators (WT-DG), in EDS.
See Electrical distribution system
(EDS), renewable DG in
Wind energy (WE), 135–136
artificial intelligence (AI), 221, 225–227t
electricity generation and installation
capacity
status of, 206, 209f
world's top countries' contribution in, 206,
210f
electricity production, 189, 206
uses, 8
wind power station, advantage of, 206
wind turbines, 206
applications, 9–10
horizontal axis, 8, 9f
speed-output power characteristics of, 136,
136f
vertical axis, 8–9, 9f
Wind energy generator (WEG), 132
Wind farms (WFs), 150
Wind source analysis
random power output samples, 25
wind power output equations, 25
wind speed data, 24, 24t
wind turbine generator set, technical data, 25,
25t
World Small Hydropower Development Report
(WSHPDR) 2019 report, 15–16

X

XGBoost model, 248
X-ray photoelectron spectroscopy
(XPS), 114–115

Printed in the United States
by Baker & Taylor Publisher Services